The Geographic Mosaic
of Coevolution

INTERSPECIFIC INTERACTIONS
A series edited by John N. Thompson

The Geographic Mosaic
of Coevolution

John N. Thompson

The University of Chicago Press
Chicago and London

The University of Chicago Press, Chicago 60637
The University of Chicago Press, Ltd., London
© 2005 by The University of Chicago
All rights reserved. Published 2005
Printed in the United States of America

14   13   12   11   10   09   08   07   5   4   3   2

ISBN (cloth): 0-226-79761-9
ISBN (paper): 0-226-79762-7

Library of Congress Cataloging-in-Publication Data

Thompson, John N.
    The geographic mosaic of coevolution / John N. Thompson.
        p.   cm. — (Interspecific interactions)
    Includes index.
    ISBN 0-226-79761-9 (cloth : alk. paper) — ISBN 0-226-79762-7 (alk. paper)
    1. Coevolution.   I. Title.   II. Series.
  QH372.T482   2005
  576.8'7—dc22

                                                      2004023861

Contents

Preface   vii

Acknowledgments   xi

Part 1   *The Framework of Coevolutionary Biology*

   1   The Overall Argument   3
   2   Raw Materials for Coevolution I: Populations,
      Species, and Lineages   11
   3   Raw Materials for Coevolution II: Ecological
      Structure and Distributed Outcomes   34
   4   Local Adaptation I: Geographic Selection Mosaics   50
   5   Local Adaptation II: Rates of Adaptation
      and Classes of Coevolutionary Dynamics   72
   6   The Conceptual Framework: The Geographic
      Mosaic Theory of Coevolution   97
   7   Coevolutionary Diversification   136
   8   Analyzing the Geographic Mosaic of Coevolution   163

Part 2   *Specific Hypotheses on the Classes*
         of Coevolutionary Dynamics

   9   Antagonists I: The Geographic Mosaic
      of Coevolving Polymorphisms   175
  10   Antagonists II: Sexual Reproduction
      and the Red Queen   205
  11   Antagonists III: Coevolutionary Alternation
      and Escalation   227

12   Mutualists I: Attenuated Antagonism
     and Mutualistic Complementarity   246
13   Mutualists II: The Geographic Mosaic
     of Mutualistic Symbioses   270
14   Mutualists III: Convergence within Mutualistic
     Networks of Free-Living Species   288
15   Coevolutionary Displacement   314
16   Applied Coevolutionary Biology   339

Appendix: Major Hypotheses on Coevolution   365
Literature Cited   371
Index   427

Preface

Coevolution is reciprocal evolutionary change between interacting species driven by natural selection. It is one of the most important ecological and genetic processes organizing earth's biodiversity. My goal in this book is to synthesize what we now know about the ways in which coevolution links species across space and time, connecting populations across landscapes and sometimes holding interactions together over thousands, even millions, of years. My parallel goal is to suggest where we can next make the greatest gains as we study coevolution in natural and fragmented environments and even try our hand at manipulating the coevolutionary process.

Together with many others I have been working for over thirty years toward the development of a framework for the science of coevolutionary biology. I have synthesized our collective progress in two previous books: *Interaction and Coevolution* (1982) and *The Coevolutionary Process* (1994). The first book confronted the problem of how different forms of interaction impose different selection pressures on interacting species. It appeared at a time when coevolutionary studies were still mostly at a stage of describing adaptations and counteradaptations of interacting taxa. My purpose then was to explore ways of moving beyond those descriptions to reach an understanding of coevolutionary selection that transcends taxonomic boundaries.

The Coevolutionary Process synthesized what we had learned in the twelve years since 1982. In that book I suggested how selection on specialization and the geographic structure of species could partition "diffuse coevolution" into more specific coevolutionary processes. My intent was to bridge the gap between studies of coevolution within local communities and studies of diversification in interacting phylogenetic lineages, thereby creating a more hierarchical view of the structure of coevolution. By emphasizing the geographic structure of interactions, I argued that much of the coevolutionary process occurs above the level of local populations but below the level of the fixed traits of species. That is, much of the coevolutionary process falls between what were then the traditional approaches of evolutionary ecology and genetics on the one hand and systematics on the other. I called that overall view the geographic mosaic theory of coevolution, but at the time all I could do was develop the general arguments and suggest the kinds of theoretical and empirical studies needed to explore the structure and dynamics of coevolution within that framework.

Since the mid-1990s, coevolution has come into its own as new studies have taken increasingly rigorous approaches to the structure and dynamics of the coevolutionary process. The major advance has come from confronting the genetic, ecological, geographic, and phylogenetic structure of real species. Rather than treating local coevolution as indicative of the overall structure of coevolving species, we now have a solid set of theoretical and empirical studies that begin with the fact that almost all species are collections of genetically differentiated populations. Mathematical models of the coevolutionary process have begun to formalize how geographic selection mosaics, coevolutionary hotspots, and trait remixing interact to create pattern and dynamics. These models create coevolutionary dynamics different from those envisaged for isolated interactions within local communities. There is still plenty to do, but the models have begun to show how coevolutionary hotspots may develop within geographic landscapes and how local maladaptation may sometimes occur as an outcome of the coevolutionary process. We also now have a growing set of empirical studies that have analyzed the same interspecific interaction in multiple populations across broad geographic landscapes. These studies include evaluations of the scale of geographic selection mosaics, the structure of coevolutionary hotspots, and the effects of trait remixing as gene flow, local extinction, and random genetic drift shape geographic patterns. The study of local matches and mismatches in the traits of coevolving species has become an important component of studies evaluating ongoing coevolutionary dynamics.

We also now have multiple studies that have analyzed the current coevolutionary structure of species within a phylogenetic context. These studies suggest that only a small subset of coevolving traits may eventually scale up to become fixed traits of species. A few traits often become the focus for coevolution, but those traits may vary among populations. We now understand that evaluation of the overall importance of coevolution to an interaction requires a thorough analysis of the geographic mosaic of coevolving traits.

Together, these ecological, genetic, mathematical, and phylogenetic studies are creating a view of coevolutionary dynamics that was not possible a decade ago. It is a view of coevolution as an ongoing ecological process that has fundamental importance for the maintenance of genetic diversity and the organization of biodiversity across landscapes worldwide. This book, then, is not about coevolution as a slow and stately process molding species through sustained directional selection over long periods of evolutionary time. That view creates a caricature of the coevolutionary process and boxes it into a nonecological framework. By that view, ongoing coevolutionary dances— meanderings, if you will—become meaningless fine adjustments, because they are often nondirectional and do not lead to major new events in the history of life. Instead, this book is explicitly an exploration of workaday coevolution, the relentless dynamics of the coevolutionary process that keep the players in the evolutionary game as they respond and counterrespond to each other, population by population, across landscapes. By the time I had finished the book, I realized that it is also a statement about why an evolution-free approach to ecology, parasitology, epidemiology, biological control, agriculture, forestry, wildlife biology, and fisheries biology is never justifiable as we attempt to manage a rapidly changing earth. Ecological time scales are also evolutionary time scales.

The Geographic Mosaic of Coevolution develops a conceptual framework for coevolutionary biology in two stages. Part 1 sets forth the overall framework. It begins with an analysis of the fundamental properties of species that provide the raw material for long-term coevolution across constantly changing landscapes. It then progresses to an evaluation of what we have learned about local coadaptation as the basic module of coevolutionary change. Once that background is in place, the remaining chapters of part 1 explore how the geographic mosaic of coevolution reshapes these local modules over space and time as interacting species diversify across landscapes. These chapters include a formal development of the geographic mosaic theory of coevolution. They also include an analysis of the longer-term phylogenetic patterns that result from the geographic mosaic of coevolution. Part 1 ends with a discussion of forms of evidence in analyses of coevolution.

Part 2 then evaluates specific hypotheses that follow from the geographic mosaic of coevolution. In particular, these chapters evaluate how the geographic mosaic of coevolution maintains genetic polymorphisms, creates multispecific networks of antagonistic trophic interaction, shapes levels of resistance and virulence, contributes to the dynamics of sexual reproduction, favors convergence of traits in mutualistic symbioses and mutualistic networks of free-living species, and molds competitive interactions across large geographic scales. Part 2 ends with a discussion of the developing science of applied coevolutionary biology.

Almost all the work I evaluate in these chapters comes from studies published in the decade after *The Coevolutionary Process* appeared in print. The number of studies of coevolution has increased so much in recent years that it is impossible to cite within a single book the entire history of work on particular topics. *The Coevolutionary Process* included an extended discussion of the history of coevolutionary theory from Darwin to the early 1990s, and that book remains in print. Consequently, I have restricted most citations in this book to papers published since 1994. In some cases, however, I have reached back into the older literature to provide a context for current arguments, views, and results.

These chapters do not provide an encyclopedia of coevolutionary models and examples. Instead, I use a wide range of empirical and theoretical studies to develop four points. First, we have in hand a developing conceptual framework that can help us organize our understanding of the structure and dynamics of coevolution. Second, coevolution is much more of an ongoing, highly dynamic process than we had previously thought. Third, coevolutionary dynamics are important for our understanding of the organization of communities even when they do not lead to long-term directional change. Last, a thorough understanding of the coevolutionary process is increasingly important as we face up to major societal concerns ranging from the rapid evolution of pathogens to the conservation of biodiversity.

Acknowledgments

I am indebted to the many colleagues who have generously shared thoughts, models, and results on the coevolutionary process. I am especially grateful to the following colleagues for discussions or responses to emails during crucial stages in the writing of this book, comments on sections of the manuscript, or preprints that helped me make this book as up to date as possible: Scott Armbruster, Jordi Bascompte, Fakhri Bazzaz, Craig Benkman, May Berenbaum, Giacomo Bernardi, Paulette Bierzychudek, Brendan Bohannan, Jacobus Boomsma, Paul Brakefield, Edmund D. Brodie Jr., Edmund D. Brodie III, Judie Bronstein, James Brown, Jeremy Burdon, Mark Carr, Scott Carroll, Patrick Carter, Yves Carton, Keith Clay, Gretchen Dailey, Peter de Jong, Paul Ehrlich, Niles Eldredge, James Estes, Stanley Faeth, Brian Farrell, Steven Frank, Laurel Fox, Douglas Futuyma, Sylvain Gandon, Sergey Gavrilets, Gregory Gilbert, Douglas Gill, Susan Harrison, Alan Hastings, Edward Allen Herre, Michael Hochberg, Robert Holt, David Jablonski, Jeremy Jackson, Pedro Jordano, Richard Lenski, Bruce Lieberman, Curt Lively, Jonathan Losos, Bruce Lyon, Marc Mangel, Mark McPeek, Kurt Merg, William Miller III, Martin Morgan, Jens Nielsen, Sören Nylin, Takayuki Ohgushi, Jens Olesen, John Pandolfi, Ingrid Parker, Matthew Parker, David Pfennig, Naomi Pierce, Grant Pogson, Don Potts, Peter Price, Peter Raimondi, O. J. Reichman, David Reznick, Kevin Rice, Victor Rico-Gray, Joan Roughgarden, Douglas Schemske, Dolph Schluter, Daniel Simberloff, Douglas Soltis, Pamela Soltis, Victoria Sork, Maureen Stanton, Sharon Strauss, Alan Templeton, David Tilman, James Trappe, Michael Turelli, Geerat Vermeij, Sara Via, Thomas Whitham, Christer Wiklund, and Arthur Zangerl. I am very grateful to Richard Gomulkiewicz and Scott Nuismer for ongoing and stimulating collaborations on formal mathematical models of the geographic mosaic of coevolution.

I thank Jeremy Burdon and Stanley Faeth for their many helpful comments on the outline for the book; Craig Benkman, Edmund Brodie III, Scott Nuismer, and Peter Thrall for their tremendously helpful comments on the entire manuscript; and colleagues in my laboratory at UCSC over the past year—Catherine Fernandez, Samantha Forde, Jason Hoeksema, Phillip Hoos, Katherine Horjus—for their insightful discussions and comments on the penultimate draft. I am indebted to the past and current graduate students, postdoctoral fellows, research associates, sabbatical visitors, and technical assistants who have kept our ongoing laboratory meetings on the coevolutionary process so intellectually challenging over the years. During the gestation and writing of this book they have included David Althoff, Paulette Bierzychudek, Ryan Calsbeek, Bradley Cunningham, Catherine Fernandez, Samantha Forde, David Hembry, Jason Hoeksema, Phillip Hoos, Katherine Horjus, Niklas Janz, Kurt Merg, Scott Nuismer, James Richardson, and Kari Segraves.

I thank Catherine Fernandez for carefully drafting the figures, Abby Young for help with final preparation of the manuscript, and Barbara Norton for copyediting. Christie Henry has provided unflagging editorial insight, encouragement, and guidance. As always, I am deeply grateful to my wife, Jill Thompson, for her suggestions and support throughout the long process of writing a book like this one.

I am also very grateful to the organizations that have provided funds for my research and collaborations with others over the past decade, including the National Science Foundation, the National Center for Ecological Analysis and Synthesis (NCEAS), the Packard Foundation, the American Society of Naturalists, Washington State University, and the University of California, Santa Cruz. NCEAS has provided multiple forms of support, including a sabbatical year of work on coevolution, a workshop on rapid evolutionary change, a working group on mathematical models of coevolution, and a separate working group on evolutionary rates that brought together paleobiologists and population biologists. I thank all my colleagues who shared their thoughts in these meetings and working groups.

Part 1

The Framework
of Coevolutionary Biology

1   The Overall Argument

This book uses the unifying framework of the geographic mosaic of coevolution to confront the major challenges in coevolutionary research: how species coevolve as groups of genetically distinct populations, how coevolving interactions can be locally transient yet persist for millions of years, and how networks of species coevolve. It is one thing to understand that a local interaction between a pair of populations eventually reaches genetic equilibrium under constant selection. It is quite another to understand how interspecific interactions are sometimes held together across millennia as species expand, contract, and diversify across complex and ever-changing landscapes. Local interacting pairs of populations are genetically linked to other populations of the same species, and these geographically variable interactions are embedded within even broader interaction networks.

The geographic and network complexity of interactions enriches the coevolutionary process. There is no more reason to expect a priori that multispecific interactions prevent coevolution than there is to expect that multiple influences of the physical environment—temperature, salinity, water, and light availability—prevent the evolution of populations. Whether researching evolution in general or coevolution in particular, all populations are confronted with multiple selection pressures and evolutionary processes. The scientific problem is to understand how species evolve in the midst of multiple conflicting selection pressures, and how species coevolve across complex landscapes amid interactions with multiple other species.

Background

Much of evolution is coevolution—the process of reciprocal evolutionary change between interacting species driven by natural selection. Most species survive and reproduce only by using a combination of their own genome and that of at least one other species, either directly or indirectly. Species evolve to a large degree by co-opting and manipulating other free-living species or by acquiring the entire genomes of other species through parasitic or mutualistic symbiotic relationships. The evolution of biodiversity is therefore largely about the evolution of interaction diversity.

As the science of coevolutionary biology has matured, we have been returning to a Darwinian appreciation of the entangled bank and an ecological approach to evolution that was largely put on hold during much of the twentieth century amid the excitement of the discovery of genes and the subsequent growth of population genetics and molecular biology. (See Thompson 1994 for a history of coevolutionary biology.) Those genetic and molecular tools have now become part of a renaissance in coevolutionary research, because they have begun to uncover the role of coevolution in the genomic and geographic complexity of life. Even hypotheses of speciation are increasingly based upon ecological and genetic processes driven directly by evolving interspecific interactions, whether by competition or by parasites that manipulate host reproduction.

Not all interactions are tightly coevolved. Nevertheless, as we learn more each year about the evolutionary ecology and genetics of species, we are finding that coevolution is a pervasive and ongoing influence on the organization of biodiversity. We now know that interactions between species can evolve and coevolve within decades. Appreciation of the speed of coevolution is increasing as the disciplines of ecology and evolutionary biology have encompassed studies of pathogens and parasites and incorporated molecular approaches. The traditional study organisms of ecology—plants, insects, rocky-intertidal invertebrates, fish, amphibians, reptiles, birds, and mammals—are now being complemented by studies of a wider array of invertebrates, fungi, bacteria, and viruses. During the past twenty years, the bacterial genus *Wolbachia* and similar intracellular symbionts have moved from being seen as interesting but esoteric causes of reproductive isolation or male sterility in a few insect species to becoming recognized as potential major causes of differentiation in the life history and population structure of a diverse mix of invertebrates. A universe of previously unknown and rapidly evolving interactions is opening as new molecular and ecological tools allow us to probe a wider range of the diversity of life.

We are also beginning to understand better the profound effects of coevolution on human societies. Human history is partly a history of coevolution with the parasites and pathogens that have shaped the spread of our species and our cultures worldwide. The story of human agriculture is to a great degree the story of human-induced coevolution between crop plants and rapidly evolving parasites and pathogens. In recent years, the fields of epidemiology, agriculture, aquaculture, forestry, and conservation biology have all become increasingly attuned to the importance of ongoing coevolution and its effects on our lives. We have, in fact, made manipulation of the coevolutionary process a central part of our human repertoire. We spend billions of dollars a year on antibiotic development, and we are working toward engineering genes to help us fight our battles with parasites, using gene against gene and parasite against parasite.

These efforts are a continuation in different forms of the coevolutionary process that has molded the organization of life on earth. There is now little question that coevolution has shaped many of the major events in the history of life. Even a short list of these events encompasses most species. The eukaryotic cell originated from coevolved symbiotic interactions that became so tightly integrated that one of the species was shaped into the organelles we now call mitochondria. The same happened again in the formation of plants, creating the organelles we now call chloroplasts. Colonization of land by plants may have been made possible through mutualistic interactions with mycorrhizal fungi. Further proliferation of plants occurred partly through coevolved interactions between flowers and pollinators and partly through coevolution with other mutualists, as well as with herbivores and pathogens. The more than seventeen thousand orchid species are thought to rely upon mycorrhizal fungi for nutrition in the early stages of development following germination, because the dustlike seeds of most orchids carry little in the way of nutritional reserves. Primary succession in terrestrial environments relies heavily upon the coevolved interactions called lichens, and subsequent succession depends in many communities upon the coevolved nitrogen-fixation symbioses between rhizobial bacteria and legumes. The very survival of many vertebrate and invertebrate species depends upon obligate coevolved symbionts that reside either within their digestive tract or in special organs, allowing them to digest plant or other tissues. In the ocean, coral reefs, which form the substrate for some of the earth's most diverse biological communities, rely upon coevolved symbioses between corals and zooanthellae and upon additional interactions between corals and algae-feeding fish, although how coevolution has shaped some of these interactions is still poorly understood. The list continues to grow.

It has taken decades for evolutionary biology to begin shifting from a restricted view of species as adapting and diversifying across "environments" to a more coevolutionary view of species as inherently dependent upon other species. It will take longer still to fully integrate interspecific interactions and coevolution into our understanding of evolutionary processes. How much of adaptation is actually coadaptation with other species? How much of population structure is due directly to coevolving interactions? How much of speciation is driven by interactions with other species? To what extent are the widespread genetic polymorphisms found in many taxa maintained by coevolution? How much has coevolution contributed to the persistence of some species across millions of years? How much of the overall organization of communities, regional biotas, continents, and oceans results directly or indirectly from the coevolutionary process?

The developing framework for coevolutionary research is allowing us to begin answering these questions. We now know that the outcomes of coevolution between a pair or group of species can differ across the geographic ranges of the interacting species. We have moved from a view of coevolution as a stately, long-term process that molds species over eons to one in which coevolution constantly reshapes interacting species across highly dynamic landscapes.

The Geographic Mosaic as the Organizing Framework of Coevolution

The goal of coevolutionary biology should be to understand how reciprocal evolutionary change shapes interspecific interactions across continents and oceans and over time. The fundamental premise of this book is that coevolution is an inherently geographic process that results from the genetic and ecological structure of species. The overall argument draws on the conclusions of two previous books (Thompson 1982, 1994) and on the empirical data and models for coevolutionary dynamics that have appeared especially over the past decade through the work of an ever-widening community of researchers. As I hope these chapters show, we now have a science of coevolutionary biology that provides a conceptual framework, specific hypotheses that follow from that framework, and predictions that can be tested within natural populations.

The framework of coevolution is built upon four fundamental attributes of species and interspecific interactions that provide the raw materials for on-

going coevolution (chapters 2 and 3). Most species are collections of genetically differentiated populations, and most interacting species do not have
identical geographic ranges. Species are phylogenetically conservative in their
interactions, and that conservatism often holds interspecific relationships
together for long periods of time. Most local populations specialize their
interactions on only a few other species. The outcomes of these interspecific
interactions differ within and among communities.

Through these attributes, interactions are simultaneously held together at
the species level even as they diversify among populations. Species become
locally adapted to other species (chapter 4), and they continue to evolve rapidly, thereby blurring the artificial distinctions between ecological time and
evolutionary time (chapter 5). These adaptations create a small set of classes
of local coevolutionary dynamics, including coevolving polymorphisms, coevolutionary alternation, coevolutionary escalation, attenuated antagonism,
coevolving complementarity, coevolutionary convergence, and coevolutionary displacement (chapter 5). The transient local coevolutionary dynamics
between any two or more species often differ among populations at any moment in time. The resulting mosaic of local adaptation and coadaptation in
interspecific interactions establishes the basic structure of the coevolutionary
mosaic.

The mosaic constantly changes as coevolving species continually adapt,
counteradapt, diverge from other populations, and occasionally undergo
speciation. The geographic mosaic theory of coevolution argues that these
broader dynamics—which go beyond local coevolution—have three components (chapter 6):

- *Geographic selection mosaics.* Natural selection on interspecific interactions varies among populations partly because there are geographic
  differences in how fitness in one species depends upon the distribution of genotypes in another species. That is, there is often a genotype-
  by-genotype-by-environment interaction in fitnesses of interacting
  species.
- *Coevolutionary hotspots.* Interactions are subject to reciprocal selection
  only within some local communities. These coevolutionary hotspots
  are embedded in a broader matrix of coevolutionary coldspots, where
  local selection is nonreciprocal.
- *Trait remixing.* The genetic structure of coevolving species also changes
  through new mutations, gene flow across landscapes, random genetic
  drift, and extinction of local populations. These processes contribute

to the shifting geographic mosaic of coevolution by continually altering the spatial distributions of potentially coevolving alleles and traits.

Through this tripartite process, coevolution produces identifiable ecological and evolutionary dynamics across landscapes (chapter 6). Populations
differ in the traits shaped by an interaction. Coevolved traits are well matched
between species in some communities but sometimes mismatched in others.
Most locally coevolved traits do not scale up to produce long-term directional change in the traits of interacting species. Traits shaped by coevolving
interactions ratchet in particular directions and become fixed in a species
only through occasional selective sweeps across all populations of interacting
species or through diversifying coevolution that creates new species (chapter
7). Most of the time, coevolution moves species around in genetic and ecological space without any sustained direction. These ongoing dynamics provide us ways of analyzing coevolution by using eleven forms of evidence and
drawing on approaches from multiple subdisciplines (chapter 8).

The various classes of local coevolutionary dynamics fit within the
broader geographic mosaic of coevolution, and part 2 develops specific hypotheses with predictions for further research (chapters 9–15). How the geographic mosaic molds these classes depends upon the mode of interaction
among species. Within antagonistic trophic interactions, predation, grazing,
and parasitism have different effects on the structure of coevolutionary selection. The geographic structure of interactions can sustain coevolving polymorphisms between parasites and hosts, while generating a mix of habitats in
which traits are matched or mismatched (chapter 9). Under some conditions
multispecific coevolution between parasites and hosts favors optimal allelic
diversification in these polymorphisms (chapter 9), and it may favor the
maintenance of sexual reproduction (chapter 10). Predators and grazers, and
some parasites, often actively choose among multiple victim species, creating
mosaics of coevolving networks through geographic differences in relative
preference and the process of coevolutionary alternation, sometimes coupled
with escalation (chapter 11).

The continuum in forms of interaction from mutualistic symbioses to
mutualism between free-living species is as fundamental to the geographic
mosaic of coevolution as is the continuum from parasitism to grazing and
predation. Mutualisms coevolve through a combination of complementarity
of traits (e.g., nutritional requirements of hosts and mutualistic symbionts;
shapes of flowers and hummingbird bills) and convergence of traits within
networks (e.g., convergent floral traits among species). The importance of

coevolutionary convergence as part of the process differs among the forms of mutualism. Local adaptation sometimes favors the evolution of attenuated antagonism within symbiotic interactions (chapter 12). Those interactions that create reciprocal fitness benefits coevolve through selection toward mutualistic monocultures and complementary symbionts (chapter 12), creating geographic mosaics (chapter 13). Local adaptation among free-living mutualists also favors complementarity among interacting species, but it also favors convergence of unrelated taxa. These two classes of coevolutionary dynamics contribute to predictable network structure even as species composition changes across landscapes (chapter 14).

The remaining major outcome, coevolutionary displacement, results from a geographic mosaic in the intensity and form of interaction among species that share similar resources or habitats (chapter 15). Species may become displaced in traits or habitat use through competition, either alone or combined with other forms of interaction, and guilds of species may become displaced in similar ways across landscapes. Local character displacement in a pair of species is therefore only one component of the overall geographic mosaic of coevolutionary displacement.

The developing framework for coevolutionary biology provides an increasingly solid basis for the establishment of a science of applied coevolutionary biology (chapter 16). Selective breeding and genetic modification of crops and livestock for resistance against parasites is a form of human-induced coevolution that has some similarities to natural coevolution but also many differences from it. The development of antibiotics and vaccines has created, in effect, surrogate genes and a process of surrogate coevolution that we are now trying to manage. More broadly, our worldwide modification of landscapes and transport of species over vast distances is creating interactions with geographic configurations that differ in some ways from anything most species have experienced in the past.

Confronting all these challenges requires a three-pronged approach in the development of coevolutionary biology. We must continue to refine our understanding of the geographic mosaic of coevolution across a broad range of landscapes and forms of interaction. We must understand the differences between natural coevolution and processes such as surrogate coevolution. And we must use more effectively the precious and free research and development that exists in the earth's remaining wilderness areas. No amount of money can ever replace the valuable information about the structure and dynamics of coevolution contained within long-coevolved interactions. The geographic mosaic of coevolution across wilderness landscapes is our touchstone for understanding the dynamics we are trying to manipulate across most of the earth's landscapes. Through these combined approaches, coevolutionary biology should become one of the most important sciences for helping us maintain the long-term health of our societies and the world that we now increasingly manage.

2 Raw Materials for Coevolution I
Populations, Species, and Lineages

The central problem of coevolution is to understand how interactions between species are shaped by reciprocal natural selection and persist across space and time even as they undergo constant and often rapid coevolutionary change. In an idealized interaction, one population of one species coevolves with one population of another species within a single local environment. Under intense reciprocal selection, the interaction coevolves either to a state of equilibrium or to local extinction of one of the species. Coevolution in real species, however, involves multiple interconnected populations, distributed across complex environments subject to ongoing major physical events such as El Niño and North Atlantic oscillations, ice ages, periods of global warming, and erratic volcanism that can change worldwide weather patterns for years on end. As environments change, so do the geographic ranges of species, bringing different populations of coevolving species into contact while shifting other populations outside the range of the interaction.

The resulting coevolutionary changes continue to reshape interactions over years, decades, and centuries. Somehow, in the midst of all this constant coevolutionary change, some interactions persist for millions of years. As these coevolving species diversify, they in turn create descendant lineages that interact in a similar way. Any realistic scientific framework for the coevolutionary process must therefore confront the temporal, geographic, and phylogenetic structure of species and interactions. It must explain the process by which short-term coevolutionary change and long-term persistence are interrelated.

The raw materials both for short-term and long-term coevolution consist of four fundamental attributes of the biology of species (Thompson 1999d), which are explored in this chapter and the next.

- Most species are collections of genetically differentiated populations, and most interacting species do not have identical geographic ranges.

- Species are phylogenetically conservative in their interactions, and that conservatism often holds interspecific relationships together for long periods of time.
- Most local populations specialize their interactions on a few other species.
- The ecological outcomes of these interspecific interactions differ within and among communities.

The first two attributes emphasize the malleable yet conservative structure of species and are explored in this chapter. The second two attributes emphasize the dynamic yet bounded ecological structure of interspecific interactions and are developed in the next chapter. Together, these components of interspecific interactions create a template upon which natural selection both drives and constrains coevolving relationships among species across multiple temporal and spatial scales. In effect, these components create the conditions that shape short-term coevolutionary dynamics and make long-term coevolution possible.

Most Species are Collections of Genetically Differentiated Populations, and Most Interacting Species Do Not Have Identical Geographic Ranges

The single clearest result from the past thirty years of research in population biology and molecular ecology is that most species are collections of populations that differ genetically from each other. Populations differ at the molecular level at DNA positions that are selectively neutral and at positions under strong selection. They differ at the phenotypic level in traits that shape their adaptations to the physical environment and their interactions with other species. Through a combination of non-panmictic breeding among populations, random genetic drift, selection on particular alleles, and geographic differences in interspecific interactions, species become collections of small evolutionary experiments. Over time, these experiments expand, contract, diversify, and anastomose across continents and oceans.

Within regions, populations may form metapopulations, with each local deme potentially differing in genetic and ecological structure from other demes (Hastings and Harrison 1994; Husband and Barrett 1996; Hanski 1999, 2003; Hastings 2003; Smith, Ericson, and Burdon 2003). Different configurations of metapopulations can lead to different genetic dynamics over time through differences in patterns of gene flow and the dynamics of extinction and recolonization (Hanski and Gilpin 1997). Similarly, different spatial structures can lead to different patterns of population dynamics (Murdoch, Briggs, and Nisbet 2003), which can feed back on the genetic dynamics by altering the temporal patterns of extinction, recolonization, random genetic drift, and natural selection. The regional structure of most species is therefore likely to be in constant genetic flux.

Over larger geographic areas, more stable genetic differences among populations create lineages of populations that have provided the basis for the development of the field of phylogeography (Avise 1994, 2000). Even some species known for long-distance migrations show geographic structure, because individuals within these species often return to their natal areas to breed (Dingle 1996). Together, local metapopulation dynamics and the broader geographic structure of most species guarantee that spatial structure will influence the coevolutionary dynamics of almost all interspecific interactions.

MOLECULAR DIFFERENTIATION

As data on DNA sequences and polymorphisms continue to accumulate for more taxa, the evidence is pointing toward more rather than less genetic differentiation among populations than we previously suspected for many species. The extreme of true panmixis throughout the range of any moderately wide-ranging species seems increasingly unlikely. Even panmixis over moderately large subregions of species seems uncommon for many taxa.

Populations can come to differ from one another simply because they are finite in size and individuals do not have equal likelihood of mating with one another among the populations. Until recently, the most commonly used measures of population-genetic differentiation were those that use, directly or indirectly, the variance in allele frequencies among populations, such as Wright's F-statistic F_{ST} (Wright 1951, 1965) or Nei's G_{ST}. (Nei 1973). F_{ST} measures the relative proportion of total genetic variation that is found among populations rather than within populations. It is essentially a measure of average population-genetic differentiation and provides no information of the spatial configuration of that differentiation (Rogers 1988; Epperson 2003). Nevertheless, calculation of F_{ST} has often provided a simple index of whether populations show some degree of differentiation across the spatial scale of a particular group of populations under study. Major reviews of population subdivision in plants and animals using these measures have usually shown some degree of subdivision in most species (Hamrick and Godt 1990; Bohonak 1999). Sometimes substantial subdivision may occur at scales of a few kilometers or less, whereas in other species substantial gene flow occurs over large scales. This overall lack of panmictic structure in many species creates part of the raw material for the geographic mosaic of coevolution. We are only starting to understand, however, how the scale of population subdivision indexed by these measures affects coevolutionary dynamics within regions.

At larger spatial scales, many molecular studies of terrestrial and freshwater taxa show evidence of substantial subdivision of populations, creating breaks among faunistic or floristic regions. For example, comparative phylogeographic studies of fish species in the southeastern United States have shown major molecular differences between populations in rivers draining into the Atlantic Ocean and those in rivers draining into the Gulf of Mexico. This pattern holds for spotted sunfish (*Lepomis punctatus*), three other sunfish species (*Lepomis* spp.), mosquito fish (*Gambusia* spp.), largemouth bass (*Micropterus salmoides*), and bowfin (*Amia calva*) (Bermingham and Avise 1986; Walker and Avise 1998) (fig. 2.1).

Similar studies in Europe have identified three broad patterns in post-Pleistocene recolonization from southern European refugia (Hewitt 2001) (fig. 2.2). For taxa such as meadow grasshoppers and alders, the Pyrenees and the Alps were major barriers, and much of Europe was recolonized from the Balkans. Other taxa, such as hedgehogs and oak species, are parapatric along a north–south line through Europe, suggesting recolonization from multiple refugia to the west and east. For brown bears and shrews, the Pyrenees seem to have been less of a barrier than the Alps, and much of central Europe was colonized by populations from the Iberian peninsula and the Caucasus.

Analyses of comparative phylogeographic structure are now appearing for an increasingly wide range of taxa and regions (Soltis et al. 1997; Bernatchez and Wilson 1998; Moritz and Faith 1998; Althoff and Thompson 1999; Ditchfield 2000; Stuart-Fox et al. 2001; Brunsfeld et al. 2002; Calsbeek, Thompson, and Richardson 2003). These studies are making it possible to compile composite phylogeographic analyses of entire ecoregions. The current conclusions, however, are best viewed as working hypotheses, requiring much infill of species and populations. For example, as recently as 2002 only fifty-five species or species complexes were available for a comparative phylogeographic analysis of California animals and plants, when the analysis was restricted to species studied in multiple populations within California using molecular markers or DNA sequences (Calsbeek, Thompson, and Richardson 2003). That analysis showed several major molecular breaks within the California Floristic Province common to multiple animal taxa (e.g., the transverse range of southwestern California) but less evident structure among

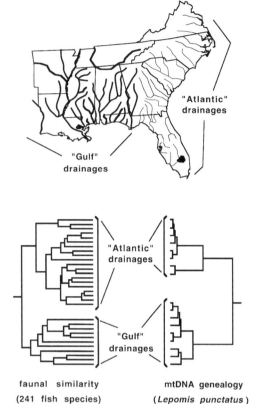

Fig. 2.1   Major river drainages of the southeastern United States, the patterns of faunal similarity among fish taxa across the region, and the pattern of mitochondrial DNA differentiation in one species, the spotted sunfish (*Lepomis punctatus*). After Walker and Avise 1998.

faunal similarity
(241 fish species)

mtDNA genealogy
(*Lepomis punctatus*)

plant species. Molecular-clock analyses of the animal species suggest major patterns of genetic differentiation (i.e., deep phylogeographic splits) starting around five to seven million years ago, with additional splits during the Pleistocene (fig. 2.3).

Some neighboring large regions show very different patterns of molecular differentiation. In both Europe and the North American Pacific Northwest, northern populations of some species show relatively little regional differentiation in DNA sequence and molecular markers compared with more southern populations that were less affected by Pleistocene ice sheets (Brown et al. 1997; Soltis and Soltis 1999; Avise 2000; Hewitt 2001; Calsbeek, Thompson, and Richardson 2003). In many cases these northern populations have resulted from rapid post-Pleistocene expansion, which has allowed little time for molecular differentiation at neutral DNA positions.

For example, the moth *Greya politella* has relatively few mitochondrial

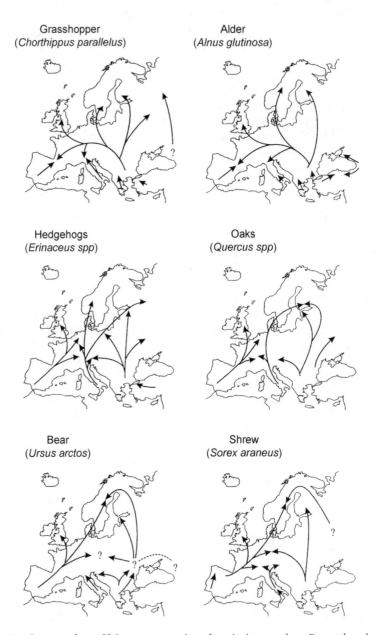

Fig. 2.2 Patterns of post-Pleistocene expansion of species into northern Europe based upon phylogeographic analyses. After Hewitt 2001.

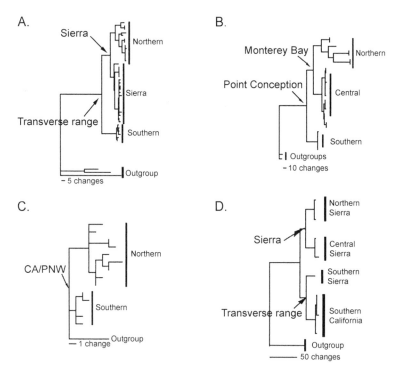

Fig. 2.3   Examples of geographic patterns in molecular differentiation in four taxa within California: (A) rubber boas, *Charina bottae*; (B) intertidal copepods, *Tigriopus californicus*; (C) the prodoxid moth *Greya politella*; and (D) mountain yellow-legged frogs, *Rana muscosa*. Arrows indicate nodes corresponding to major geographic boundaries within California. After Calsbeek, Thompson, and Richardson 2003.

DNA haplotypes in Idaho, Washington, and Oregon relative to populations in California (fig. 2.4). All the northern populations so far tested for cytochrome oxidase I and II share the same one or two haplotypes. The southern Oregon populations are genetically very similar to those found farther north, differing only by a few base substitutions. In contrast, Californian populations show greater regional differentiation in cytochrome oxidase haplotypes, even though only parts of that region have been sampled (Brown et al. 1997). Similar patterns of shallow molecular differentiation at neutral markers in the Pacific Northwest have been found in the few other insect, plant, and vertebrate species that have been studied across the same habitats (e.g., Althoff and Thompson 1999; Segraves et al. 1999; Soltis et al. 1997;

Fig. 2.4   Molecular differentiation in cytochrome oxidase I and II among populations of the moth *Greya politella* in the western United States. Populations in the Pacific Northwest (Washington, Idaho, and Oregon) show less molecular differentiation among populations (haplotypes W1–2) than populations in California (haplotypes C1–7). Haplotypes are more closely related within regions than between regions (i.e., W1–2 group together, and C1–7 group together). After Brown et al. 1997.

Nielson, Lohman, and Sullivan 2001; Janzen et al. 2002; Good et al. 2003; Thompson and Calsbeek 2004).

There are still too few studies of marine taxa to make any general conclusions about geographic patterns of differentiation in marine environments as compared with terrestrial and freshwater environments. Some intertidal species show strong phylogeographic structure (Burton 1998), but until recently marine species were considered to be fundamentally different from most terrestrial organisms, because many marine species have pelagic larvae that drift in ocean currents for extended periods of time. Panmixis over large regions does, in fact, seem to occur in some species such as plaice (*Pleuronectes platessa*) and turbot (*Scophthalmus maximus*) (Hoarau et al. 2002), but examples of restricted gene flow are accumulating for species in all the major oceans (Palumbi 1994; Terry, Bucciarelli, and Bernardi, 2000). In some cases, the molecular differentiation is between major regions, such as occurs in crown-of-thorns starfish (*Acanthaster planci*) populations between the Indian and Pacific oceans (Benzie 1999). Other species, such as the widespread scleractinian coral *Plesiastrea versipora,* show molecular evidence of geographic structure in some regions of the Pacific Ocean but not in others (Rodriguez-Lanetty and Hoegh-Guldberg 2002). Still others, such as Atlantic cod (*Gadus morhua*), show differentiation within regions. Analysis of natural selection on the pantophysin locus in cod has suggested that coastal and arctic populations have undergone recent diversifying selection (Pogson 2003).

Some oceanic species or species complexes showing restricted regional gene flow are associated with island habitats (Taylor and Hellberg 2003), which may favor reduced pelagic larval periods and restricted gene flow into the open ocean. Yet other species such as superb blackfish (*Embiotoca jacksoni*) and the predatory snail *Nucella caniliculata* lack pelagic stages (Bernardi 2000; Sanford et al. 2003). Superb blackfish is subdivided into groups of populations along the California and Baja California coasts, showing major breaks north and south of the Big Sur/Monterey Bay region, near Santa Monica in southern California, and near Punta Eugenia in Baja California (fig. 2.5). Populations on the northern Channel Islands also differ in their haplotypes from those on the southern Channel Islands (fig. 2.5). Similarly, *Nucella caniliculata* shows genetic differences among populations along the west coast of North America, although less pronounced and at larger geographic scales (Sanford et al. 2003). The populations show significant isolation by distance between California and the Pacific Northwest.

Overall, marine taxa show a wide range of geographic scales at which populations are genetically differentiated. Even studies of phytoplankton are showing more population differentiation or cryptic speciation than previously suspected (Goetze 2003). As with terrestrial and freshwater species, there is therefore a potential for marine species and species complexes to differ geographically in coevolution with other species. If it turns out, however, that geographic patterns of population differentiation differ fundamentally in terrestrial and marine environments, those results will be crucial for our understanding of how the coevolutionary process has shaped the earth's biodiversity in different major environments.

MOLECULAR DIFFERENTIATION COMPARED
WITH PHENOTYPIC DIFFERENTIATION

Even with high levels of gene flow among populations, regional genetic differentiation may still be possible, if natural selection is stronger than the effects of gene flow. Consequently, estimates of gene flow and population

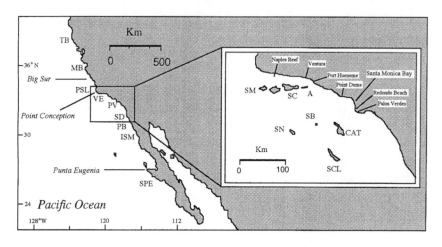

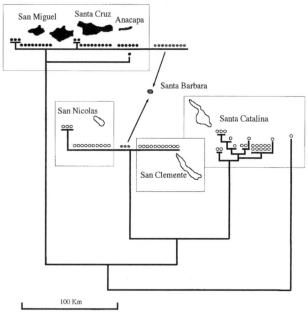

Fig. 2.5 Molecular differentiation at the mitochondrial control region among populations of the black surfperch, *Embiotoca jacksoni,* among the Channel Islands off California. The top panel shows the position of sampling sites off the Channel Islands. The bottom panel shows the phylogeographic structure of Channel Island populations. Each circle is a sample. The northern populations (black circles) are significantly differentiated from the southern populations (white circles). From Bernardi 2000.

subdivision using molecular markers alone are insufficient as a template for understanding the geographic mosaic of coevolution. In fact, our current estimates of molecular differentiation among populations often do not match patterns of phenotypic differentiation. At one extreme, microsatellite analyses may show so much fine-scale molecular diversity that they swamp the spatial scale of local adaptation. At the other extreme, DNA sequences of several genes and genome-wide analyses of restriction fragment length polymorphisms (RFLPs) or amplified fragment length polymorphisms (AFLPs) often underestimate the more fine-scaled geographic structure of species that can result from regional differences in natural selection acting on populations. Low levels of molecular regional differentiation sometimes found in these studies do not automatically suggest a lack of differentiation in traits under selection across the geographic range of a species. In fact, there is now strong evidence of differentiation across landscapes in phenotypic traits of some species that show little molecular differentiation across the same regions. For example, color patterns differ among butterfly fish in the Pacific Ocean despite the high levels of gene flow indicated by studies of allozymes and mitochondrial DNA (McMillan, Weigt, and Palumbi 1999).

This is where studies of phylogeography and evolutionary ecology meet, showing the need for caution in both kinds of analysis. The patterns found in most molecular data show the combined history of gene flow, mutation, and random genetic drift in neutral genes, and those processes alone can create geographic structure given enough time (Charlesworth, Charlesworth, and Barton 2003). In some regions, such as the North American Pacific Northwest, there has been little time for neutral molecular differentiation among populations of many species. Populations are now living in regions that were under ice less than eighteen thousand years ago. Phylogeographic data therefore provide important information on the large-scale geographic breaks among populations, but they probably underestimate the amount of geographic differentiation in the phenotypic traits under selection in evolving interspecific interactions.

The saxifragaceous herb *Heuchera grossulariifolia* in the North American Pacific Northwest provides a clear example. This species is restricted to the mountains of northern Idaho and adjacent western Montana and has only a small number of chloroplast DNA and restriction fragment length differences across its geographic range (Segraves et al. 1999), suggesting relatively little phylogeographic structure in comparison to some other species that have been studied in other regions. Nevertheless, *H. grossulariifolia* has repeatedly produced autopolyploid populations, presumably since the end of

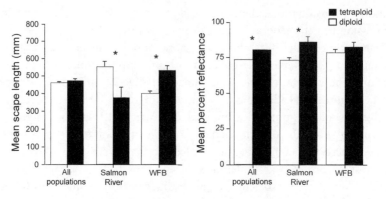

Fig. 2.6   Mean scape (= floral stalk) length and mean percent reflectance of the sepals in diploid and tetraploid plants of *Heuchera grossulariifolia* in northern Idaho and western Montana. The plants of different ploidy show little differentiation in DNA sequence or restriction fragment length polymorphisms, but they differ in multiple phenotypic traits. The phenotypic traits of two of the multiple suggested origins of polyploid populations are graphed here: the Salmon River and the West Fork of the Bitterroot (WFB) River. Also shown are composite means for all populations studied. After Segraves and Thompson 1999.

the Pleistocene (Wolf, Soltis, and Soltis 1990; Segraves et al. 1999), and these polyploid populations differ from sympatric diploid populations and from one another in a wide range of phenotypic traits, including flowering time, length of floral stalks, and floral size and color (Segraves and Thompson 1999) (fig. 2.6).

More important for the geographic mosaic of coevolution, diploid and tetraploid plants differ in the pattern of attack by herbivores. The moth *Greya politella* attacks tetraploids, whereas its sympatric congener, *G. piperella,* attacks diploids, where the plants of these two ploidy levels are sympatric (Thompson et al. 1997; Nuismer and Thompson 2001; Janz and Thompson 2002) (fig. 2.7). Moreover, *G. politella* attacks a higher proportion of tetraploids than diploids in three regions in which tetraploidy appears to have arisen independently, according to phylogeographic analyses (Thompson et al. 1997; Segraves et al. 1999). Hence, some yet unknown aspect of polyploidy per se, rather than the specific genotypic structure of any one polyploid event, seems to have caused a higher level of attack on tetraploids by *G. politella.*

Similar differences in the use of diploid and tetraploid plants occur among floral visitors. The most remarkable example is in a population of

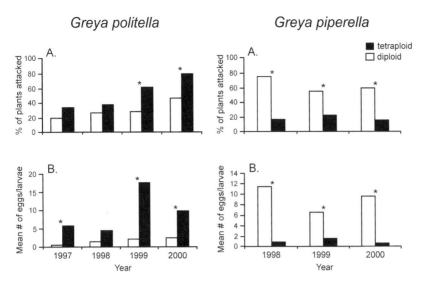

Fig. 2.7   Attack of sympatric diploid and tetraploid *Heuchera grossulariifolia* plants by the moths *Greya politella* (left two panels) and *Greya piperella* (right two panels). For both species (A) shows the percentage of plants attacked and (B) shows the mean number of eggs or larvae per floral capsule. The asterisk (*) indicates a significant difference in level of attack on diploid and tetraploid plants. After Nuismer and Thompson 2001.

H. grossulariifolia along the Salmon River in Idaho, where queens of the bumblebee *Bombus centralis* preferentially visit tetraploid flowers and the workers preferentially visit diploids (Segraves and Thompson 1999). Hence, even though there is only weak phylogeographic differentiation among these plant populations, as assayed using current phylogeographic techniques, there are strong differences in phenotypic traits and interspecific interactions resulting from ploidy differences and the effects they have on phenotypes.

Weak molecular differentiation but strong population differences in interspecific interactions occur in the same region of the Rocky Mountains in the interactions between braconid parasitoids in the genus *Agathis* and the *Greya* moths they attack (Althoff and Thompson 1999, 2001). The parasitoids show little molecular differentiation across the deeply divided river drainages in which they occur, and they show little evidence of isolation by distance. Nevertheless, the wasps differ among populations in ovipositor length, which reflects regional differences in the plant tissues they probe in search of hidden *Greya* larvae. The parasitoids also differ behaviorally in how they search for host larvae hidden within plant reproductive and vegetative tissues.

IMPLICATIONS

Ongoing genetic differentiation among populations is an inherent part of the evolutionary biology of most species, which is becoming more evident each year as more taxa are studied in greater detail. Populations are the basic units of evolutionary and coevolutionary dynamics, and they form the fundamental structure of the geographic mosaic of coevolution. In a brave attempt at an initial estimate of just how many populations occur worldwide, Hughes, Daily, and Ehrlich (1997) calculated it to be somewhere between 1.1 billion and 6.6 billion. Their estimate used as an index the number of molecularly separable populations found within well-studied species over defined geographic areas. They extrapolated those estimates to the overall geographic ranges of those and other species. The results, of course, were highly biased taxonomically, reflecting real biases in the taxa for which data are available. Their analysis, however, provided an initial estimate that can be refined as more detailed studies are published for a wide range of species. It does not really matter if this initial estimate is off even by half an order of magnitude. Their estimates highlight the complex populational structure of the earth's biodiversity.

There is, however, one important caveat. We do not know the extent to which free-living microbial populations undergo extensive geographic differentiation. Although work on diseases has commonly shown strong geographic differentiation in many parasitic microbial taxa, there are still too few detailed studies of free-living microbial taxa to make any general statements about these organisms. Part of the problem comes from disagreements over species limits. By some views the earth has only a small number of free-living microbial eukaryotic species that often have worldwide distributions unrestricted by geographic barriers (Finlay 2002). By other views, free-living microbial eukaryotes are more diverse than previously thought, based upon recent DNA sequence comparisons (Coleman 2002). Similarly, molecular tools are revealing a rich diversity of bacteria (Horner-Devine, Carney, and Bohannan 2004). Whether these molecularly differentiated populations represent geographically differentiated populations within species or separate species matters less than the question of whether they represent different lineages in their functions within ecosystems and their coevolving interactions with other taxa. The technical challenges remain daunting. Even mesocosm studies on microbial diversity across environmental gradients (e.g., cattle tanks that mimic small ponds) are limited in the diversity they can assess, because the taxa are identified through cloning and DNA sequencing (Horner-Devine et al. 2003). The overall issue of the geographic diversity of microbial taxa will become increasingly important as microbial biology continues its trajectory toward incorporation into mainstream ecology, evolutionary ecology, and coevolutionary biology.

The constantly changing geographic ranges and relative abundances of species add to the raw material for the geographic mosaic of coevolution. In the eastern Pacific, anchovy and sardine populations have oscillated several times in relative abundance over multidecadal periods during the past century, producing ripple effects on the geographic distributions of multiple other species (Chavez et al. 2003). During the same time period multiple terrestrial species have undergone major shifts in geographic distribution in North America and Europe (Parmesan et al. 1999; Hill et al. 2001). Human intervention complicates any simple interpretation of these changing geographic ranges, but the changes themselves illustrate the elasticity of geographic ranges that form part of the template for the geographic mosaic of coevolution. As species' ranges expand and contract, peripheral populations sometimes differ in their genetic structure from more central populations, due to differences in the combined effects of natural selection, gene flow, and random genetic drift (Volis, Mendlinger, and Orlovsky 2000; Jones, Gliddon, and Good 2001; Ball-Ilosera, Garcia-Marin, and Pla 2002).

The dynamics of species' ranges are even greater over longer time periods. Paleontological data show tremendous shifts in the geographic ranges of species and genera over geologic time (Lieberman and Eldredge 1996; Kaustuv, Jablonski, and Valentine 2001; Ricklefs and Bermingham 2001; Rode and Lieberman 2002). Periods of mass extinction have been followed by quirky patterns of reinvasion within and among biogeographic regions (Jablonski 1998). Pleistocene glaciation events scoured landscapes with ice, fragmented forests, lowered seabeds, altered worldwide climates, and repeatedly changed species distributions (Davis and Shaw 2001). Probably no species on earth has the same geographic distribution that it had only fifteen thousand years ago. Consequently, at any moment in time some populations of most widespread species are probably coming into contact with novel populations of other species. Each of these events is a potentially new coevolutionary experiment.

Species Are Phylogenetically Conservative in Their Interactions, and That Conservatism Often Holds Interspecific Relationships Together for Long Periods of Time

CONSERVED TRAITS AND TETHERED LINEAGES

As species diverge into genetically differentiated populations, their adaptations remain constrained by the genetic architecture they inherit from their ancestors. The traits used by species in their interspecific interactions are jury-rigged from their ancestral traits, biasing adaptation in particular directions. As a result, the members of each species are phylogenetically constrained to eat, compete against, and defend themselves against a minuscule fraction of the earth's biological diversity (Ehrlich and Raven 1964; Futuyma, Keese, and Funk 1995; Futuyma and Mitter 1996). In leaf fossils, leaf-mines and other damage to plants caused by some insect taxa are almost identical to leaf-mines and damage caused by those insect taxa on the same plant genera today (Labandeira 2002). Some of these documented similarities between fossil and extant interactions are tens of millions of years old (Labandeira et al. 1994; Wilf et al. 2000). Phylogenetic lineages also often show similarities among species in their life histories and population dynamics (Price 2003), thereby adding to the adaptive conservatism on which coevolving interactions are shaped. The conservatism imposed by phylogenetic continuity therefore provides the opportunity for ongoing coevolution between lineages over long periods of geological history.

As we have learned more about the phylogeny of species interactions in recent decades, it has become evident that simply opting out of an interaction is not a commonly viable option for most taxa. Large phylogenetic jumps that allow species to interact with taxa very different from those their ancestors encountered are relatively uncommon. The most viable option, other than extinction, is coevolution with the small subset of taxa that one's ancestors also coevolved with. These closely related species often differ phenotypically and ecologically from one another in relatively small ways (e.g., Harvey 1996; Hedderson and Longton 1996; Silvertown, Franco, and Harper 1997; Clayton et al. 2003). For example, some of the major features of fleshy fruits used to attract birds and mammals are shared among related plant species and genera (table 2.1) and cannot be interpreted as results of direct and recent selection on each individual species (Herrera 1995; Jordano 1995; Herrera 2002). Some characteristics, such as protein levels and energy per gram of dry mass, vary considerably among species within a genus, whereas

Table 2.1 Hierarchical analysis of differences among plant taxa in characteristics of fleshy fruits

	% of phenotypic variation occurring		
	≥Family	Genus	Within genus
Fruit			
Fruit length	28	24	48
Fruit diameter	23	29	48
Fruit fresh mass (FFM)	38	35	27
Pulp dry mass (PDM)	39	28	33
FFM/PDM	4	43	53
Seed			
Seed total dry mass	46	20	34
Individual seed dry mass	58	30	12
Number of seeds	94	1	5
Water and energy			
Percent water	27	26	47
Energy per fruit	29	32	39
Energy/g dry mass	25	0	75
Nutrients (%)			
Lipids	33	36	31
Protein	8	32	60
Nonstructural			
carbohydrates	22	0	78
Minerals	38	37	25
Fiber	5	50	45

Source: From Jordano 1995.

Note: The table shows the percentage of the range of difference found in each character that occurs among plant families, genera, and species.

other characteristics, such as the number of seeds per fruit and the mean dry mass of individual seeds, vary relatively little among congeners. Individual plant species differ from one another in many small ways that can affect their interactions with frugivores, but species within a plant genus often interact with similar groups of frugivore species. That does not mean in any way that there is little variation on which natural selection can act. It simply means that evolution of interactions is more constrained along some trajectories than others.

This phylogenetic conservatism is as important as coadaptation in making coevolution a central force in the organization of biodiversity. Species remain tethered together, allowing repeated bouts of coevolutionary change— often in slightly different ways in different populations. When species are able to shift their interactions to other species, they often do so by shifting onto species that are phylogenetically close to the species used by their ancestors. For a parasite of mammals, the evolutionary options rarely include shifting to a frog as its primary host. For insects that feed on conifers, incorporation of orchids in their diets is unlikely. The ancestral-descendant sequence of populations that make up a phylogeny imposes historical structure on patterns of specialization and the organization of biodiversity as lineages respond to new ecological opportunities (Futuyma and Mitter 1996).

Phylogenetic conservatism also shapes the geographic template of interspecific interactions by limiting the range of habitats available to species. The distribution of clades of *Enallagma* damselflies across North America is among the most carefully studied examples. Most lakes throughout eastern North America harbor several *Enallagma* species that share similar defenses against predators (McPeek 2000) (fig. 2.8). The lakes either have fish or large dragonflies as the top predators, and the damselfly species differ in their ability to defend themselves against these two predator taxa. Damselflies that use crypsis to avoid predators coexist with fish, whereas those that actively swim away from attacking predators coexist with dragonflies (McPeek 1998). These behavioral differences are also associated with morphological and physiological differences among damselfly species. Damselfly larvae use their caudal lamellae to generate thrust while swimming (fig. 2.9), and species that inhabit lakes with dragonflies characteristically have larger caudal lamellae than species that inhabit lakes with fish. These species also have higher levels of arginine kinase per unit of tissue, which replenishes the pool of ATP to muscle tissues during swimming (McPeek 1999).

Phylogenetic analysis of diversification in mitochondrial genes has suggested that the distribution of *Enallagma* damselflies among lakes resulted primarily from two radiations (fig. 2.10), a relatively old diversification in the southeastern United States and a more recent one in New England (McPeek and Brown 2000). *Enallagma* originated in lakes with fish, and crypsis was the

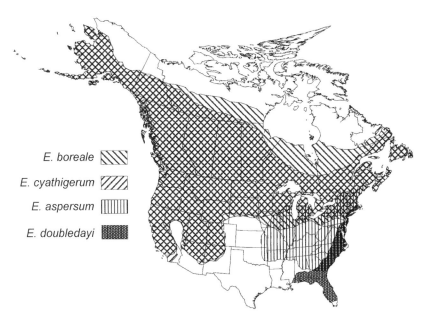

Fig. 2.8   Geographic distribution of four *Enallagma* damselfly species that inhabit lakes in which dragonflies are the top predators. After McPeek and Brown 2000.

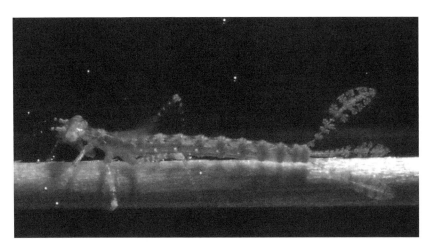

Fig. 2.9   Larval form of the damselfly *Enallagma vesperum* showing the caudal lamellae. This species occurs in lakes in which fish are the top predators. Photograph courtesy of Mark A. McPeek.

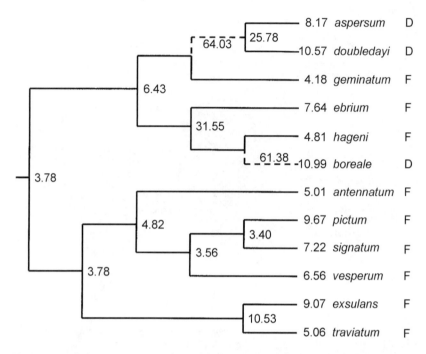

Fig. 2.10 Evolutionary contrast analysis of the lateral area of the caudal lamella of final in-star *Enallagma* damselfly larvae. The mean lateral area for each species is shown at the tip of each branch next to the name of the species. D indicates that the species inhabits lakes with dragonfly predators; F indicates that it inhabits lakes with fish predators. Standard evolution-ary contrast values are shown for each branch. The broken lines highlight two hypothesized independent shifts to dragonfly lakes. After McPeek and Brown 2000.

ancestral defense mechanism for predator avoidance. Shifts to lakes with dragonflies have occurred twice, each time within only one of the two pri-mary clades of *Enallagma* (McPeek and Brown 2000). Hence, the current dis-tribution of damselflies among lakes results from a combination of two dif-ferent periods of diversification and differential colonization of lakes by one of the subclades, followed by additional local adaptation within the bounds of the fundamental niche conservatism found within each clade. The results for *Enallagma* damselflies therefore illustrate the combined phylogenetic and geographic background that commonly forms the raw material for the geo-graphic mosaic of coevolution.

OPPORTUNITY WITHIN CONSERVATISM

Phylogeny, however, does not impose a straitjacket on the evolutionary structure and dynamics of interactions, which is why the coevolution of species rarely shows evidence of unflagging parallel speciation of interacting taxa. Genetic correlations and developmental processes make it easier for natural selection to change traits in some ways than in others, but artificial selection experiments have shown repeatedly that new mutations coupled with intense selection can shift the traits of populations in novel ways (Beldade and Brakefield 2002; Beldade, Koops, and Brakefield 2002). Coevolution is inherently a genetic and ecological process that mixes phylogenetic conservatism with new opportunity across complex landscapes. Moreover, some forms of interaction (e.g., dispersal mutualisms between frugivores and fleshy fruits) inherently coevolve toward interspecific networks of phylogenetically unrelated species, creating much opportunity for new interactions within the broader constraints.

The combination of phylogenetic conservatism and new opportunity is apparent in the diversification of prodoxid moths. The family Prodoxidae includes the yucca moths, whose interactions with yuccas have become one of the most commonly cited textbook examples of coevolution. Over the past ninety-five million years, this ancient family has radiated into groups of species that feed on different plant families (fig. 2.11). There is, however, considerable phylogenetic conservatism in the use of plant species by closely related prodoxid species. The genus *Greya,* near the base of the clade, includes a subclade specialized to feeding on the Apiaceae and another subclade on the Saxifragaceae (Thompson 1987b; Davis, Pellmyr, and Thompson 1992; Pellmyr, Leebens-Mack, and Huth 1998; Althoff and Thompson 1999; Nuismer and Thompson 2001). The moths that feed on Apiaceae are restricted to one subfamily within that plant family, and the moths that feed on Saxifragaceae are restricted to a small group of closely related genera within the *Heuchera* group. Hence, there has been much phylogenetic conservatism during diversification.

Nonetheless, the prodoxid phylogeny shows some major shifts in interactions as well. *Lampronia* is a complex genus that has been recorded on four plant families. Even more impressive, the most derived group of genera (*Mesepiola, Prodoxus, Parategeticula,* and *Tegeticula*) has shifted completely from dicotyledonous plants onto monocots (Nolinaceae and Agavaceae). Among the monocot-feeders, yucca moths in the genera *Tegeticula* and

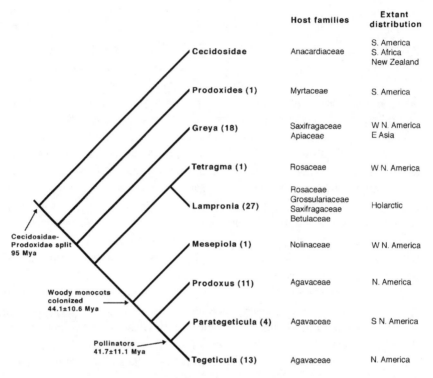

Fig. 2.11   Phylogeny of the moth family Prodoxidae, showing shifts onto different host plant families and the origin of the interaction between yucca moths (*Parategeticula* and *Tegeticula*) and yuccas (Agavaceae). Numbers in parentheses are the number of species within that insect genus or family. From Pellmyr 2003.

Parategeticula have diversified in North America into a diverse group of species that are restricted to the plant genus *Yucca* and are the sole pollinators of their host plants. Most of these moths actively pollinate the yucca flowers into which they lay their eggs, collecting pollen from flowers and carrying it in specialized tentacles that are highly derived components of the proboscis (Pellmyr and Krenn 2002).

Many of the morphological and behavioral characteristics of yucca moths and their interactions with yuccas are not unique, further reinforcing the observation of phylogenetic conservatism in interspecific interactions. The entire family is composed of species that feed as internal parasites of angiosperms, and many species throughout the family oviposit into flowers, including *Greya* and *Tetragma*, near the base of the clade. Consequently, whole suites of traits have been transported wholesale as the moths have occasionally colonized new plant genera or families. Conservatism provides a structure to the diversification of these interspecific interactions and subsequent coevolution with new taxa.

How the combination of phylogenetic conservatism and ecological opportunity shapes the structure of interaction webs remains one of the least understood aspects of community ecology and assembly, despite a long tradition of studies in historical biogeography, historical ecology, and evolutionary ecology. Part of the problem has been that, until recently, there were too few robust phylogenies of co-occurring taxa to evaluate how patterns of phylogenetic diversification shape the community structure and organization of regional and worldwide biotas. Similarly, there were few statistical approaches that allowed clear predictions against null models of community assembly. Those methodological constraints, however, are disappearing, allowing more explicit links between phylogeny and community organization (Ricklefs and Schluter 1993; Futuyma and Mitter 1996; Losos 1996; Thompson 1997; McPeek and Brown 2000; Tofts and Silvertown 2000; Webb 2000; Silvertown, Dodd, and Gowing 2001; Webb et al. 2002; Cavender-Bares and Wilczek 2003; Losos et al. 2003; Gillespie 2004). These approaches are beginning to make their way into the mainstream of community ecology. They are enhancing our understanding of the phylogenetic structure of local communities and the network structure of interspecific interactions by showing why particular subsets of phylogenetic lineages may occur in some regions but not in others.

Conclusions

Most species that have been studied in detail show a strong geographic structure to their divergence across landscapes. That structure allows for some geographic continuity in interspecific interactions as local populations of interacting species coevolve over time. Continuity in interspecific interactions is maintained over longer periods of time by the phylogenetic conservatism of species lineages. Descendant species within lineages often interact with the same genera or families as their immediate ancestors. The combined phylogeographic and phylogenetic conservatism of taxa holds interactions together in a way that allows the possibility of long-term coevolution in many interspecific interactions. To be sure, novel interactions are also always forming worldwide, and that fuels the dynamic structure of coevolution. But the novelty is built upon a backbone of species and interactions that show geographic and phylogenetic conservatism.

3   Raw Materials for Coevolution II
Ecological Structure and Distributed Outcomes

Species are not only phylogeographically and phylogenetically conservative in their interspecific interactions. They are ecologically specialized as well. Each population encounters only a tiny fraction of the earth's biodiversity. At the extreme, coevolution produces intricately specialized and mutually dependent interactions such as those between figs and fig wasps (Bronstein and Hossaert-McKey 1996; Herre 1996) and yuccas and yucca moths (Pellmyr, Leebens-Mack, and Huth 1996; Addicott 1998; Pellmyr and Krenn 2002), which are often viewed as showy exceptions within the overall structure of biodiversity. They are exceptions, however, only in that these species show obligate reciprocal specialization throughout their geographic ranges and are parts of lineages that have repeated the same coevolutionary theme multiple times. At the population level, where evolution meets ecology, most other interactions are much more ecologically variable. Interactions fit within broader networks that vary in ecological structure and outcome across landscapes, permitting coevolution only within a subset of local communities.

Variation in the number of interacting species and ecological outcome therefore provides further raw material for the geographic mosaic of coevolution. This chapter explores how that variation is structured in ways that provide the opportunity for the evolution of geographic variation in natural selection on interspecific interactions.

Most Local Populations Specialize their Interactions on a Few Other Species

As coevolutionary biology has developed, perceptions of the importance of coevolution have been shaped partially by changing views on how many species are involved in coevolving interactions. If local populations interact with a few species in ways that affect Darwinian fitness, then it is easy to understand how coevolution could shape an interaction. The challenge is to understand how coevolution proceeds when local populations interact with a broad range of other species, and many of them exert selection on an interaction.

Most empirical and theoretical studies of coevolving interactions over the past half century have concentrated on interactions between pairs of species. Although that is still the mainstay of coevolutionary studies, almost all detailed studies of coevolution now involve some evaluation of how pairwise interactions coevolve within a broader community context of multispecific interactions. As we learn more about the structure both of local interaction networks and of selection within those networks, the results are suggesting that local ecological specialization provides plenty of opportunity for coevolution between pairs of species or networks of species.

The proportion of extreme specialists and generalists in species interactions differs among lineages and habitats, but there are many more specialists than generalists in any real sense of those words. Until recently, that has not been the prevailing view in ecological approaches to coevolution. That view, however, is changing. We now understand that many species appear to be generalists only when viewed at the species-wide rather than at the population level. A long list of known food items for a predator species is not the same as a population-level list of prey species that are important as selective agents on a local predator population. In addition, ecological research is now overcoming its historical bias toward studying relatively large, long-lived organisms such as vertebrates and perennial plants. Most species are very small and short-lived, and most are specialized symbionts on other species: phytophagous insects and nematodes, parasitoids, the myriad of tiny marine taxa, and the wide range of fungal and bacterial pathogens and mutualistic symbionts that interact with eukaryotic organisms.

There are, of course, species that routinely interact with many other species, and some forms of coevolution directly favor the evolution of multispecific networks. The lifestyles of predation, grazing (eating parts of many victims without killing them), and mutualism among free-living species generally favor coevolution of multispecific networks. Even then, the number of species in an interaction at the local level is rarely large. In fact, highly generalist species within lineages are often turning out to be collections of sibling species or genetically differentiated populations that differ in the species with which they interact most frequently (Thompson 1994).

The search for the processes that produce differences in specialization within lineages remains one of the most important problems in biology, because it is fundamental to our understanding of how biodiversity is organized locally and globally and of how species respond to environmental change. We do not yet have systematic estimates of the mean number of other species with which a local population interacts. If restricted, however, to the species likely to impose natural selection on local populations, the number is often likely to be in the tens rather than the hundreds. The number will generally be higher for large, long-lived organisms, such as many vertebrates, which have a diverse complex of gut microorganisms, and large angiosperms, which can harbor dozens of fungal endophytes in their leaves and an equally complex mix of fungi and bacteria on their roots. But these taxa constitute only a small proportion of extant species.

Even if a species interacts on average with hundreds of other species, a local population will interact with only a subset of that number. An even smaller subset of those species will impose strong selection on the population. The few available estimates of how often individuals within a population must confront multiple enemies indicate that most local interactions are not highly diffuse. For example, during a ten-year study of a goldenrod species (*Solidago altissima*), most plants surviving for longer than five years were attacked at least once by seventeen *Solidago*-feeding insect herbivores found within the local communities, and those herbivores were partitioned among different plant parts (Maddox and Root 1990). A similar ten-year study of the pattern of attack on a local population of the long-lived herb *Lomatium dissectum* (Apiaceae) found that the population was attacked by five major herbivore species and a *Puccinia* rust species (fig. 3.1). Individual plants were attacked each year on average by 1.6 enemies, and all plants surviving throughout the ten years had been attacked by at least two enemies (Thompson 1998a) (fig. 3.2). These numbers are undoubtedly underestimates, because they exclude most underground enemies, pathogens that do not produce obvious external symptoms, and generalist herbivores that may attack individuals rarely, at time scales beyond the study. Also, this same plant species must simultaneously contend with competitors, pollinators, and other mutualists. Nonetheless, the numbers suggest that the network of interactions that potentially impose selection on this plant population is small rather than large, especially since not all these species are likely to impose conflicting selection pressures on the plant population.

Which enemy species is most important may also vary with the age and size of individuals, allowing natural selection to turn genes on and off during development and creating the potential for ontogenetically structured coevolution with multiple taxa. Currently, we do not have any analyses of how

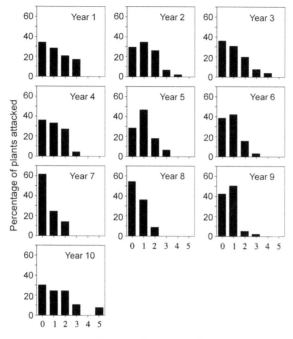

Fig. 3.1   Percentage of *Lomatium dissectum* plants attacked by 0–5 herbivore and pathogen species during each year of a ten-year study. All the species attack either leaves or flowers. Additional species attack the roots. After Thompson 1998a.

ontogenetic shifts in species interactions shape the overall coevolution of a species and its interactions with other species. The accumulating evidence, however, suggests that species can genetically partition their coevolution with multiple taxa. The simple fact that butterflies change from plant-chewing larvae to nectar-sucking adults points to the ability of natural selection to partition specialization to other species into different stages or ages of life histories.

The Outcomes of Interspecific Interactions Differ within and among Communities

DISTRIBUTED OUTCOMES

Each interaction between two or more species has a potential fitness consequence for the participants through its ecological outcome. Just as genetic

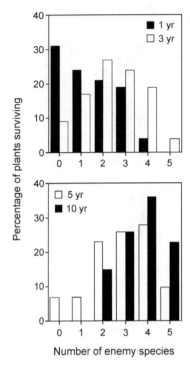

Fig. 3.2   The number of herbivore and pathogen species attacking *Lomatium dissectum* plants that survived one, three, five, or ten years. Values are the percentage of plants that remained unattacked (0 enemy species) or were attacked at least once by 1–5 enemy species over one, three, five, or ten years. After Thompson 1998a.

variation among individuals is the raw material for evolution, variation in the ecological outcome of interactions is a major part of the raw material of coevolution (Thompson 1988b, 1999d). The ecological importance of coevolution is not in the details of which traits are countered by which other traits, but rather in the ways in which coevolution alters the ecological outcomes of interactions and the structure of biodiversity through these coevolving traits. Much of what drives the ongoing coevolutionary process is therefore the continued alteration of the distribution of ecological outcomes (i.e., distributed outcomes) in interactions between species (fig. 3.3).

Within populations, individuals differ in genetic composition, age, size, group structure, and positions along physical gradients, all of which affect the outcomes of interactions with other species (Thompson 1988b; Travis 1996; Thompson 1999d; Harley 2003). That, in turn, may favor enemies or mutualists that preferentially interact with individuals possessing particular traits or combinations of traits. Studies of insects and other invertebrate herbivores, for instance, have shown time and again the remarkable ability of

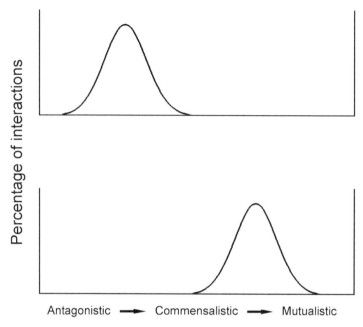

Fig. 3.3   Distribution of ecological outcomes in an interspecific interaction. In this hypothetical example, the interaction can vary in outcome from antagonistic to mutualistic. The
relationship between the two panels can be read in several ways: the distributed outcomes of
an interaction in different interacting populations of a pair of species; change in the distributed outcomes over time through natural selection; or the distributed outcomes of the same
interacting populations placed in different environments.

many species to choose preferentially among plant individuals and plant
parts that differ sometimes only slightly in traits (Brody and Waser 1995; Linhart and Thompson 1995; Price and Ohgushi 1995; Abrahamson and Weis
1997; Craig et al. 2000). Through natural selection on both participants,
those preferences translate in future generations into further modification of
the distributed outcomes within populations.

In some cases, differences in ecological outcome may occur among individuals growing in different environments only meters apart. In the marine
intertidal zone along the coast of Washington State, *Halichrondria panicea*
sponges co-occur with the erect coralline alga *Corallina vancouveriensis* and
benefit from the association through protection from physical desiccation.
This commensalistic outcome, however, occurs only at one level of the inter-

tidal zone. At lower intertidal levels, the algae have no effect on the sponges. In addition, the commensalistic effect on the sponges depends upon the presence of the herbivorous chitons. If the chitons are removed, then the coralline and other algae exclude the sponges and prevent them from growing into a dense canopy. Hence, within a space of a few meters within a community the outcome of this sponge/alga interaction can vary between commensalism and antagonism, thereby creating complex distributed outcomes that may vary among communities (Palumbi 1985).

In other cases, local variation in ecological outcome may arise from genetic differences among individuals that, in turn, have ripple effects on other interactions. The important long-term analyses of pinyon pine (*Pinus edulis*) at Sunset Crater in Arizona provide the most complete study ever undertaken to analyze these ripple effects (Gehring and Whitham 1991, 1994; Brown et al. 2001). The volcanic ash and cinder soils of Sunset Crater contrast with the sandy loams of the neighboring woodlands. Pinyon pines grow on both soils, but they are subject to very different abiotic and biotic conditions. Growth rates are lower and death rates during drought are much higher on cinder soils. Most plants on cinder soils also suffer much attack by stemboring moths (*Dioryctria albovittella*) that destroy the terminal shoots and turn these plants into shrubs. A few pines on cinder soils, however, are resistant to the moths, and they retain their treelike architecture. The resulting differences in architecture are due directly to attack by the moths, as has been shown in studies that have experimentally prevented attack on susceptible plants (fig. 3.4).

The ripple effects of moth attack on other interactions with pinyon moths are unexpectedly wide. Susceptible trees produce fewer cones than resistant trees, and the harvest rates by birds are greater on old, resistant trees than on susceptible trees. Susceptible trees also show a reduced coarse root structure and reduced colonization of fine roots by mycorrhizal fungi. When moths are removed, ectomycorrhizal numbers increase to levels comparable to those in resistant trees (Gehring and Whitham 1995). There are yet other direct and indirect interactions among the species that interact with pinyon pine (Gehring, Cobb, and Whitham 1997), and the overall result is a set of distributed outcomes in the ecological interactions between pinyon pine and its enemies and mutualists on Sunset Crater. Those outcomes fall into two classes, based upon resistance to the bud moths, with additional variation resulting from tree age and the indirect interactions between the tree's enemies and mutualists.

Moth Susceptible   Moth Resistant

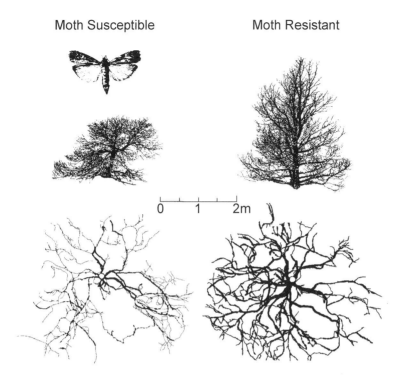

Fig. 3.4   Aboveground and belowground architecture of pinyon pines (*Pinus edulis*) on cinder soils at Sunset Crater, Arizona. Most trees are susceptible to attack by stem-boring moths (*Dioryctria albovittella*) that destroy the terminal buds (top), resulting in a reduced underground root structure (below). From Brown et al. 2001.

INTERACTION NORMS

The distribution of outcomes can differ among populations through differences in the genetic structures of populations, the physical or community context in which an interaction takes place, or the demographic structure of the interacting populations. Even two populations with identical population-genetic structures can differ in the distribution of ecological outcome among environments (see caption of fig. 3.3). Most genotypes differ phenotypically among environments, due to differences in enzyme activity, nutritional levels, and the switching on and off of various genes. The result is a reaction norm, which reflects the range of phenotypes produced by a genotype. Reaction norms extend to interspecific interactions as "interaction norms"

(Thompson 1986c, 1988b). The outcomes of an interaction between two genotypes will differ among environments through environmentally induced differences in physiologies and life histories of the interacting species and the demographic and community context in which those genotypes occur. Also, in some species a genotype will produce a particular phenotype only in environments in which another particular species is present, as occurs in some *Daphnia* species that develop defensive morphologies only in environments in which their predators occur in significant numbers (Agrawal 2001).

As a result of interaction norms, natural selection on a coevolving interaction between two or more species is likely to differ to some extent across landscapes even without major genetic differences among populations. In the example of pinyon pines, susceptible trees growing on sandy-loam soil adjacent to Sunset Crater do not suffer the high level of moth attack found on trees on the crater's cinder soils. Supplementing the cinder-soil susceptible trees with water and nutrients increases resin production in the trees and decreases attack by moths (Cobb et al. 1997; Brown et al. 2001). Pinyon pines and bud moths therefore show an interaction norm that is mediated by water and nutrient availability.

Only a few other studies have begun to analyze in detail the structure of interaction norms or distributed outcomes of interacting species within natural populations. Recent analyses of interactions between grasses and endophytic fungi, however, are an exemplar and have been designed specifically to analyze the pattern of variation in outcome among interacting genotypes in different environments. All plants appear to harbor at least one endophytic fungal species, and some woody plants may harbor hundreds (Saikkonen et al. 1998; Arnold et al. 2000). Much of the work on ecological outcome in these interactions has focused on interactions between grasses in the family Poaceae and fungi in the family Clavicipitaceae, which appears to be a monophyletic group of species that has diversified along with these grasses (Clay and Schardl 2002). These interactions can alter not only the population structure and dynamics of these species but also local patterns of species diversity within communities (Clay and Holah 1999; Rudgers, Koslow, and Clay 2004).

The interactions between the grasses and these endophytes have generally been considered to be mutualistic, because studies of some plant species have shown that endophytic infection can increase resistance to herbivores and pathogens through production of novel defenses such as alkaloids, improve competitive ability through increased drought tolerance, and enhance uptake of nutrients (Bush, Wilkinson, and Schardl 1997; Malinowski and Belesky 1999; Clay and Schardl 2002). Moreover, some of these fungi are vertically transmitted only through their host's seeds as asexual propagules (Clay and Schardl 2002); such conditions are thought often to favor the evolution of reduced antagonism with hosts and, potentially, mutualism (Law 1985; Ewald 1994; Frank 1994a).

Three kinds of evidence, however, suggest that these interactions remain evolutionarily complex in outcome. First, the percentage of plant individuals in a population harboring the endophytes varies among species. In some plant populations, infection levels are close to 100 percent, and that result has been reported for a wide number of species studied (Clay 1998; Clay and Schardl 2002). One survey of over five hundred perennial ryegrass populations in Europe, however, reported infection in only about two-thirds of the populations, and fewer than half the plants were infected in most populations that contained the endophytes (Lewis et al. 1997). In some cases, low infection rates may result from taking point samples of populations not yet under equilibrium for local levels of infection, but there is some evidence that variation in infection rates may be common in some of these interactions (Schulthess and Faeth 1998; Saikkonen et al. 1999).

Second, the potential benefits of the interaction may vary among plant and fungus species and populations. Although endophyte-infected individuals of some grass species have higher resistance against herbivores and pathogens, that benefit does not hold for all plant species that have been tested (Saikkonen et al. 1999; Tibbets and Faeth 1999). In fact, Faeth (2002) has argued that mutualism may be the exception rather than the rule among populations of endophyte-infected species, partly because the outcome of the interaction can vary widely with plant genotype and the growth conditions in which the interaction occurs.

The clearest evidence of this variation comes from experiments on Arizona fescue (*Festuca arizonica*) (Faeth and Fagan 2002; Faeth and Sullivan 2003). This grass occurs in semiarid ponderosa pine communities in the southwestern United States and is commonly infected with a species of *Neotyphodium*. The endophyte is transmitted asexually through its host's seeds. When plants were cloned in the greenhouse, grown for a year, and then put into a field setting either as infected or uninfected plants, they differed among genotypes in how endophyte infection affects plant growth and seed production (fig. 3.5). The experiment allowed attack by small herbivores but excluded the effects of large herbivores or competitors. Over the two years

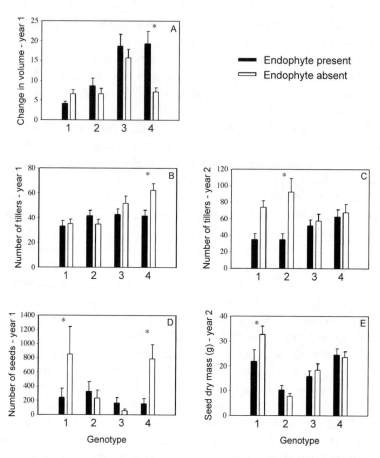

Fig. 3.5   Effect of infection by the endophyte *Neotyphodium* sp. on growth in four genotypes of the grass Arizona fescue (*Festuca arizonica*) during two years. Change in plant volume at the end of year 1 (panel A) was calculated as multiples of initial volume. The asterisk (*) indicates significant differences. After Faeth and Sullivan 2003.

subsequent to transplantation into the field, most of the four tested genotypes had lower growth or seed production in the field when infected than when uninfected (Faeth and Sullivan 2003). Each plant genotype, however, responded in a unique way to infection, and the effects of the infection on plant growth varied with water and nutrient availability (Faeth and Fagan 2002). Hence, the overall ecological outcome of this interaction between a native grass and its fungal endophyte depends upon plant genotype, water and nutrient availability, and patterns of natural herbivory. Other factors such as endophyte genotype, interspecific competition, and herbivory by

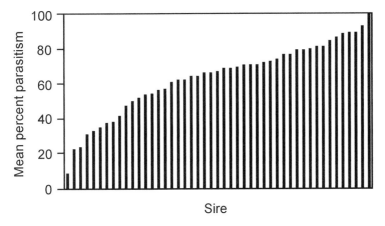

Fig. 3.6   Mean percentage of aphids (N = 15 aphids) parasitized by a single parasitoid wasp during a twenty-four-hour trial, shown in rank order. The aphids were from a clone known to show some resistance to parasitoid attack. Each bar is the mean of all female progeny of one male (sire) mated to two to five females, resulting in forty-seven paternal half-sib families. After Henter 1995.

large mammals could add to the complexity of ecological outcomes within and among natural communities in this interaction.

ANALYZING VARIATION IN ECOLOGICAL OUTCOMES

Few studies have tested the outcome of an interaction among enough families to get an overall sense of the distribution of outcomes within a population for at least one of the species. Quantitative genetic analysis of introduced pea aphids (*Acyrthosiphon pisum*) and the parasitoid wasp *Aphidius ervi* provide the clearest data currently available. Pea aphids were first observed in North America in the 1870s and are major pests of clover, alfalfa, and some other legumes. The parasitoid *A. ervi* was introduced as a biological control agent in the late 1950s and 1960s and is now the dominant pea aphid parasitoid in temperate North America (Henter 1995).

Parasitoids collected from alfalfa in Tompkins County, New York, show a tremendous range of variation in their ability to attack one pea aphid clone known to exhibit resistance to attack by this parasitoid (fig. 3.6). This continuum of variation is expressed among paternal half-sib families, which indicates that there is additive genetic variance for the ability of these parasitoids to overcome resistance in their hosts. Narrow-sense heritability estimated from these experiments is 26 percent. Because heritability estimates can

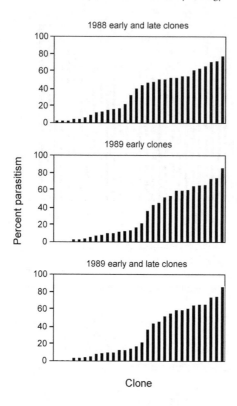

Fig. 3.7   Susceptibility of pea aphid clones to parasitism by the wasp *Aphidius ervi*. Clones were collected from an alfalfa field, propagated in the laboratory to obtain thirty aphid nymphs per clone, and then challenged for twenty-four hours with a single parasitoid from a laboratory colony derived from the same alfalfa field. Values are the means of four to five replicates for each clone. The three panels show the results for three sets of collections. After Henter and Via 1995.

vary among environments, these laboratory estimates may be higher or lower than the actual levels of heritability found within field populations. They suggest, however, a potentially wide range of genetically determined variation in these parasitoids to overcome resistance deployed by their aphid hosts.

This population of pea aphids shows a similarly wide variation in the ability to resist parasitoid attack. Some pea aphid clones collected from one of the alfalfa fields are almost completely susceptible to parasitism when challenged with parasitoids from a genetically variable laboratory colony derived from the alfalfa-field population. Other clones are highly resistant, and the resistance within the population varies continuously between these extremes. The same overall distributions of outcome have been obtained among aphid clones collected from the field in different years and different months, propagated in the laboratory, and then challenged with the wasps from the laboratory colony (fig. 3.7).

Even though these aphid colonies show similar distributions of outcome among years and seasons in this study, it is likely that the distributions vary

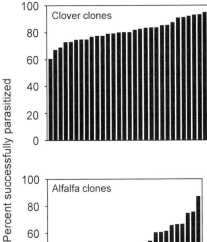

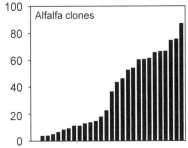

Pea aphid clones, rank order

Fig. 3.8   Percentage of pea aphid clones successfully parasitized by the parasitoid wasp *Aphidius ervi*. The wasps were from a mixed colony composed of parasitoids from aphids on clover and alfalfa. All values are means of four replicates per clone. Clones on clover did not differ significantly, but clones on alfalfa did differ significantly. After Hufbauer and Via 1999, based on data from Hufauer and Via 1999 for clover and Henter and Via 1995 for alfalfa.

over time through the effects of local selection, random genetic drift, and gene flow. In fact, a later study of pea aphid resistance to the same parasitoids on clover in Tompkins County found very different patterns of genetic variation for resistance (Hufbauer and Via 1999). Although it is not possible to separate the effects of space and time in these studies, the results suggest that the distributions of resistance in these aphids can differ greatly over short distances or short spans of time.

Pea aphids and these parasitoids have been evolving in upstate New York for less than fifty years, and there appears to be strong spatial structure in the distributed outcomes wherever these interactions occur. Studies of additional clover fields in New York and Maryland have indicated that parasitism rates of aphids on clover are generally higher than for aphids on alfalfa (Hufbauer 2001) (fig. 3.8). Despite these differences among crops, however, there is no evidence that the parasitoids have developed differential adaptation to the clover-feeding aphid clones and the aphid-feeding aphid clones. The habitats are agricultural monocultures that differ in spatial distribution from year to year, potentially swamping the potential for longer-term local adaptation. Nevertheless, these shifting agricultural landscapes create a mosaic of parasitism levels on pea aphid populations in eastern North America.

The studies of pea aphids and parasitoids, together with the studies of Arizona fescue and its endophytes and related studies of other taxa (e.g., Ebert, Zschokke-Rohringer, and Carius 1998), emphasize that "mean" outcomes mask the distributed outcomes that characterize most interactions. As species spread across landscapes, the mean and distribution of ecological outcomes in most interspecific interactions vary among populations. This variation has been given specific names for some forms of interaction—for example, conditional mutualism (Cushman and Beattie 1991; Bronstein 1994a)—but the importance is the same for all forms of interaction. Distributed outcomes and interaction norms are an integral part of the raw material for the geographic mosaic of coevolution.

Natural selection can shift distributed outcomes by favoring individuals that interact only with particular genotypes in the other species. It can also shift distributions by favoring individuals that restrict their interactions to particular ages, sizes, or social compositions in the other species. A highly harmful parasite that attacks pre-reproductive hosts can impose strong selection on the evolution of defenses in the host population. The next round of natural selection on the parasite population could favor parasites that attack only the genetically least defended pre-reproductive individuals. It could just as well, however, favor parasites that attack late-reproductive or post-reproductive individuals that have lessened defenses due to senescence. If natural selection favors parasites that attack post-reproductive individuals, the interaction itself could begin to shift toward commensalism. Attack on post-reproductive hosts will generally have fewer consequences for selection on host populations, unless post-reproductive individuals are crucial to the success of kin groups. In addition, selection could favor symbionts, whether parasitic or mutualistic, that infect hosts only in particular environments. The possibilities for creating differences among populations in the mean and distribution of outcomes are almost endless, leading almost inevitably to a geographic mosaic structure in any interspecific interaction that persists over long periods of time.

Conclusions

This and the previous chapter suggest that the raw materials for coevolution consist of four fundamental attributes of the biology of species. The most

fundamental is the fact that most species are collections of genetically differentiated populations, making each population a small evolutionary experiment (Wade and Goodnight 1998) and each local interspecific interaction a
potential coevolutionary experiment. Replicated many times across landscapes, natural selection on these experiments has the potential to "explore"
multiple coevolutionary options. Many of these experiments undoubtedly
result in the local extinction of one or all the participating species, but other
coevolving populations of the same species persist. In addition, interactions
are held together over the long term partly because most species are phylogenetically conservative in their interactions with other species, and they
probably interact with only a small group of species that have the potential to
affect the fitness of individuals. As those interactions persist, interaction
norms and distributed outcomes continue to create new possibilities for further coevolutionary change. All that is required for the next step in the geographic mosaic of coevolution is for natural selection to take the raw materials and create varying patterns of local coadaptation.

4   Local Adaptation I

Geographic Selection Mosaics

The raw materials for coevolution—genetically differentiated populations, phylogenetic conservatism, ecological specialization, and variable outcomes—provide the opportunity for natural selection to continually reshape
interspecific interactions across geographic landscapes. In fact, geographic
variation in natural selection must occur almost inevitably in any widespread
interaction. Different mutations arise in different populations, creating population differences in the genes under selection. Even without new mutations, two interacting species exhibiting the same initial distributions of
genotypes in different environments could coevolve in different ways. The
reason is that the way in which the fitness of a genotype of one species depends upon the distribution of genotypes in another species could vary with
community context or among physical environments. This environment-
dependent effect of how the distribution of genotypes of one species affects the
fitness of a genotype in another species creates a geographic selection mosaic.

Coevolution is therefore an inherently hierarchical process. It begins with
the coevolutionary dynamics of locally interacting populations, gathers
geographic complexity as interacting species affect each other's fitnesses in
different ways in different environments, and partitions some of that complexity more permanently over time as populations diversify into separate
species. Understanding coevolution as a major process organizing biodiversity therefore requires a conceptual framework that takes into account the hierarchical genetic and ecological structure of interacting populations, species, and lineages. Ongoing local coevolutionary adaptation is the starting
point for the dynamics, because it creates the basic template for the geographic mosaic of coevolution.

This chapter and the next explore the structure and dynamics of local
adaptation as an essential building block of the geographic mosaic of coevolution. I begin by evaluating studies that have demonstrated local adaptation

in interspecific interactions. These studies illustrate the expanding range of approaches to the study of local adaptation and the multiple geographic scales at which local adaptation has now been observed. In the next chapter I then discuss the speed at which local adaptation can occur and the major outcomes of local coadaptation that develop between pairs of species.

Geographic Variation in Outcomes and Traits

Almost all widespread species that have been studied in detail for their adaptations to other species show geographic differences in ecological outcomes or traits linked to interactions (Thompson 1988b, 1994; Travis 1996; Brodie, Ridenhour, and Brodie 2002; Burdon, Thrall, and Lawrence 2002; Benkman 2003; Zangerl and Berenbaum 2003). This variation sets the stage for the geographic mosaic of coevolution. It is only one part of the process, but it is a crucial part. The geographic pattern of defense and counterdefense in *Drosophila melanogaster* and its parasitoids illustrates the mosaic structure, which is becoming evident as more interactions are studied in multiple localities. *Drosophila melanogaster* shows strong differences among European populations in the ability to thwart attack by the braconid wasp *Asobara tabida* (Kraaijeveld and Godfray 1999) (fig. 4.1). Populations of the parasitoid occur throughout most of Europe. Adults inject their eggs into *Drosophila* larvae, which then defend themselves by encapsulating the parasitoid eggs with hemocytes that fuse and harden (Gillespie, Kanost, and Trenczek 1997). If encapsulation is successful, the parasitoid eggs do not hatch. When challenged with a test strain of the parasitoid, populations of *D. melanogaster* in France, northern Italy, and a few other regions show relatively high levels of defense, whereas other populations show lower levels.

Similarly, populations of the parasitoid differ geographically in their ability to resist encapsulation when challenged with a test strain of *D. melanogaster* (fig. 4.2). Northern populations of *A. tabida* commonly attack *D. subobscura* rather than *D. melanogaster,* which may at least partially explain the lower ability of these populations to resist encapsulation by *D. melanogaster* (Dupas, Carton, and Poiriè 2003). Nevertheless, there is no simple geographic pattern of matching levels of defense and counterdefense. Local host populations tested against sympatric and allopatric parasitoids, rather than against a test strain, yield some differences in these geographic patterns, but the reference strains appear to capture the broad geographic picture of defense and counterdefense (Kraaijeveld and Godfray 2001). The reference

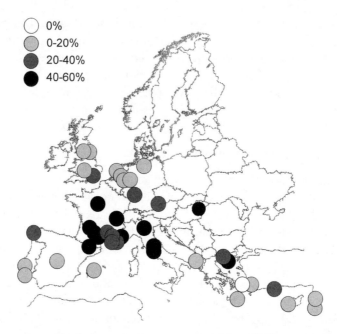

Fig. 4.1   Geographic differences in the ability of *Drosophila melanogaster* populations to encapsulate eggs of a test strain of the braconid wasp parasitoid *Asobara tabida*. Values are the percentage of parasitoid eggs encapsulated when attacked by a test strain of the parasitoid. After Kraaijeveld and Godfray 1999.

strains may act as effective surrogates to local populations in these experiments, because resistance to this parasitoid appears to be under simple genetic control. Differences in resistance to *A. tabida* among isofemale lines and among at least some *D. melanogaster* populations can be explained by allelic differences at a single locus, with resistance dominant to susceptibility (Orr and Irving 1997; Benassi, Frey, and Carton 1998). The overall results from these challenge experiments suggest major populational differences in the ability of the hosts and parasitoids to interact with particular genotypes of the other species.

Another parasitoid, *Leptopilina boulardi*, attacks *D. melanogaster* within Europe only along the Mediterranean. Resistance of *D. melanogaster* to this parasitoid shows a geographic pattern that does not simply reflect high attack along the Mediterranean and low attack elsewhere (fig. 4.3). Instead, fly populations show a patchwork distribution of defense when challenged with a test strain of the parasitoid. The causes of this pattern of geographic variation

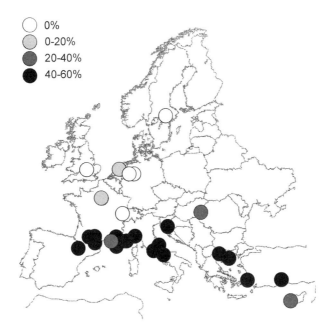

Fig. 4.2   Geographic differences in the ability of a braconid parasitoid, *Asobara tabida*, to resist encapsulation by its host, *Drosophila melanogaster*, when challenged with a test strain of the host. The values are the percentage of eggs that escape encapsulation by the test strain of the host. After Kraaijeveld and Godfray 1999.

are still being evaluated. Since *D. melanogaster* is attacked by several parasitoid species throughout Europe, adaptation to one parasitoid species could potentially affect adaptation to other species. The geographic pattern of cross-resistance to these two parasitoids, however, shows no correlation across Europe (Kraaijeveld and Godfray 1999). Moreover, although selection for defense against *L. boulardi* increases resistance to *A. tabida*, selection for defense against *A. tabida* does not increase resistance to *L. boulardi* (Fellowes, Kraaijeveld, and Godfray 1999; Fellowes and Godfray 2000). Hence, selection on *D. melanogaster* driven by *A. tabida* cannot explain the geographic pattern of resistance to *L. boulardi*.

The geographic mosaic pattern of defense and counterdefense in these two species may result at least partially from the interaction of major genes in these two species (Carton and Nappi 1997; Dupas, Frey, and Carton 1998; Dupas, Carton, and Poiriè 2003). Populations of *D. melanogaster* and *L. boulardi* show strong regional and continental differences in encapsulation and

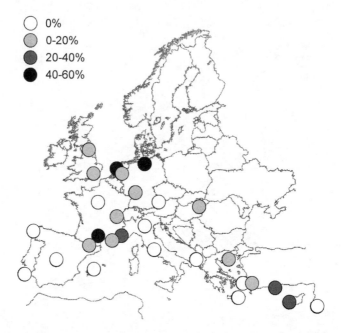

Fig. 4.3   Geographic differences in the ability of *Drosophila melanogaster* populations to encapsulate eggs of a test strain of the hymenopteran parasitoid *Leptopilina boulardi*. Values are the percentage of eggs encapsulated when attacked by a test strain of the parasitoid. After Kraaijeveld and Godfray 1999.

ability to escape encapsulation, respectively (fig. 4.4A). These large-scale geographic differences in outcome reflect geographic differences in the frequencies of resistant flies and of parasitoids able to overcome host resistance (fig. 4.4B, C). Hence, at least some of the mosaic pattern of defense and counterdefense in this interaction may result from broad geographic differences in the frequencies of host and parasitoid alleles controlling the outcome of parasitoid attack.

These studies of *D. melanogaster* and its parasitoids are particularly useful, because they show that interactions can differ greatly in patterns of defense and counterdefense across the same landscapes. In this case, a single host species differs in its geographic pattern of defense against two parasitoid species, and the two parasitoid species differ from each other in their geographic patterns of counterdefense. The resulting mosaic sets the stage for the ongoing geographic dynamics of coevolving species.

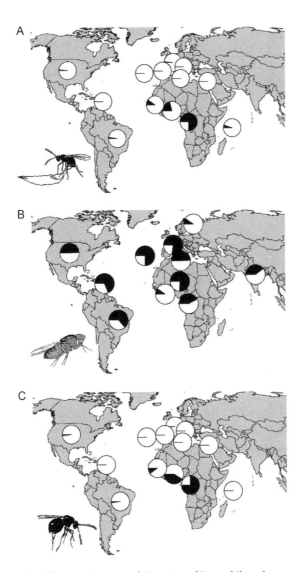

Fig. 4.4 Geographic differences in encapsulation rates of *Drosophila melanogaster* attacked by the parasitoid *Leptopilina boulardi*. The black portions of the pie diagrams show (A) the percentage of parasitoid eggs encapsulated in sympatric populations of the flies and parasitoids, (B) the geographic distribution of a known resistance phenotype in *D. melanogaster*, as estimated by the percentage of flies in each population able to encapsulate a reference strain of the parasitoid, and (C) the geographic distribution of a known virulence phenotype of *L. boulardi*, as estimated by the percentage of instances in which the parasitoids from each population were encapsulated by a reference strain of the flies. After Dupas, Carton, and Poiriè 2003.

Species Often Become Locally Adapted to Particular Populations of Other Species

Studies of an increasingly wide range of species have now provided strong evidence of local adaptation to other species. Plant species commonly differ among populations in many traits, including their defenses against herbivores and the rewards they offer to mutualists (Linhart and Grant 1996). Herbivore populations differ in their counterdefenses to particular plant species and in their degree of specialization to one or a few local plant taxa (Fox, Waddell, and Mousseau 1994; e.g., Singer and Thomas 1996; Berenbaum and Zangerl 1998; Thompson 1998b; de Jong and Nielsen 1999; Nylin and Janz 1999; Janz 2003; Singer 2003). Parasitic animals, fungi, plants, and microorganisms differ geographically in life histories, morphology, and physiology in ways that suggest adaptation to their local host populations (e.g., Webster and Woolhouse 1998; Burdon and Thrall 1999; Koskela, Salonen, and Mutikainen 2000; Lively and Dybdahl 2000). The list continues to grow and encompasses all forms of interspecific interaction.

DEMONSTRATING LOCAL ADAPTATION

Studies demonstrating local adaptation to other species are of several types: comparison of how a local population interacts with sympatric and allopatric populations of other species (e.g., Ebert 1994; Kaltz and Shykoff 1998; Thrall, Burdon, and Bever 2002), comparison of populations where the interaction occurs to those where the interaction does not occur (e.g., Reznick et al. 1997; O'Steen, Cullum, and Bennett 2002), and comparison of how the traits or outcome of an interaction differ among populations in a way consistent with local natural selection (e.g., Benkman, Holimon, and Smith 2001; Brodie, Ridenhour, and Brodie 2002; Sanford et al. 2003; Zangerl and Berenbaum 2003). Not all studies using these protocols have found evidence for local adaptation (Strauss 1997b; Kaltz et al. 1999; Lammi, Siikamaki, and Salonen 1999; Mutikainen et al. 2000; Hufbauer 2001; Lajeunesse and Forbes 2002), and some interactions may, in fact, not show local adaptation. Most studies that fail to demonstrate local adaptation, however, caution against over-extrapolation of the results. Studies often use a single protocol, and each protocol has limitations in interpretation. Because the point of these studies is to assess the geographic structure of the interaction, any single protocol can only corroborate or falsify the particular geographic structure tested by that experiment (e.g., performance of parasites on sympatric and allopatric hosts separated by ten, one hundred, and one thousand kilometers).

Studies that combine populations at geographic distances greater than the scale of local adaptation will mask the pattern of local adaptation. Similar masking can occur if a study is conducted at too small a scale. There may be little relationship between adaptation and geographic distance within a metapopulation, if populations are connected by high levels of gene flow (Thrall, Burdon, and Bever 2002). The interacting species may coevolve within each region, but the coevolutionary adaptation may be evident only at the overall metapopulation level rather than at the local population level. The problem in all analyses of local adaptation is to understand the geographic scale at which local adaptation may occur.

Use of only a single experimental protocol can also mask local adaptation if populations differ in overall investment in the interaction. For example, assessing how a parasite species performs on sympatric as compared with allopatric host populations can fail to provide evidence of local adaptation across all populations if host populations differ greatly from each other in their overall level of defense or in their range of defenses (Thrall, Burdon, and Bever 2002). Through local adaptation, some populations may harbor defenses against a wider range of enemy genotypes than other populations. A population with the wider range of defenses may therefore, by chance, also fare equally well against sympatric and some allopatric populations of the parasite in challenge experiments.

In addition, failure to demonstrate local adaptation can also result from protocols that do not take into account time lags in response to selection in multispecific interactions or ephemeral patterns of apparent maladaptation that accompany fluctuating polymorphisms (Kaltz and Shykoff 1998). Single samples within a locality may capture an interaction at a moment when a new beneficial mutation has appeared in a local population of one species but a beneficial countermutation has yet to appear in the other species. Even in the absence of new mutations, fluctuating selection on genes will create periods of mismatching in coevolving species. Consequently, interpretation of local adaptation requires analyses at multiple spatial and temporal scales.

Our current knowledge of local adaptation in interspecific interactions is highly biased toward terrestrial and freshwater species (Kaltz and Shykoff 1998; Thompson 1998c; Hoekstra et al. 2001), because few marine species or species complexes have been studied in a way that would allow evaluation of the form or scale of local adaptation (Duffy 1996; Dietl 2003; Sanford et al. 2003; Sotka, Wares, and Hay 2003). Studies of intertidal and nearshore marine communities have mostly evaluated adaptation to physical conditions, regional differences in species composition, the ecological structure of interactions, and the ecological "strengths" of interaction between particular species. Ecological interaction strengths, however, cannot be translated directly into information on selection intensities and local adaptation. Some strong ecological interactions may impose only weak selection on populations, because they cause little differential mortality or reproduction among genotypes. Nevertheless, these studies of regional differences in ecological structure coupled with the growing number of studies showing molecular differentiation among marine populations are beginning to foster interest in work on local adaptation in marine environments. For example, populations of the snail (or whelk) *Nucella caniliculata* not only show molecular differentiation among populations along the west coast of North America, they also differ between California and Washington State in their tendency to prey upon the intertidal mussel *Mytilus californianus* (Sanford et al. 2003). Whelks reared from eggs in common environments retain their geographic differences in the tendency to drill into *M. californianus* shells. The great majority of California whelks drill into the offered snails, but less than 10 percent of the Oregon snails and less than 5 percent of the Washington State snails attempt to drill into this species.

MOSAICS OF LOCAL ADAPTATION

Most studies of local adaptation still analyze only one side of an interaction (e.g., parasites but not hosts; prey but not predators), but these studies often provide some evidence for the ways in which local adaptation may contribute to coevolution between species. Even species that we think about as interacting with a wide range of other species are often much more genetically specialized in their interactions at local scales when they are studied in more detail population by population. Geographic differences in adaptation of the Malayan pit viper *Calloselasma rhodostoma* to its prey are a good example (Daltry, Wüster, and Thorpe 1996). Snakes are often considered to be somewhat general predators, even though there has long been evidence of genetic differentiation in food choice and prey-handling abilities among some snake populations (Arnold 1981; Brodie and Brodie 1990). At the species level, the Malaysian pit viper appears to be a highly generalized predator of vertebrates. It is broadly distributed throughout Southeast Asia and has been recorded feeding on amphibians, reptiles, birds, and mammals. Like most pit vipers, it uses venom to subdue and digest its prey. The species-level list of known prey taxa, however, masks the local adaptation to different prey found among Malayan pit viper populations.

Among thirty-six populations analyzed in Vietnam, Thailand, Malaysia, and Java, some populations eat almost exclusively reptiles and others eat mostly birds and mammals (Daltry, Wüster, and Thorpe 1996). Amphibians make up about a quarter of the diet in some populations but are not eaten at all in other populations. Venom varies among the snake populations in ways that broadly match these differences in diet, and the differences in venom production appear to be genetically controlled (Daltry, Wüster, and Thorpe 1996). Hence, this snake species is not a highly generalized predator that randomly attacks vertebrate species. It is a collection of populations each locally adapted to a smaller range of vertebrate prey.

In some cases, mosaics of local adaptation in species interactions result directly from the presence of particular defense or counterdefense genes in host or prey populations in some regions but not in others. Detailed genetic and chemical analyses of the interaction between the cruciferous plant *Barbarea vulgaris* and the flea beetle *Phyllotreta nemorum* have characterized the geographic structure of chemical defense and counterdefense in these species across Denmark. *Barbarea vulgaris* is a common European plant, and a subspecies, *B. vulgaris* ssp. *arcuata,* occurs in Denmark in two forms, called G-type and P-type. G-type plants are morphologically and chemically distinguishable from P-type plants, and they are unsuitable as hosts for most populations of the flea beetle. Usually, the leaf-mining larvae begin to form mines on G-type plants but then stop feeding and die (Nielsen 1996).

Some Danish flea beetles, however, have overcome the chemical defenses of G-type plants and can complete development on those plants (see fig. 4.5). Genetic crossing studies have shown that the ability to feed on G-type plants differs genetically among flea beetle populations, perhaps suggesting independent origins of this counterdefense. Initial genetic crosses of beetles at Ejby suggested that the ability involves at least two sex-linked genes (X and Y) and one or more autosomal genes (Nielsen 1997a, 1997b; de Jong and Nielsen 1999; Nielsen 1999), but recent additional crosses suggested more complicated genetic interactions (de Jong and Nielsen, personal communication). At Kvaerkeby, the ability is conferred by a single autosomal gene, although some additional modifier genes may also contribute to the counterdefenses (de Jong et al. 2000). The autosomal and sex-linked genes confer the ability of larvae to develop on the plant as well as the ability of adult beetles to feed on the plant (Nielsen 1996).

Individuals having the ability to survive on G-type plants occur at low frequency in most but not all localities sampled throughout Denmark (de Jong and Nielsen 1999). At least one population, however, includes large numbers

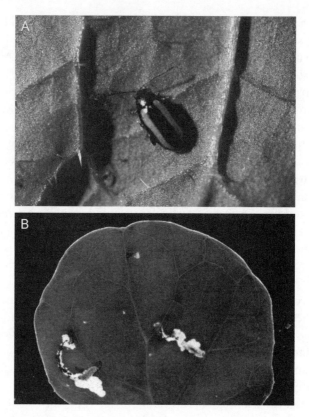

Fig. 4.5   (A) Adult flea beetle *Phyllotreta nemorum,* which consists in Denmark of populations that differ in their ability to survive on the G-type of *Barbarea vulgaris* ssp. *arcuata.*
Photograph from Jens Nielsen, courtesy of G. Brovad. (B) Damage to leaves of G-type *Barbarea vulgaris* by resistant and susceptible larval genotypes. Larvae of resistant genotypes are
seen within the two relatively large areas of the leaf they have damaged. Larvae of the susceptible genotypes died after damaging at most very small areas of leaf, visible as light dots. Photograph courtesy of Jens Nielsen.

of individuals capable of feeding on G-type plants. The selection pressures
driving these differences are still under study, but part of the explanation may
lie in the geographic patterns of host use in these beetles. *Phyllotreta nemorum* is not restricted to *B. vulgaris* as a host, and local populations of the
beetle differ in the combination of crucifer species they use as hosts.

The ability to complete development on G-type *B. vulgaris* is genetically
independent of the ability to feed on other plants (Nielsen 1999). Hence, the
differences found among *P. nemorum* populations are not due simply to genetic correlations in their ability to survive on different local hosts. Instead,

the differences are more likely due to geographically variable selection across Danish landscapes, strongly favoring the ability to survive on G-type *B. vulgaris* in some regions but not in others. Selection for coadapted gene complexes favoring adaptation to local hosts may play a role (de Jong and Nielsen 2002). Geographic differences in selection in this interaction may therefore be driven by differences among regions in the occurrence and frequency of G-type *B. vulgaris* plants, the availability of alternative hosts for the beetles, and the genetic basis of the counterdefenses in the beetles.

Similar geographic differences in local adaptation in Denmark and other parts of Europe are evident in the interactions between lycaenid butterflies in the genus *Maculinea* and the ants whose nests they parasitize. Many lycaenid butterflies form mutualistic relationships with ants (Pierce 1987; Fiedler and Saam 1995; Fiedler 1996; Pierce et al. 2002). In some of these interactions the ants even herd the butterfly larvae, moving them onto host plants to feed for part of the day, milking special glands on the larvae, and then moving them off the plants to places safe from predation during other parts of the day. As the interactions between lycaenids and ants have diversified, some butterfly species have exploited these interactions and become parasites of ant nests. *Maculinea* butterflies are the best-known examples, and the interactions between these species and their ant hosts have been studied intensively throughout Europe, partly because changing agricultural practices have driven many populations close to extinction (Thomas 1995). Five *Maculinea* species exploit the ant genus *Myrmica* in Europe (Thomas et al. 1989), and the relationships are geographically complex. The larvae feed initially on plant tissue, but later instars move inside ant nests, where they eat larval ants. The butterfly larvae probably escape attack within the nests by chemically mimicking their host ants, as has been shown in one *Maculinea* species (Akino et al. 1999). This chemical matching may be a major cause of the high level of host specificity found in these butterflies. For instance, *Maculinea alcon* is a widespread species and, in Denmark, consists of a group of genetically distinct populations that differ in the *Myrmica* species they primarily use as hosts (Als, Nash, and Boomsma 2002). There is no simple clinal pattern to host use across Denmark, but Als, Nash, and Boomsma (2002) suggest that the pattern exhibits a geographic mosaic in local adaptation to *Myrmica* species.

Over even larger geographic scales, evidence of local adaptation has come from combining at least two kinds of analysis: the traits of populations in regions where the interaction occurs as compared with regions where the interaction is absent, and the outcomes of interactions between sympatric and allopatric populations. For instance, colonies of *Leptothorax acervorum* ants in Russia and Germany are attacked by the slave-making ant *Harpagoxenus sublaevis* (Foitzik, Fischer, and Heinze 2003). Host workers in Russian populations of *L. acervorum* sting invading slave makers from allopatric populations but not sympatric populations. In contrast, German populations of *L. acervorum* do not sting either sympatric or allopatric slave makers. German populations post more guards at their nests than do Russian populations, and they bite the invading slave makers. English populations, which fall outside the range of the interaction, do not post guards and sting invading slave makers when challenged under experimental conditions. These results illustrate how different populations may mix traits in different ways, creating a strong geographic mosaic in the interaction.

In addition to large-scale geographic differences in local adaptation of interacting species, some interspecific interactions may evolve local adaptation at very fine scales. Red oak (*Quercus rubra*) subpopulations in different habitats within a four-hectare study area of a Missouri oak and hickory forest differ in resistance to local herbivores, and those differences remain when plants from each subpopulation are transplanted to the other plots in a reciprocal transplant experiment (Sork, Stowe, and Hochwender 1993). At similar scales, non-random gene flow between populations of *Euphydryas editha* butterflies separated by only two hundred meters created subpopulations in the 1980s and 1990s that differed in host plant preference. These subpopulation differences resulted from creation of a novel habitat following logging, which favored non-random colonization of that habitat by insects that differed from individuals in nearby undisturbed habitat in their host preference (Singer and Thomas 1996). Although these kinds of subpopulation differences are unlikely to be stable over long periods of time, the selection mosaics they create provide fuel for ongoing coevolutionary dynamics.

At the absolute extreme of local adaptation, there is now evidence that some parasites become adapted to individual long-lived hosts. The known examples include parasitic taxa as different as bacteria that are pathogenic on animals and insects that feed throughout their entire larval lives on an individual plant (Ebert, Zschokke-Rohringer, and Carius 1998; Mopper and Strauss 1998; Van Zandt and Mopper 1998). The rapid evolution of HIV within individual humans highlights the potential importance of this kind of local adaptation in most interactions between parasites and relatively long-lived hosts. Whenever a parasite clone or population can live through many generations on a single host individual, such extreme local adaptation is always possible. It adds another level to the hierarchical structure of co-evolutionary selection.

FROM LOCAL ADAPTATION TO REGIONAL ADAPTATION

Between the extremes of local adaptation and species-wide adaptation is the middle ground of regional adaptation in interspecific interactions. Because local populations are often connected by gene flow to surrounding populations, coevolving interactions have the potential to exhibit metapopulation dynamics. Through gene flow, local extinction, and local recolonization, traits of locally coevolving species often occur together in different combinations or with different means and variances. A small but growing number of long-term studies have compared the metapopulation structure and dynamics of traits in coevolving species (Antonovics et al. 1994; Burdon, Ericson, and Muller 1995; Antonovics, Thrall, and Jarosz 1998; Burdon, Thrall, and Brown 1999; Burdon and Thrall 2000; Burdon, Thrall, and Lawrence 2002; Ericson, Burdon, and Müller 2002). These studies suggest that a continually shifting geographic mosaic of traits is inevitable within any interaction showing a metapopulation structure.

Beyond metapopulations, interspecific interactions may also show broader patterns among biogeographic regions. The beginnings of evolutionary ecology in the 1950s and 1960s provided tantalizing suggestions of geographic patterns in the strength of biotic interactions and, possibly, particular kinds of biotic interactions (review in Thompson 1994). In recent years, evolutionary approaches to community ecology and regional biogeography have begun to explore how coincident patterns in pairwise interactions or assemblages may vary across ecoregions and how defenses, counterdefenses, and mutualisms in multiple species may differ among regions (Ricklefs and Schluter 1993; Vermeij 1994; Steinberg, Estes, and Winter 1995; Coley and Barone 1996; Losos et al. 1998; Sotka, Wares, and Hay 2003).

Studies of geographic variation in species interactions in marine environments, for instance, suggest major biogeographic patterns in both the Pacific and Atlantic Oceans. Along the Atlantic coast of North America, northern populations of salt-marsh grasses are more palatable than southern populations of the same species. This result holds fairly consistently across ten salt-marsh species tested with a wide range of invertebrate herbivores (Pennings, Siska, and Bertness 2001). Along the same latitudinal gradient, predation levels on a number of marine animals are generally higher south of Cape

Hatteras (Vermeij 1978; Lankford, Billerbeck, and Conover 2001). Species spanning these regions differ geographically in their adaptations to other species. Populations of the crab *Libinia dubia* differ in defenses against predators, and populations of the amphipod *Ampithoe longimana* differ in feeding preference and tolerance to chemical defenses in seaweed species (Stachowicz and Hay 2000; Sotka and Hay 2002). The geographic differences in both species are related to corresponding geographic differences in the occurrence of the chemically defended tropical seaweed genus *Dictyota*, which occurs only south of Cape Hatteras (fig. 4.6). *Libinia* crabs decorate themselves with algae throughout their geographic range, but south of Cape Hatteras they specialize on *Dictyota* in camouflaging their carapace (Stachowicz and Hay 2000). Although *Ampithoe* ranges widely along the eastern seaboard of North America, populations south of Cape Hatteras show genetically higher preference and tolerance for *Dictyota* (fig. 4.6). Phylogeographic studies show that the northern and southern populations of *Ampithoe* form separate monophyletic clades (Sotka, Wares, and Hay 2003).

Defenses in the bryozoan *Bugula neritina* also differ between these regions, although recent molecular work has suggested that the northern and southern species are now so separated that they are actually cryptic species (Mcgovern and Hellberg 2003). Interestingly, the differences in defense—as measured by palatability to predators—in the *B. neritina* complex between these two regions are due to differences in bacteria they harbor rather than to differences in chemical compounds produced directly by the bryozoans themselves (Mcgovern and Hellberg 2003). Analyses on *B. neritina* in the Pacific Ocean along the coast of California have similarly provided evidence for a *B. neritina* species complex that differs geographically in the same bacterial species (Davidson and Haygood 1999).

Over even larger geographic scales, the kelp forests of the North Pacific Ocean may differ from those in temperate Australasia in levels of chemical defense against herbivores, and the difference may be caused by the presence of large predators in the North Pacific but not in temperate Australasia (Steinberg, Estes, and Winter 1995). In the North Pacific, predation by sea otters historically kept herbivore populations at low levels, and chemical defenses against herbivores is relatively low in these kelp forests. In contrast, herbivory on kelp forests in temperate Australasia has been historically higher, and these kelp exhibit higher levels of chemical defense. Australasian herbivores tested for response to phlorotannins are unaffected by these compounds, whereas North Pacific herbivores are deterred, whether the regional

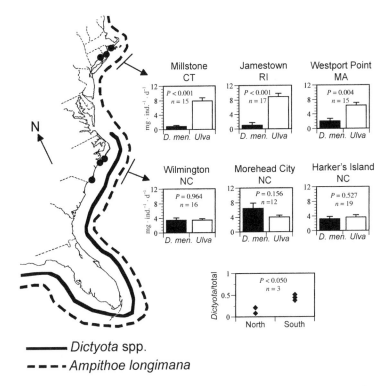

Fig. 4.6   Regional differences in preference for seaweed taxa by the amphipod *Ampithoe longimana*, using consumption rate as an index of preference. The tropical seaweeds in the genus *Dictyota* have high levels of chemical defense, and experiments on *D. menstrualis* show that only populations of *Ampithoe* that coexist with *Dictyota* have a high preference and tolerance for this genus. The bottom panel indicates that southern populations consumed proportionately more *D. menstrualis* tisssue than northern populations. After Sotka and Hay 2002.

source of the compounds is the North Pacific or Australasia (Steinberg, Estes, and Winter 1995). For now this geographic pattern in kelp and herbivores and its predator-based explanation remain a working hypothesis, but the results of these studies suggest a potential for large biogeographic differences among kelp forests in the coevolutionary structure of interactions.

If biogeographic regions generally differ in the patterns of selection they impose on interactions (Coley and Barone 1996), then geographic differences in selection on interspecific interactions are an almost inevitable component of coevolution in any widespread pairwise interaction. As community context changes across regions, so will the structure of selection on coevolving taxa. Coincident selection on defense, counterdefense, or mutualism among multiple species within regions may therefore also provide a basis for regional differences in the organization of multispecific coevolutionary networks.

Human Malaria: An Exemplar of Geographically Variable Selection and Local Coadaptation

The geography and genetics of human malaria exemplify how the geographic structure of interactions may diversify across landscapes through local adaptation. Human malaria shows strong geographic patterns in prevalence, disease severity, the complex of species involved, and the genes that act as defenses. It is also the best-studied disease for analysis of geographic structure. It therefore provides a glimpse into what we can expect from the structure of other coevolving interactions once they have been studied in more genetic and ecological detail across their geographic ranges.

Malaria is a major cause of death in humans. It afflicts annually almost 500 million people worldwide, resulting each year in more than a million deaths (Sachs and Malaney 2002). It is broadly but patchily distributed among tropical, subtropical, and some warm temperate environments (fig. 4.7). Prior to major eradication efforts in the 1950s and 1960s, it had a much wider worldwide distribution. Now, malaria is mostly a disease of the tropics, with 90 percent of malarial deaths occurring in sub-Saharan Africa (Sachs and Malaney 2002).

Malaria results from infection by *Plasmodium falciparum, P. malariae, P. ovale,* or *P. vivax* parasites, which are transmitted by multiple mosquito species that vary in occurrence worldwide. These species are members of a genus that has radiated over millions of years into taxa that attack reptiles, birds, rodents, and primates (Evans and Wellems 2002; Paul, Ariey, and Robert 2003). *Plasmodium falciparum,* which causes the most deadly form of human malaria, may have shared a common ancestor with a closely related chimpanzee parasite several million years ago (Escalante, Barrio, and Ayala 1995; Evans and Wellems 2002). Although estimates vary, *Plasmodium falciparum* appears to have diverged from its most recent common ancestor about 100,000–180,000 years ago (Mu et al. 2002). Since then it has maintained large effective population sizes (Hughes and Verra 2001) but has genetically diverged among continents, as shown by genome-wide analysis of molecular markers (Wooton et al. 2002). *Plasmodium falciparum* therefore appears to

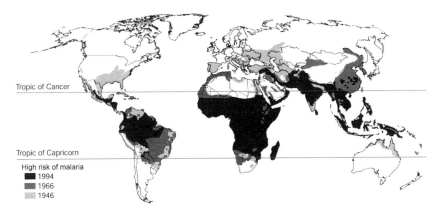

Fig. 4.7   Global distribution of malaria, showing its historical patchwork distribution in 1946 and more restricted distributions at two subsequent times following eradication efforts during the second half of the twentieth century. After Sachs and Malaney 2002.

have had a long history of adaptation to humans, and it may have undergone periods of marked expansion. One such period may have occurred ten thousand years ago, but the interpretation of the data suggesting this expansion remains controversial (Hartl 2004). *Plasmodium falciparum* may not have been present in the Western Hemisphere until the New World was colonized by Europeans (Evans and Wellems 2002). The effects of those introductions on New World human populations that had not coevolved with *Plasmodium* (and other European pathogens) were devastating (Diamond 1997).

The most efficient vector of *Plasmodium* to humans is *Anopheles gambiae,* which is the major vector in sub-Saharan Africa. It is part of a group of species in the subgenus *Cellia,* which also includes other major vectors such as *A. arabiensis* and *A. funestus. Anopheles gambiae* itself is a complex of species that differ geographically (fig. 4.8) and diverged from *A. funestus* about five million years ago. Since that time, *A. gambiae* and *A. funestus* have become fixed for at least seventy chromosomal inversions (Sharakhov et al. 2002). Often such inversions are associated with ecological adaptation (Torre et al. 2002), and these two species are known to differ in a number of ecological characteristics, including breeding site preferences and seasonal abundance.

Even within *A. gambiae* sensu stricto, populations differ markedly in chromosomal inversions, molecular markers, and ecological characters (Torre et al. 2002). Because the chromosomal inversions are associated with adaptation to different environments, they suggest divergence mediated by natural selection. The species also shows divergence into two molecular

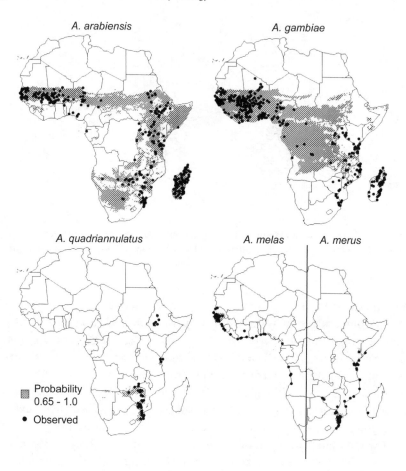

Fig. 4.8   Geographic distribution of five species within the *Anopheles gambiae* mosquito complex in Africa as predicted by temporal Fourier-processed satellite data and elevation coverage. *Anopheles gambiae* in this figure refers to *A. gambiae* sensu stricto. After Rogers et al. 2002.

forms, M and S. Because this divergence is likely associated with neutral molecular substitutions, it suggests historical barriers to gene flow between these forms (Torre et al. 2002). The chromosomal forms and molecular forms show a close correspondence to each other in some parts of Africa but not in others, suggesting that a complex mix of natural selection for local adaptation and gene flow among populations has shaped patterns of divergence within this species.

Although *A. gambiae* shows a close association with humans throughout its geographic range, the M form is especially associated with human environments and breeding sites created by human activities. The S form is more common in temporary breeding sites created by rain (Torre et al. 2002). The M form may therefore have resulted from creation of new environments by humans, allowing *A. gambiae* to extend its range into new habitats. *Anopheles gambiae* sensu stricto is therefore a set of diverging populations that seem to have evolved during human expansion and in the environments created by that expansion.

The worldwide geographic structure of malaria includes many other local variants of the interaction between humans and anopheline mosquitoes. Altogether, more than sixty species, species groups, and subspecies of anopheline mosquito transmit *Plasmodium* parasites to humans, adding to the worldwide geographic complexity of this three-way interaction between host, vector, and parasite (Budiansky 2002). Unlike *A. gambiae,* most species are inefficient vectors. This large worldwide list of inefficient vectors, however, provides important insight into how multispecific selection mosaics develop in interspecific interactions. As an interaction diversifies across landscapes, populations evolve different degrees and forms of specialization and adaptation to each other. Those geographic relationships then become increasingly complex over time as populations speciate and sometimes come back together into secondary contact. In sub-Saharan Africa, for example, *A. gambiae, A. arabiensis,* and *A. funestus* all inhabit the same villages and even, in some regions, the same houses (Sharakhov et al. 2002).

The genetics of resistance to malaria adds to the geographic mosaic of local selection and adaptation in these interactions. Resistance to malaria is influenced by at least fourteen different genes, including globin, HLA, and ABO blood group genes (Hill 1998). The genes vary in effectiveness against malaria, and their frequencies vary geographically, creating complex worldwide patterns in the genetics of malarial resistance. For example, sickle-cell trait (which in homozygous individuals manifests as sickle-cell anemia) and glucose-6-phosphate dehydrogenase (G6PD) are both common in malarial regions. They produce human disorders, but they also confer some protection against severe malaria. Sickle cell reaches its highest frequencies in Africa and in parts of the Mediterranean, Middle East, and India (Weatherall and Provan 2000). G6PD deficiency reaches its highest frequencies in parts of Africa, the Mediterranean, the Middle East, Asia, and Papua New Guinea (Ruwende and Hill 1998).

G6PD is an X-linked gene that may be the most polymorphic locus in humans. More than three hundred alleles have been identified, and at least eighty-seven have been recorded at frequencies greater than 1 percent within populations (Ruwende and Hill 1998). Many of these variants cause G6PD deficiency, which is the most common enzyme pathology in humans, affecting over four hundred million people (Ruwende et al. 1995; Ruwende and Hill 1998). Linking G6PD deficiency to protection against malaria has not been as straightforward as for sickle-cell trait, but field and molecular studies over the past decade have suggested a strong link between this deficiency and selection imposed by malaria. Two studies, one in Kenya and the other in Gambia, have shown that the common form of G6PD deficiency (G6PD A—) results in a 46–58 percent reduction in severe malaria both in female heterozygotes and in male hemizygotes. Because G6PD deficiency provides protection to hemizygous males as well as heterozygous females, it has the potential to reach fixation within populations where it is favored. (If protection were limited to heterozygous females, the allele would stabilize at an intermediate frequency that would depend upon the relative fitnesses of the different genotypes.) Allele frequencies in populations within many regions, however, are 5–40 percent, suggesting that counterbalancing selection associated with this enzyme deficiency has probably constrained the broader spread of this deficiency (Ruwende et al. 1995).

Molecular studies using restriction fragment length polymorphisms (RFLP) and microsatellites have indicated that the "A-" and "Med" mutations, which are two of the most common causes of the deficiency, arose within the past 12,000 years. These two mutations arose independently at different times: the A-mutation between 11,760 and 3840 years ago and the Med mutation between 6640 and 1600 years ago. Their rate of increase within and among populations has been too rapid to be explained by random genetic drift (Tishkoff et al. 2001). Instead, the spread of these defenses appears to have been associated with an increase in conditions favoring the disease: warmer and more humid conditions in Africa 12,000 to 7,000 years ago and the spread of human cultures during the past 12,000 years (Tishkoff et al. 2001). Greek and Egyptian documents suggest that milder forms of malaria— possibly caused by *P. malariae* and *P. vivax*—were present before 1500 years ago, but severe malaria did not occur around the Mediterranean until then. Consequently, these alleles may represent defensive responses to *P. falciparum* as it spread into the Mediterranean. The rate of spread, however, may have been mediated by multiple factors, including the development of trade routes and even the movement of Alexander the Great's army (Tishkoff et al. 2001). The result has been a geographic mosaic in the particular forms of adaptation to malaria.

Just as humans differ in their resistance to *Plasmodium,* so do their vectors. Mosquitoes infected with *Plasmodium* can suffer reduced longevity or fecundity, but most studies so far have been in the laboratory. The first study within a natural population showed that *A. gambiae* females infected with *P. falciparum* had a 37.5 percent greater chance of dying during nighttime feeding activities than noninfected individuals (Anderson, Knols, and Koella 2000; Schwartz and Koella 2001). More recent studies have shown that family lines within *A. gambiae* differ in their resistance to *Plasmodium falciparum* (Niaré et al. 2002). Crossing experiments among these lines and genome-wide scanning of pedigrees have detected segregating resistance alleles. The geographic distribution of resistance is not yet known, but geographic variation for resistance seems likely now that it is known that the mosquitoes are genetically variable for resistance traits. Any such geographic variation in resistance levels is likely to have multiple genetic origins, because 242 genes from 18 gene families have been implicated in innate immunity in *A. gambiae* (Christophides et al. 2002).

As new molecular and populational approaches refine our understanding of malaria, they continue to reveal ever-greater evidence for geographic complexity in these interactions. Few interspecific interactions will ever be studied in as much geographic and genomic detail as is now becoming possible for human malaria. This interaction will therefore be an increasing source of our understanding of the geographic mosaic of coevolution and how human culture is reshaping the coevolutionary process.

Conclusions

Geographic differences in selection and the resulting local adaptation in interspecific interactions create the genetic and ecological template for the geographic mosaic of coevolution. They produce innumerable local coevolutionary experiments that fuel the ongoing geographic dynamics of evolving interactions. Local adaptations are linked hierarchically, creating differences that are apparent among local populations, metapopulations, and broader collections of populations from different biogeographic regions. Natural selection is ongoing at all these levels and can occur sometimes very quickly along any one of several potential trajectories. These rates and trajectories are explored in the next chapter.

5   Local Adaptation II

*Rates of Adaptation and Classes
of Coevolutionary Dynamics*

Local adaptation in interspecific interactions is observable within human lifetimes (Thompson 1998c). Local coevolution may even increase the rate of local evolutionary change over that observed in adaptation to physical environments, because each local evolutionary change in one species can impose selection on the other species in an interaction. Unlike in adaptation of populations to the physical environment, the adaptive peaks for locally coevolving populations keep shifting specifically in response to recent evolutionary changes in other species. Local adaptation to enemies and mutualists is therefore a relentless process, favoring genotypes that can keep up with the rapid pace of change in the selective landscape. This process can favor local specialization to a small set of other species. The result is local coevolution along a set of outcomes that have been identified through empirical studies and mathematical models of a wide range of interactions.

In this chapter I consider the rates of local adaptation in interspecific interactions and the classes of coevolutionary dynamics produced by local coadaptation. These local dynamics form the final components of the template for the geographic mosaic of coevolution. Only some of the examples I discuss show actual coevolutionary change, because many studies have analyzed only one of the species in an interaction. All the examples, however, show the potential for rapid coevolutionary change in interspecific interactions.

Ongoing Local Adaptation to Other Species Can Occur Quickly, Blurring the Artificial Distinction between Ecological Time and Evolutionary Time

RAPID EVOLUTION OF INTERACTIONS UNDER CONTROLLED CONDITIONS

I will define rapid evolution as evolution over time scales that can influence the population dynamics of species and the dynamics of community assembly and organization. For most species that definition essentially means evolution over the time scales of a few decades or less, which is the time scale at which most studies of ecological dynamics take place. Rapid evolution of traits and outcomes has now been shown in laboratory or mesocosm studies of all forms of interaction, sometimes altering ecological outcomes in only a few generations. These studies include, for example, increased resistance to parasites or parasitoids (e.g., Bohannan, Travisano, and Lenski 1999; Fellowes, Kraaijeveld, and Godfray 1999), changes in interspecific competitive ability (e.g., Joshi and Thompson 1995, 1996; Goodnight and Craig 1996), and altered preferences for hosts or ability to survive on hosts (e.g., Agrawal 2000). Although these analyses use a wide range of experimental protocols, many of them have been designed explicitly to exclude coevolutionary change in order to study how one species evolves adaptation to another in the absence of a coevolutionary response.

Some of the most carefully controlled studies of local adaptation have been of parasites, including pathogens. A common way of evaluating rapid evolution and the potential for local adaptation in microparasites is through serial passage experiments. Experimental populations are propagated under one or more sets of specific conditions and compared with control populations that have been maintained in the ancestral state. The hosts used in these experiments usually have little or no genetic variation. This experimental design allows a clear analysis of rapid evolution and local adaptation against a constant host background. These experiments commonly show evidence of rapid adaptation of the parasites to their new hosts. Compared with ancestral control populations, the experimental populations rapidly increase their ability to cause damage or death in the host (Ebert 1998). Such increases are assayed in a variety of ways, including percentage of hosts that die, dosage needed to kill 50 percent of the hosts, and feeding rates on host tissue. These increases in "virulence" are generally accompanied by concomitant decreases in the same measures on the former host.

The design of these experiments usually favors genotypes that proliferate fastest, because the most numerically abundant genotypes have the highest probability of being transferred to the next hosts (Ebert 1998). Hence, the increases in virulence may be partly a result of the experimental design. The interpretations demand additional caution, because the potential for coevolution has been explicitly excluded from most experimental designs using serial transfer. The lack of any coevolutionary response by the hosts removes the barriers that might otherwise limit or slow adaptation of the parasite to its new host.

Many of most highly replicated experiments on coevolution have evaluated interspecific competition in *Drosophila*. Various versions of these experiments have been conducted during the past several decades, carefully controlling for population-genetic structure, population sizes, and environmental influences. These studies often show that relative competitive abilities of *Drosophila* can vary with environmental conditions (Hedrick and King 1996). Moreover, the experiments show that interactions can evolve quickly in different ways in different replicates, even when the replicates begin with the same genetic structures and environmental conditions. The subset of these studies that have allowed coevolution show that genetically variable populations can evolve in ways that both suppress populations of competitor species and resist competition by those competitors (e.g., Joshi and Thompson 1995).

For example, *Drosophila melanogaster* and *D. simulans* can both impose strong competition on each other, and populations of both species can evolve rapidly when placed in competitive environments. In one set of replicated experiments, a population of each species was brought to genetic equilibrium over multiple generations on normal growth medium and then subdivided into three environments that imposed interspecific competition (Joshi and Thompson 1995, 1996, 1997). In environment I, the populations competed on normal growth medium. In environment II, the populations experienced a novel environment, which was their normal growth medium with 4 percent ethanol added. In environment III, the populations were provided with a two-resource environment with normal growth medium on one half and 4 percent ethanol added on the other half. Each experiment was replicated three times to evaluate the repeatability of competitive coevolutionary trajectories under the same genetic and environmental conditions. A simultaneous experiment tracked evolution and overall density of each species in the same environments but in the absence of interspecific competition. This experimental structure made it possible to separate evolution in response to interspecific competition from evolution imposed by the physical environments

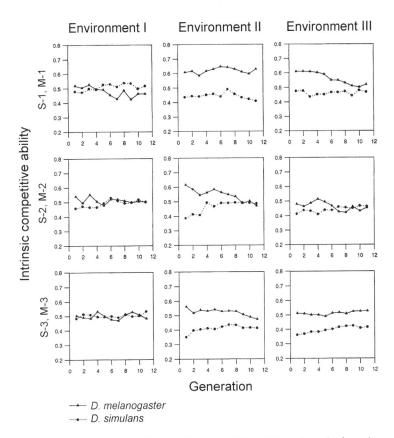

Fig. 5.1 Evolutionary changes in interspecific competitive ability in nine mixed-species laboratory cultures of *Drosophila melanogaster* and *D. simulans*. Values plotted are the decrement in population numbers of one species on the other species, proportional to the potential numbers of the second species in that environment. Three replicate two-species cultures (S-1 M-1, S-2 M-2, S-3 M-3) were tested in each of three environments: (I) no ethanol in the growth medium, (II) 4 percent ethanol in the medium, and (III) a two-resource environment with no ethanol in one half and 4 percent ethanol in the other half. Each generation was started with a total of fifty males and fifty females from each species. After Joshi and Thompson 1995.

and intraspecific competition alone. This particular experiment, however, differs from coevolution in many natural populations, because neither competitor was allowed to become locally extinct during the experiment.

Under these conditions, the interaction evolves along multiple short-term trajectories (fig. 5.1). In all the tested environments, *D. melanogaster* is initially the stronger competitor, and that is especially true in the novel environments containing 4 percent ethanol. Over time, however, the relative competitive abilities of the two species change, and the trajectories differ even among replicate populations in the same environment. If we imagine these as nine populations distributed across a variable landscape, this set of experiments provides a picture of the kinds of geographic selection mosaics that are likely to shape coevolutionary dynamics across natural landscapes. Addition of other evolutionary processes, such as gene flow, random genetic drift, and local extinction, would add further evolutionary complexity to the geographic mosaic of coevolution between these interacting species.

RAPID EVOLUTION OF INTERACTIONS IN NATURAL COMMUNITIES

Studies on the heritability of traits and the strength of natural selection within natural populations suggest that the rapid evolutionary changes found within laboratory studies are not artifacts of the controlled conditions that focus selection on a few traits. There is, however, disagreement on typical strengths of selection within natural populations (Kingsolver 1995; Hendry and Kinnison 1999; Conner 2001; Hoekstra et al. 2001; Kingsolver et al. 2001). Most measurements continue to be on morphological traits despite the fact that Endler (1986) noted this strong bias in the mid-1980s. Nevertheless, some measured potential levels of selection and rates of evolution are high. For example, the alpine flower *Polemonium viscosum* has a corolla that flares 12 percent wider in populations pollinated solely by bumblebees in comparison with populations pollinated by a wide variety of floral visitors (Galen 1996). Heritability estimates coupled with pollinator selection experiments indicate that a difference of that magnitude could arise within a single generation following bumblebee-mediated selection.

In addition, scores of interspecific interactions from natural populations have been shown to have evolved over the past two centuries, and the examples span a wide range of taxa and forms of interaction (Lively 1993; Thompson 1998c; Burdon and Thrall 1999; Carroll et al. 2001; Lively and Jokela 2002; O'Steen, Cullum, and Bennett 2002; Zangerl and Berenbaum 2003). For many years, the most commonly cited example of rapid evolution of defenses was the wing color of peppered moths (*Biston betularia*), which evolved first toward the melanic form following the Industrial Revolution in Europe and North America and then back toward the peppered form following implementation of new pollution standards. Experiments over the past half century on the causes of this change have suggested that differential predation is a likely major cause of the change, although further experiments are needed to understand how predation and other selective factors may interact (Majerus 1998; Cook and Grant 2000). The remarkable speed of these changes in Britain, continental Europe, and North America highlights the rate at which traits at least partially associated with species interactions can change repeatedly over short spans of time (Brakefield and Liebert 2000; Cook and Grant 2000; Grant and Wiseman 2002).

In some species with very fast generation times, natural selection may even fine-tune defenses against enemies in different years within local populations. The result is fluctuation in the proportion of individuals with different suites of traits. An eight-year study of a population of the freshwater copepod *Diaptomus sanguineus* found that the timing of diapause evolves rapidly among years, driven by natural selection imposed by predatory fish (Ellner et al. 1999). The date in spring on which females switch from producing clutches that hatch immediately to those that remain in diapause until autumn is under genetic control. The major selective agent on the timing of diapause is predation by sunfish, which varies from year to year. Such changes in timing of diapause must surely also affect the species on which the copepods feed. The result is a potential selective cascade of rapid evolutionary effects, tugging populations first one way and then another within and across environments.

The most remarkable of all studies of rapid evolutionary change are the long-term studies on two species of Darwin's finch on the Galápagos Islands over the past thirty years (Boag and Grant 1981; Grant and Grant 2002). These studies have set the standard for evaluating the evolutionary processes that drive rapid change in interspecific interactions within natural populations. Evolution has been studied only in the birds, so the coevolutionary dynamics are unknown. These studies, however, show how rare events with strong selective effects on populations can have long-term effects on traits associated with interspecific interactions.

During the past several decades, medium ground finches (*Geospiza fortis*) and cactus finches (*G. scandens*) have changed repeatedly in mean beak size, beak shape, and body size on the island of Daphne Major. The changes have been driven primarily by fluctuating selection on the interaction between the birds and the plants on which they feed. Fluctuating selection has, in turn, been driven by climatic fluctuations associated with El Niño and La Niña events that create cycles of wet and drought years.

The distribution of seed sizes available to the finches differs between wet years and drought years. In drought years, the seeds available during the dry season are from plants that make relatively large, hard seeds. Only birds with large beaks are capable of breaking open these seeds, leading to selective death of birds with small beaks. As a consequence, a sharp rise in beak size, beak shape (associated with ability to crack large seeds), and body size occurred in *G. fortis* in the late 1970s in response to severe drought (fig. 5.2). Beak size and body size, but not beak shape, also increased in *Geospiza scandens* during the same time period.

A prolonged wet season in 1983, associated with an El Niño event, created a different set of selective conditions and favored introgressive hybridization between the two species. That event had a greater overall effect on *G. scandens* than on *G. fortis,* because the sex ratio of *G. scandens* was biased toward males at that time. Socially subordinate female *G. scandens* died disproportionately as their principal dry-season food, *Opuntia* cactus seeds and flowers, was smothered by rampantly growing vines (Grant and Grant 2002). *Geospiza fortis* was less affected because it depends upon the small seeds of other plants. It retained a near-equal ratio of males and females during that period. This period was followed by two years of drought. When the finches began to breed again in 1987, all *G. scandens* females were able to mate with *G. scandens* males, but two *G. fortis* females mated with *G. scandens* males. These interspecific offspring later mated with *G. scandens* males and females, resulting in introgression of *G. fortis* genes into the *G. scandens* population. The introgression contributed to subsequent shifts in mean beak size, beak shape, and body size in *G. scandens.*

The links among climatic fluctuations, selection on traits associated with interspecific interactions, sex ratio, and introgressive hybridization point to the inevitability of continuing rapid local evolution. So many constantly changing factors shape local populations that it is unlikely populations could remain viable for very long without continuing evolution, generation after generation. These studies have also shown that, once the mean values for important traits have changed within a population, subsequent reversal in the direction of selection does not necessarily return a population to its previous state, at least over these time periods (Grant and Grant 2002).

Rapid divergence of experimental native populations has provided further evidence that interspecific interactions can continue to evolve on the time scale of decades within natural communities. Among the longest term and best controlled are the replicated studies on rapid evolution of defensive behavior in Trinidadian guppies (*Poecilia reticulata*). The guppies occur in freshwater streams on the island of Trinidad. Guppies in some streams are subject to high levels of predation, whereas guppies in other streams are relatively free of predators. The suite of predators differs among streams and

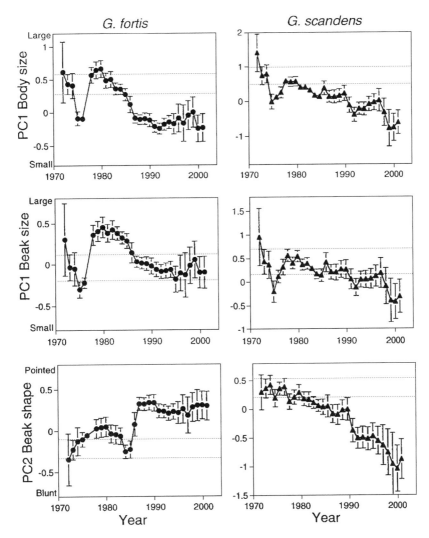

Fig. 5.2 Changes in the means of morphological traits over time in two Galápagos finches, *Geospiza fortis* and *G. scandens*. The horizontal dotted lines on each graph are the 95 percent confidence intervals of the estimates from the 1973 samples. Values outside those intervals indicate that the population changed significantly in that trait over time. After Grant and Grant 2002.

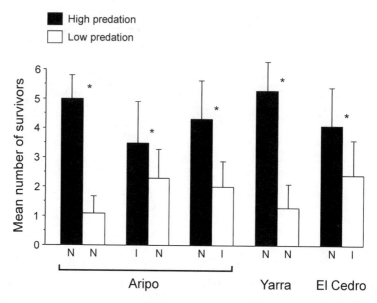

Fig. 5.3   Escape ability of wild caught guppies (*Poecilia reticulata*) collected from high-predation and low-predation populations from three different areas in Trinidad, when placed in pools with the pike cichlid *Crenicichla alta*. N = natural populations of guppies. I = introduced populations. The asterisk (*) indicates significant differences. Figure drawn from results in O'Steen, Cullum, and Bennett 2002.

includes fish and invertebrates. The high-predation populations differ from the low-predation populations in multiple traits, including escape behavior from predators (Endler 1980, 1995; Magurran et al. 1995; Reznick, Butler, and Rodd 2001). Guppies were experimentally introduced into high-predation and low-predation environments in the mid-1970s and early 1980s, and the evolution of traits in these populations has been followed ever since. Since the 1980s, populations in the high-predation environments have evolved improved escape ability, and those in low-predation environments have rapidly lost their escape ability (fig. 5.3). These changes have occurred in only 26–36 generations, and the results have been consistent across the natural and introduced populations that have been tested (Reznick et al. 1997; O'Steen, Cullum, and Bennett 2002).

These carefully designed studies of Trinidadian guppies are important for our understanding of the dynamics of coevolution, because they suggest that some traits associated with interactions can increase rapidly and also decrease rapidly in response to different selective regimes within natural communities. That is, there is no obvious asymmetry in the rate at which traits associated with interspecific interactions are acquired or lost. That conclusion, however, is highly tentative, because there are very few studies that have analyzed the rates of both gain and loss of traits in an interaction once it has appeared within a population and risen through selection above levels maintained by recurrent mutation. If asymmetries in rates occur in some traits or in some forms of interaction, they would strongly shape coevolutionary dynamics. In fact, at least some kinds of traits may be lost only slowly once the interaction that initially favored them is lost. Some Central American plant species have very large, hard fruits that, until the reintroduction of horses, were beyond the ability of contemporary native mammals to handle and disperse. Janzen and Martin (1982) argued that these fruits are adapted to dispersal by large mammals that became extinct in the neotropics more than ten thousand years ago. Future coevolutionary studies will need to confront more fully the problem of differential rates of gain and loss of traits important to interspecific interactions.

RAPID EVOLUTION OF INTERACTIONS WITH INTRODUCED SPECIES

When species are introduced into new regions, they often leave behind their old enemies and mutualists. They are, however, immediately surrounded by a network of new potential interactions that can impose strong selection not only on the newly introduced species but also on the native species. Accumulating studies of the evolution of interactions between introduced and native species suggest that novel interactions can evolve and diversify rapidly as they spread across geographic landscapes.

As with many studies of interspecific interactions, most of these analyses are of how one species evolves in an interaction rather than of how the interacting species coevolve. Nevertheless, they illustrate how geographic selection mosaics and local evolutionary adaptation can develop rapidly as introduced species encounter various combinations of native species in different communities. Because so many insects and plants have been moved among continents during the past several hundred years, many of the examples involve these taxa. In some cases rapid evolution has occurred in the native species, as in the case of North American *Rhagoletis pomonella* flies. These fruit-feeding flies are specialized to native hawthorns, but some populations have formed distinct host races on introduced apples since the mid- to late 1800s (Feder et al. 1997; Berlocher 2000; Berlocher and Feder 2002). Similarly, Western anise swallowtail butterflies (*Papilio zelicaon*) in a part of Cali-

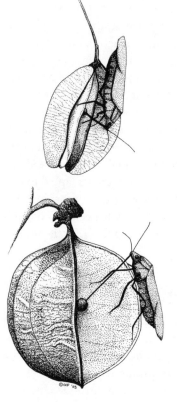

Fig. 5.4   Soapberry bugs feeding on introduced goldenrain tree (*Koelreuteria elegans*) (top) and native balloon vine (*Cardiospermum corindum*) in Florida by piercing the hollow fruit wall to reach a seed. The left side of both fruits shows the outer fruit wall, which is removed on the right side. Drawing by Catherine C. Fernandez.

fornia now dominated by introduced fennel exhibit higher preference for fennel than other populations, which prefer native host species (Thompson 1993; Wehling and Thompson 1997; Thompson 1998b).

In some cases, rapid evolution has favored different traits in different geographic regions, fueling the rapid development of geographic differences in interactions between insects and plants. Soapberry bugs (*Jadera haematoloma*) feed on the seeds of plants in the Sapindaceae and have rapidly evolved longer piercing mouthparts in some regions of the southern United States and shorter mouthparts in other regions in response to the introduction of new sapindaceous species into these environments (Carroll and Boyd 1992; Carroll, Dingle, and Klassen 1997). Soapberry seeds are encased in a hollow fruit. To successfully feed on a seed, a soapberry bug must have mouthpart lengths that can span the distance from the outside of the fruit to the seed (fig. 5.4). That distance varies among soapberry species. In Florida, the native host species is balloon vine (*Cardiospermum corindum*), which has a wide gap between fruit and seed, and the soapberry bugs have mouthparts that match that distance. In contrast, the introduced host plant in Florida is goldenrain tree (*Koelreuteria elegans*), which was introduced from Southeast Asia in the 1950s. It has a flat pod and therefore a narrow gap between the fruit coat and seeds. As populations of soapberry bugs have colonized and adapted to goldenrain trees, they have evolved shorter mouthparts, while other dimensions of body size have remained largely unchanged. Comparison of the lengths of mouthparts in museum collections and current populations shows that the mouthpart lengths of some soapberry populations now fall outside the range found prior to the introduction of new sapindaceous species (Carroll and Boyd 1992).

Reciprocal crossing experiments have shown that the differences in beak length among populations on the two hosts are genetically based, as are differences in development time and other measures of survival, growth, and reproduction (Carroll, Klassen, and Dingle 1998; Carroll et al. 2001). Moreover, populations on their normal host do better by most performance measures than populations on the alternative host. These results suggest that local adaptation to the introduced host involves multiple traits. The crossing experiments have shown that the genetic architecture of adaptation varies among these traits (Carroll et al. 2001), suggesting that adaptation has involved selection on multiple loci.

The geographic selection mosaic on soapberry bugs across the southern United States has produced an inverse pattern of rapid local adaptation in mouthpart length in other populations (Carroll and Boyd 1992). In Louisiana and other parts of the south-central United States, the native sapindaceous plants have narrow gaps between fruit coats and seeds, whereas the introduced species have relatively wide gaps. Mouthpart lengths in soapberry bug populations that feed on the introduced plants have evolved accordingly. Rapid evolution in opposite directions in different geographic regions illustrates the speed with which some species can fine-tune their local adaptation to other species and create geographic differences in the structure of interspecific interactions.

Rapid evolution is perhaps more easily observed in introduced species than in native species, because selection may initially be strongly directional as species adapt to new abiotic as well as new biotic environments (e.g., Ohgushi and Sawada 1997). Despite the growing number of examples of rapid evolution involving introduced taxa, only recently has rapid evolution become one of the working hypotheses for the dynamics of invasive species and their interactions with native taxa. Successful invasion into new communities, however, often may be as much an evolutionary process as an ecological one (Thompson 1998c; Sakai et al. 2001). Some mathematical models of species invasions suggest that the rate of spread of invasive species may be governed by the rate of adaptation of invaders to their new spatially varying landscapes (García-Ramos and Rodríguez 2002). Spatial complexity can have many causes, but geographically varying species interactions are likely to be a major part of it. Incorporation of local adaptation and coadaptation in invasive species into studies of community dynamics is one of the next frontiers in community ecology and invasive species biology.

A growing number of examples of rapidly evolving interactions involve viral, fungal, and bacterial pathogens, and the most ecologically visible examples often involve introduction of new species or genotypes into new regions. Myxoma virus was introduced into Australia in the 1950s as a biological control agent against rabbits and evolved rapidly toward decreased virulence within only a few years as rabbit populations plummeted and selection on virulence apparently changed (Fenner and Kerr 1994; Saint, French, and Kerr 2001). Such rapid change is expected in viruses, because analyses of nucleotide substitution rates have shown that viruses can continue to maintain high evolutionary rates over thousands of years. Influenza A virus that attacks humans continually undergoes divergence at rates that are readily observable within a single century. The virus produces a rapidly evolving glycoprotein coat called hemagglutinin, and the evolution of new glycoprotein coats is associated with new human pandemics of the disease (Suzuki and Nei 2002). Phylogenetic analyses indicate that different subtypes of hemagglutinin genes originated about two thousand years ago and have undergone continued divergence ever since. Human and swine hemagglutinin sequences have shown an almost constant rate of divergence from an amino acid sequence dated to approximately 1849 (fig. 5.5). Different subtypes of influenza A evolve at different rates, and those rates differ from the evolutionary rate of influenza B, but all these rates are remarkably high (fig. 5.6).

Why Is Rapid Evolution Still Generally Perceived as Uncommon?

Our increased appreciation for rapid evolution is developing from four sources: a renaissance of new approaches and perspectives that link evolutionary ecology with molecular approaches, an infusion of perspectives from evolutionary epidemiology into ecological disciplines, long-term studies of

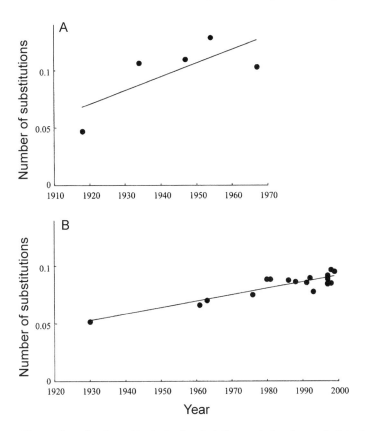

Fig. 5.5   The number of amino acid substitutions in influenza A virus hemagglutinin, in (A) human and (B) classical swine sequences. Hemagglutinin is the major envelope glyco-protein produced by the virus. The x axis shows the year of isolation or divergence, and the y axis shows the number of amino acid substitutions from a sequence dated at approximately 1849. After Suzuki and Nei 2002.

local interspecific interactions, and analyses of geographic differences in in-terspecific interactions. Yet, there is still a general perception that rapid evo-lution and local adaptation are unusual events or occur at time scales irrele-vant to our understanding of ecological dynamics. Despite the burgeoning number of examples of rapid evolution, even now few studies of ecological dynamics include rapid evolution among the hypotheses to be tested in eval-uating the causes of those dynamics.

Even as the perspective is changing, three views continue to cloud an appreciation of ongoing, rapid evolution. The first is that, until recently,

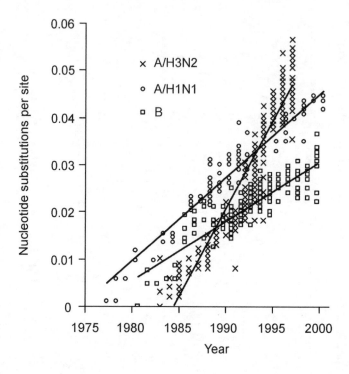

Fig. 5.6   Divergence of nucleotide sequences of influenza A subtypes H3N2 and H1N1 and influenza B along phylogenetic trees rooted in 1977, 1984, and 1980 respectively. After Ferguson, Galvani, and Bush 2003.

some—although certainly not all—of the theory of population genetics was based upon the assumption that natural selection is weak within populations. Historically, some population-genetic models have assumed that local selection is on the order of the inverse of population size and that the effects will therefore be apparent on a time scale that equates approximately with population size (Neuhauser et al. 2003). Under those conditions, evolutionary dynamics would not contribute to short-term ecological dynamics. As evidence of rapid evolution in natural population continues to accumulate, the fallacy of assuming that selection is almost always weak is becoming more evident.

The assumption of weak selection in many population-genetic models is accompanied by a concomitant traditional emphasis on equilibrium genetic states. Until recently, the goal of many population-genetic models has often been to understand the equilibrium values of alleles. Under weak selection, those equilibrium states are sometimes not reached for thousands of generations, which reinforces the view that evolution is a slow process. Focusing discussion on the final equilibrium values diverts attention from the transient dynamics. These nonequilibrium states are probably more common in local populations than equilibrium states in an ever-changing world, and the short-term dynamics found in nonequilibrium states often produce large changes in population-genetic structure over short periods of time.

The third cause of the general perception that rapid evolution is uncommon comes from a misuse of the word "evolution" to mean long-term, sustained directional change in species. During more than twenty-five years of teaching undergraduate and graduate evolution, there has never been a year when I did not have to work hard to overcome this misconception early in the course. Moreover, my own experience over the years suggests that this misconception of what constitutes evolution is more common even among professional biologists than most evolutionary biologists would think at first hearing. It is not uncommon to listen to discussions in which someone says, for example, that horseshoe crabs have not evolved over millions of years — meaning that the morphology of extant populations is similar to that found in fossils. The implication is that, in the absence of sustained directional change, there has been no evolution.

Such views reflect a fundamental misunderstanding of the evolutionary process. Evolution is simply heritable genetic change within populations, and there is nothing in the definition of the word that implies sustained directional change. Evolution is much more than sustained directional change within species, speciation, and large-scale patterns in the fossil record. Most of evolution and coevolution is a local and geographic process that continually shifts populations in one direction and then another in a constantly changing world. Rapid local adaptation keeps players in the local evolutionary game as they respond to the vagaries of their local physical and biotic environments. Much of evolution is therefore an ecological process that results in no net long-term changes. Most evolutionary changes are like the changes in positions of boats held by loose lines to the dock as the surrounding sea rises and falls. The boats are constrained by their loose tethers, but they are able to stay afloat because they constantly change their positions in the ever-changing water levels.

Failure to appreciate this crucial ecological role for rapid evolution is perhaps the major reason many biologists continue to view evolution as a slow, long-term process with little relevance to understanding the current dynamics of populations and species. The same applies to coevolutionary change. Most of coevolution is not about relentless directional change that produces ever-increasing levels of coadaptation across all populations. Much of local coevolution is simply coevolutionary meandering, shifting populations over time as they coadapt repeatedly in different ways across dynamic landscapes (Thompson 1999a). The geographic mosaic that results is by no means inconsequential. It creates the coevolutionary process that links species across regions and allows pairs and groups of species to continue to interact across millennia.

The General Classes of Coevolutionary Dynamics

As species adapt to each other, their coevolutionary dynamics fall into seven general classes (table 5.1). These local coevolutionary dynamics are the modules on which the geographic mosaic of coevolution is built. Starting with Mode's (1958) first model of gene-for-gene coevolution, hundreds of mathematical models have explored the rates, trajectories, and equilibrium outcomes of most classes of local coevolutionary dynamics (e.g., Roughgarden 1979; Slatkin and Maynard Smith 1979; Anderson and May 1982; Taper and Case 1992; Frank 1994b; Gavrilets 1997; Gavrilets and Hastings 1998; Doebeli and Knowlton 1998; Abrams 2000a; Roy and Kirchner 2000; Bergelson, Dwyer, and Emerson 2001; Law, Bronstein, and Ferriere 2001; Switkes and Moody 2001; Nuismer, Gomulkiewicz, and Morgan 2003a). These models have given us tremendous insight into the ways in which the structure of local selection, local population dynamics, the form of interaction, and the genetic architecture of species shape local coevolution. The different classes of local coevolutionary dynamics result from different temporal combinations of the basic forms of natural selection: frequency-dependent, density-dependent, stabilizing, and directional selection (table 5.1).

These pure forms of local coevolutionary dynamics are for populations living in constant environments, and most of the pairwise models assume that reciprocal selection on local populations is constant over time. That does not imply in any way, however, that models of local coevolutionary dynamics are not useful. Rather, these models are crucial starting points for our understanding of coevolution. They show us the range of coevolutionary dynamics and outcomes that are possible under constant conditions in the absence of geographic structure in interacting species. In fact, extinction of one or both species is a common outcome in many models of local coevolution, and that, in itself, is an important result for our understanding of the conditions necessary for the long-term coevolution of species.

Table 5.1   Local coevolutionary dynamics and the primary forms of selection driving the outcomes

Local coevolutionary dynamics	Primary forms of selection
Coevolving polymorphisms	Negative frequency dependent and short-term directional
Coevolutionary alternation	Multispecific fluctuating selection
Coevolutionary escalation	Directional
Attenuated antagonism	Directional, stabilizing, and density dependent
Coevolving complementarity	Positive frequency dependent, stabilizing, and directional
Coevolutionary convergence	Positive frequency dependent, stabilizing, and directional
Coevolutionary displacement	Directional

Within these models, coevolving polymorphisms develop through selection favoring rare genotypes in at least one of the interacting species. In a common version of this form of coevolutionary dynamics, natural selection favors host genotypes that are rare because they are more likely to escape attack by parasites that are locally adapted to the most common host genotype. At the same time, selection favors parasites capable of attacking the currently most common hosts. Fluctuating polymorphisms appear to be a widespread form of coevolutionary dynamics and have been implicated in the complex immune dynamics of vertebrates, gene-for-gene coevolution between hosts and parasites, and the maintenance of sexual reproduction. Depending upon the genetic architecture of the interaction, the strength of natural selection, and time lags in response to selection, the frequency and amplitude of allele fluctuations in the interacting species can differ greatly among environments (fig. 5.7).

Predators, grazers, or parasites that actively choose their victims may drive a class of multispecific coevolutionary dynamics called coevolutionary alternation. If these species preferentially attack the least-defended local victim species, then selection may favor increased levels of defense in that victim species. The resulting changes in the relative levels of defense among victim species may then favor predators or parasites with different preference rankings among available victim species. Coevolution proceeds as preference

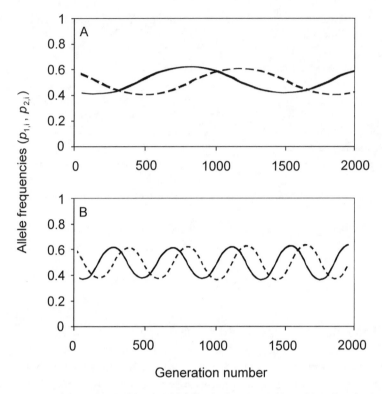

Fig. 5.7   Coevolving polymorphisms within parasite-host interactions driven by selection favoring rare genotypes within host populations. The number of generations is shown on the x axis. The graphs show fluctuation in one host allele (solid line) and one parasite allele (broken line); (A) weak selection, (B) strong selection. After Nuismer, Thompson, and Gomulkiewicz 2000.

hierarchies and relative defense levels continue to change over time. This form of multispecific coevolution is discussed in detail in chapter 11.

Coevolutionary escalation is driven by directional selection toward higher levels of investment in defense and counterdefense in antagonistic trophic interactions (fig. 5.8). This is the arms-race view of coevolution, although the phrase "coevolutionary arms race" has also been used more loosely, and less usefully, as a general term for any kind of antagonistic coevolutionary dynamics. Coevolutionary escalation occurs through sequential replacement of alleles in antagonistic species in a way that demands higher investment in defense and counterdefense in descendant populations. It may involve devel-

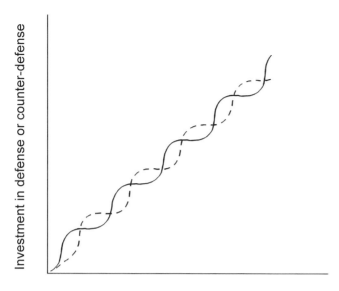

Fig. 5.8   Coevolutionary escalation in which local populations of two species (solid line and broken line) continue to increase their investment in defense and counterdefense over time. Although the figure shows this "arms race" as increasing indefinitely between a pair of species within a local population, such an unlimited arms race is not possible.

opment of qualitatively different but more expensive defenses and counter-defenses, inclusion of more kinds of defense so that overall investment increases, or deployment of a greater quantitative investment in a current defense (e.g., higher levels of a chemical defense; thicker shells). It sometimes involves genetic changes that attract mutualists that aid in defense (e.g., defense of plants by ants). True coevolutionary escalation between pairs of species, in the sense of ever-increasing reciprocal investment in defense or counterdefense, is limited at the local population level by trade-offs imposed by competing selection pressures (Thompson 1986a; Brodie and Brodie 1999a, 1999b). Unlimited coevolutionary escalation is not possible, and populations are likely to differ in the proportion of their total resources that they devote to defense or counterdefense.

This use of the word "escalation" differs from Vermeij's (1987, 1994), which restricts the word to the hypothesis that overall selection for defense has increased over evolutionary time through diffuse selection imposed by predators, competitors, and dangerous prey. That hypothesis does not invoke reciprocal selection between predators and prey. Instead, it suggests that selection by enemies has increased and diversified so much over eons that escalating defenses have become an inherent part of the background of evolutionary change. Descendant lineages have faced an ever increasing range of ever more dangerous enemies, raising the bar of minimum defense levels over time. There are, in fact, multiple ways by which overall levels of defense may increase over time within multispecific networks beyond an escalating background of danger (Dietl and Kelley 2002). The challenge is to elucidate and evaluate each of the potential mechanisms. Some paleobiologists have argued that trends in defenses may reflect the differential origin of new taxa over time or increases in variance in enemy-related traits over time rather than directional selection on the mean level of defenses within species (Jablonski and Sepkoski 1996), as would be expected if escalation were driven by diversifying selective forces across complex geographic landscapes. More recently, Vermeij (2002a) has suggested that, although he still believes that predators retain an evolutionary edge over prey in the long term, the data now suggest that predators and prey influence the direction of evolution in one another. One coevolutionary mechanism by which escalating changes could occur is the process of coevolutionary alternation with escalation in multispecific networks, which is discussed in chapter 11.

Coevolutionary polymorphism, coevolutionary alternation, and coevolutionary escalation may all involve some serial replacement of alleles over time. For example, polymorphisms coevolve by continual fluctuation in frequency among alleles already present within the interacting populations. As new mutations arise, however, they join the existing arsenal of defenses and counterdefenses. Each of these mutant alleles becomes subject to selection relative to the existing alleles. As long as some of these new mutations are continually added to the populations, the allelic structure of local populations will be in continual flux. If rarity or novelty confers as much selective advantage as higher investment through escalation, then alleles are likely to fluctuate in frequency and simply be replaced occasionally by yet rarer or more novel alleles. The underlying coevolutionary dynamics for most local antagonistic interactions between predators and prey, parasites and hosts, and grazers and victims may therefore be fluctuating polymorphisms and coevolutionary alternation with occasional allelic replacements.

Attenuated antagonism results from a complex interplay of multiple forms of coevolutionary selection. Under conditions of low host density or low parasite transmission rates, natural selection may favor more benign forms of parasites even as selection on hosts is for higher resistance. Those parasites that have less immediate effects on host survival are more likely to be transmitted to their new hosts than parasites that cause quick death of the current host. If host populations eventually expand to high density, or if environments change in ways that increase transmission rates among hosts, natural selection on the parasite could change and favor higher rather than lower levels of virulence. There is now a large set of mathematical models and experiments on selection for reduced virulence and coevolution toward attenuated antagonisms, which are discussed in chapter 12. The coevolutionary trajectories depend not only on the pairwise interaction itself, but also on the pattern of infection by multiple parasite genotypes or species and the structure of host immunity. The resulting coevolutionary dynamics provide one of the potential paths toward the evolution of mutualism in symbiotic interactions.

Mutualisms coevolve through a combination of complementarity of traits and convergence of traits within networks. Examples of coevolving complementarity include the nutritional requirements of hosts and mutualistic symbionts, and the morphological matching of hummingbird bills and floral shapes. In addition, some forms of mutualism, especially those among free-living species, coevolve as networks through convergence of additional species on these complementary traits. Examples include coevolutionary convergence in floral morphology among plants pollinated by hawkmoths and convergence in fruit traits among plants dispersed by birds rather than mammals. Forms of mutualism differ in the importance of convergence as part of the process, as will be discussed in chapters 12 through 14.

Coevolutionary complementarity and convergence in mutualisms are often driven by a complex combination of positive frequency-dependent selection, stabilizing selection, and directional selection. Complementarity generally favors individuals of the most common genotype. These individuals are more likely to benefit from the interaction than are rare genotypes, because they are more likely to complement the most common local genotype of the other species. Selection on this part of the process is therefore often stabilizing, although directional selection will result whenever a mutually beneficial novel mutation arises. Directional selection will also continually favor individuals that extract the greatest fitness gain from the interaction with the least cost. Hence, there is an inherent coevolutionary tension in all mutualisms that results from the interaction of these multiple forms of selection on the participants.

Overall, coevolving complementarity favors fixation of traits that increase

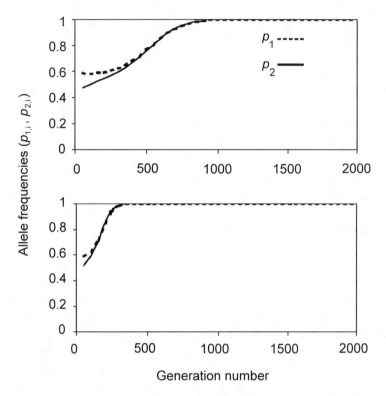

Fig. 5.9 Coevolving complementarity leading to fixation of alleles of two interacting mutualistic species. The top panel shows a slow rate of evolution to fixation of alleles under weak mutualistic selection. Bottom panel shows rapid evolution under strong mutualistic selection. The x axis shows the number of generations and the y axis shows the frequencies of the favored alleles in species 1 (broken line) and species 2 (solid line). The alternative alleles for each species (not shown) undergo a symmetrical decline to local extinction. After Nuismer, Thompson, and Gomulkiewicz 2000.

the probability of interaction between two species and maximize the complementarity (fig. 5.9). As new alleles arise that retain or increase the fitness benefit of the interaction but at reduced cost, they will be favored by selection. Such complementarity contributes to the development of a predictable network structure in mutualisms among free-living mutualistic species, because natural selection favors the spread of mutualism through convergence of species with similar traits (see chapter 14). Additional selective complexity may be embedded within mutualisms if genotypic combinations vary within communities in the degree of symmetry of fitness benefits they provide each other. These asymmetries may impose some negative frequency

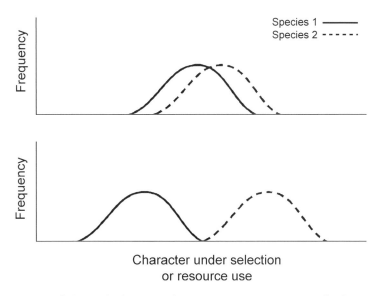

Fig. 5.10   Coevolutionary displacement of sympatric species. Divergence in the character under selection should match divergence in resource use, as indicated by the interchangeable
designations for the *x* axis. The *y* axis denotes the frequency of individuals in each species.
The upper panel shows the pattern before coevolutionary displacement, and the bottom
panel shows the pattern after displacement.

dependence within mutualisms, thereby preventing complete fixation of
genotypes (Bever 1999).

Finally, coevolutionary displacement differs fundamentally from other
classes of coevolutionary dynamics, because selection acts on both species to
decrease the chance that the species will interact (fig. 5.10). Consequently,
coevolutionary selection should decrease over time within local populations,
as the populations diverge in their use of limited resources or habitats. Most
cases of ecological character displacement invoke competition among closely
related species for a locally limiting resource or habitat, but multispecific coevolutionary divergence among unrelated taxa is possible. In addition, non-
linear competitive interactions and apparent competition driven by interactions with shared enemies can favor displacement among species (Holt and
Lawton 1994; Abrams 2000b; Abrams and Chen 2002; Morris, Lewis, and
Godfray 2004).

Recent models of most of these classes of coevolutionary dynamics have
begun to explore the more biologically realistic situation in which local selection varies in intensity or even direction among years or generations (e.g.,

Nuismer, Gomulkiewicz, and Morgan 2003). These models show a broader range of transient dynamics and longer persistence of coevolving polymorphisms than those found in most models of constant coevolutionary selection. Even small differences among populations in the initial genetic structure of an interaction can maintain large differences across future generations in the transient dynamics of a coevolving interaction. At any moment interactions are likely to be at very different points in their transient trajectories in different communities, even if the long-term trajectories are the same across all communities. Together, the results of all local coevolutionary models suggest that local coevolution of real species creates a vast universe of transient coevolutionary change as local populations coevolve along their trajectories toward equilibrium states. The models provide a theoretical underpinning for our understanding of why coevolving interactions generally develop as a geographic mosaic.

Conclusions

We now have strong evidence of rapid evolution of interspecific interactions from laboratory and field studies on native species and introduced species. Many observed evolutionary changes are probably due to fluctuating selection, but these ongoing changes are the interface of evolution and ecology. These are the evolutionary changes that keep populations viable as they react to their constantly changing enemies and mutualists. At the local level, these changes exhibit any of seven general classes of coevolutionary dynamics, which are then further modified by the overall geographic mosaic of coevolution. We turn to the analysis of that mosaic in the next chapter.

6   The Conceptual Framework
The Geographic Mosaic Theory of Coevolution

It is still possible to read in the scientific literature statements asserting that an interaction is not truly coevolved because it "breaks down" in some populations or does not occur at all in others. By that typological, nonbiological definition, an interaction is coevolved only if it shows evidence of reciprocal change in all populations. That view, now mostly past, has become increasingly untenable as we have learned more about the genetic and ecological structure of real species and the hierarchical structure of coevolutionary dynamics. Coevolution is a crucial process in the organization of biodiversity specifically because it is simultaneously flexible yet conservative in how it shapes interspecific interactions in a constantly changing world. The empirical observations outlined in chapters 2 through 5 suggest that coevolution is a dynamic and inherently geographic process.

This chapter explores coevolution as an ongoing process shaped by the geographic structure of interactions among species. It outlines and updates the geographic mosaic theory of coevolution and explores what we have learned about geographically structured interactions in recent years. The framework provides a biologically based explanation for both the ongoing rapid evolution of interactions observed in ecological studies and the long-term persistence of many interactions evident in phylogenetic and paleobiological analyses. The ramifications of the theory outlined in this chapter are explored in more detail in later chapters for specific classes of coevolutionary dynamics.

The Conceptual Framework

The geographic mosaic theory of coevolution developed as an attempt to organize what we have learned about the ecological, genetic, and phylogenetic

Table 6.1   The overall structure of the geographic mosaic theory of coevolution

Major Assumptions

Species are groups of genetically differentiated populations, and most interacting species do not have identical geographic ranges.

Species are phylogenetically conservative in their interactions, and that conservatism often holds interspecific relationships together for long periods of time.

Most local populations specialize their interactions on a few other species.

The ecological outcomes of these interspecific interactions differ among communities.

Species often become locally adapted to local populations of other species and continue to evolve rapidly.

Evolutionary Hypothesis

Geographic selection mosaics. Natural selection on interspecific interactions varies among populations partly because there are geographic differences in how fitness in one species depends upon the distribution of genotypes in another species. That is, there is often a genotype-by-genotype-by-environment interaction in fitnesses of interacting species.

Coevolutionary hotspots. Interactions are subject to reciprocal selection only within some local communities. These coevolutionary hotspots are embedded in a broader matrix of coevolutionary coldspots, where local selection is nonreciprocal.

Trait remixing. The genetic structure of coevolving species also changes through new mutations, gene flow across landscapes, random genetic drift, and extinction of local populations. These processes contribute to the shifting geographic mosaic of coevolution by continually altering the spatial distributions of potentially coevolving alleles and traits.

General Ecological Predictions

Populations differ in the traits shaped by an interaction.

Traits of interacting species are well matched only in communities.

Few coevolved traits spread across all populations to become fixed traits within species, because few coevolved traits will be favored across all populations.

Note: The major assumptions are discussed in chapters 2–5.

structure of real species into a hierarchical framework for coevolutionary analysis (Thompson 1994, 1997, 1999b, 1999e, 2001; Thompson and Cunningham 2002). It attempts a testable framework for the structure of the coevolutionary process (table 6.1), and it continues to develop as new models and data add refinements (e.g., Gomulkiewicz et al. 2000; Benkman, Holimon, and Smith 2001; Brodie, Ridenhour, and Brodie 2002; Burdon, Thrall, and Lawrence 2002; Gandon and Michalakis 2002; Nuismer, Thompson, and Gomulkiewicz 2003; Zangerl and Berenbaum 2003).

Taken as assumptions, the results outlined in chapters 2 through 5 suggest that ongoing coevolution has three attributes not evident at the level of local coevolutionary selection alone.

Geographic selection mosaics. Natural selection on interspecific interactions varies among populations partly because there are geographic differences in how fitness in one species depends upon the distribution of genotypes in another species (box 6.1). Imagine two interacting panmictic species that are both suddenly subdivided into separate populations in different environments. The environments could differ physically (e.g., temperature) or in how these two species interact with yet other species. The two species continue to interact in both environments, but how any particular genotype of the one species affects the fitness of any genotype of the other species may differ between the environments. That is, there is often a genotype-by-genotype-by-environment interaction of fitnesses of interacting species.

Coevolutionary hotspots. Natural selection on an interaction is not reciprocal in all communities in which it occurs (fig. 6.1). Instead, regions in which selection on the interaction is truly reciprocal—coevolutionary hotspots (box 6.1)—will often be embedded within a broader network of habitats and regions in which the interaction imposes selection on only one of the species or on neither species (i.e., coevolutionary coldspots). In a pairwise interaction, a coevolutionary coldspot is one in which selection on at least one of the species does not depend upon the distribution of genotypes in the other species. Since few interacting species co-occur throughout their entire geographic ranges, some regions will be coevolutionary coldspots for an interaction simply because one of the species does not occur there. Such structural coldspots are almost inevitable except for some reciprocally obligate mutualisms. Coevolution therefore does not require that interacting species have completely coincident geographic ranges. Such a view ignores the geographic mosaic of the coevolutionary process.

Trait remixing. New mutations, gene flow across landscapes, random genetic drift, and extinction of local populations further fuel the geographic mosaic of coevolution by altering the spatial distributions of potentially coevolving alleles and traits. Even if an interaction is antagonistic across all populations, local populations may differ at any moment in time in the range of defense and counterdefense alleles on which selection can act.

The geographic mosaic theory of coevolution therefore argues that coevolution is a tripartite evolutionary process that continually remolds evolv-

Box 6.1

General formulations of the conditions specifying a coevolutionary hotspot and a geographic selection mosaic for a pair of species interacting in two environments. The formulations are generalized forms of the expressions for coevolutionary hotspots and geographic selection mosaics used in Gomulkiewicz et al. (2000).

Coevolutionary Hotspot

A coevolutionary hotspot between two species is defined as the condition in which a local interaction exhibits reciprocal selection such that

$$W_{1,i} = f_{1,i}(\phi_{2,i})$$

and

$$W_{2,i} = f_{2,i}(\phi_{1,i})$$

where $W_{1,i}$ and $W_{2,i}$ represent fitnesses of particular genotypes in environment i of species 1 and 2, respectively; $\phi_{1,i}$ and $\phi_{2,i}$ represent the distributions of genotypes in species 1 and 2, respectively, and $f_{1,i}(\phi)$ and $f_{2,i}(\phi)$ are genotype-specific functions describing how fitnesses in species 1 and 2, respectively, depend on the distribution of genotypes in the other species. These expressions imply that local selection within each species typically depends on the distribution of genotypes in the other species. Although the interaction is shown here for two species, reciprocal selection can occur among more than two species.

Coevolutionary Coldspot

All local fitnesses, $W_{k,i}$, in at least one species k are independent of the distribution of genotypes in the other species in environment i.

Geographic Selection Mosaics

Geographic selection mosaics occur when

$$f_{1,i}(\phi) \neq f_{1,j}(\phi) \text{ or } f_{2,i}(\phi) \neq f_{2,j}(\phi) \text{ for some distribution } \phi,$$

where i and j are two locations harboring different populations of species 1 and 2. That is, local *fitness functions* of at least one species must differ between environments i and j. This definition includes the possibility that fitnesses in one environment are independent of the distribution of genotypes in the other species (i.e, the environment is a coevolutionary coldspot).

. . .

Note that local fitnesses affected by an interspecific interaction can vary among environments for two reasons. First the distribution of genotypes of the other species might differ between two environments (e.g., $\phi_{2,i} \neq \phi_{2,j}$), creating geographically variable fitnesses without a true geographic selection mosaic. In this case, even if, for instance, $f_{1,i}(\phi) = f_{1,j}(\phi) = f_1(\phi)$ for all ϕ (i.e., species 1 genotypes have exactly the same fitness functions in both environments), the fitnesses $W_{1,i} \neq W_{1,j}$ since $f_1(\phi_{2,i}) \neq f_1(\phi_{2,j})$ in general. This difference in fitnesses does not indicate a geographic selection mosaic, although it would contribute to the geographic structure of a coevolving interaction. Alternatively, local fitnesses might differ among environments because fitness functions themselves differ between environments, say $f_{1,i}(\phi)$ and $f_{1,j}(\phi)$. *This* is the defining property of a geographic selection mosaic. Note that in this case, it is possible for local fitnesses to vary even when the distribution of genotypes in the other species does not. For example, it is possible that $W_{1,i} \neq W_{1,j}$ even though $\phi_{2,i} = \phi_{2,j} = \phi_2$ because $f_{1,i}(\phi_2) \neq f_{1,j}(\phi_2)$ for the same distribution ϕ_2.

ing species interactions across landscapes (Thompson 1994). Geographic selection mosaics, coevolutionary hotspots, and trait remixing together add a higher level of dynamics to the coevolutionary process than occurs at the level of local populations. Through this dynamic structure, interspecific interactions coevolve across millennia over constantly changing landscapes.

Most coevolution is therefore as much an ecological process as it is an evolutionary process, creating highly dynamic, ongoing reciprocal change across landscapes without necessarily leading to major changes in the diversity of species. It is a process that connects populations and keeps interactions intact, even as some local populations become extinct. Parasites drive their local host populations to extinction in some environments even as the hosts drive their parasites to extinction in other environments. Populations of one competitor species win in warmer or drier environments as populations of the other species win in colder or wetter environments. Mutualists in different populations converge on different combinations of traits and even on different combinations of partners. Individuals with different coevolved traits move among populations, adding to the regional dynamics of the coevolutionary process.

This mixing and matching of genes, traits, outcomes, and participants

Universal coevolutionary hotspots

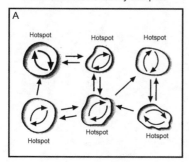

Complex mosaics

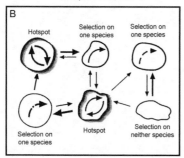

Fig. 6.1   Hypothetical examples of some major components of the geographic mosaic of coevolution between a pair of species. Interactions within local communities are shown as arrows within circles and indicate selection on one or both species. Different arrow types represent differences in how selection acts on the interaction in different communities. Arrows between communities indicate gene flow, with thicker arrows indicating more gene flow. (A) The interaction coevolves in all communities in which it occurs, although it coevolves in different ways among communities. (B) Coevolutionary hotspots occur within in a broader matrix of coevolutionary coldspots.

across landscapes becomes the true ecological structure of interspecific interactions and coevolution. The shifting geographic mosaic becomes every bit as important as the role of coevolution in the diversification of species, because it directly connects evolution with ecology, species with communities, and communities across regions and even continents. It is likely a central part of the explanation for how species persist for so long in the midst of continuing onslaughts from multiple enemy species and ongoing climatic change.

General Predictions of the Geographic Mosaic Theory of Coevolution

The geographic mosaic theory of coevolution makes three overall predictions about the ecological structure and dynamics of coevolving taxa.

Populations differ in the traits shaped by an interaction. Geographic differences in coevolved traits are an inevitable consequence of geographic selection mosaics, coevolutionary hotspots, and trait remixing. These population differences can be manifested in several ways. Coevolving species may show evidence of local adaptation only in some populations. Some populations in an interaction may show stronger evidence of reciprocal adaptation than others (i.e., coevolutionary hotspots). The overall geographic structure of a coevolving interaction may show evidence of clines or mosaics in traits molded by coevolution. The geographic edges of species ranges are especially likely to be highly dynamic zones for the evolution of novel traits in coevolving interactions, as the species reach regions that differ in abiotic conditions or community context.

Traits of interacting species are well matched in only some communities. Gene flow and metapopulation dynamics mix locally coevolved traits with coevolved traits from other populations. Local trait mismatches are therefore likely in at least some populations. The temporal dynamics of coevolution can also produce transient mismatches within local communities, as species respond sequentially to one another. Gene flow and metapopulation dynamics, however, add still more dimensions, creating clines in adaptation across landscapes. In some cases, populations may maintain coevolved traits only through gene flow with other populations.

Few coevolved traits spread across all populations to become fixed traits within species, because few coevolved traits are favored across all populations. This prediction follows from the first two. Newly coevolved traits can arise in any population throughout the range of an interaction, but only a few of these traits will spread beyond their site of origin. Coevolution is such a dynamic process that any analysis of fixed differences in traits among species is bound to underestimate the traits shaped by coevolution.

Selection Mosaics, Coevolutionary Hotspots, and Trait Remixing

CURRENT COEVOLUTIONARY CONTEXT

The accumulating examples of local adaptation in interactions (chapters 4 and 5) provide strong evidence of geographic selection mosaics across landscapes. Most of the examples, however, reflect a combination of current and past selection acting within and among populations. It has been more difficult to evaluate how selection mosaics and coevolutionary hotspots are distributed geographically at any moment in time for any interaction. Yet these

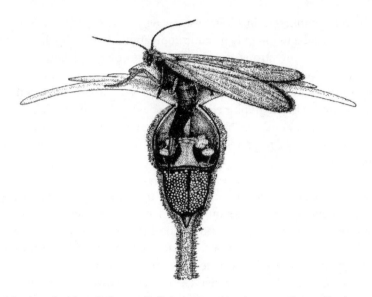

Fig. 6.2   A prodoxid moth *Greya politella* laying eggs into the ovary of a woodland star *Lithophragma parviflorum* (Saxifragaceae) while passively pollinating the flower with pollen adhering to the moth's abdomen. Drawing by Lisa Roberts.

data are crucial for refining our understanding of coevolutionary dynamics, because spatially realistic models of the geographic mosaic of coevolution require empirical examples of the current geographic selection mosaics on which to draw.

The interaction between woodland stars (*Lithophragma parviflorum*) and the moth *Greya politella* shows evidence of the potential for differences in current coevolutionary selection across landscapes. Most populations of *G. politella* in northwestern North America are restricted to feeding on this one plant species. The moths oviposit into the flowers through the corolla and, in the process, passively pollinate the flowers with pollen adhering to the base of their abdomens (Thompson and Pellmyr 1992) (fig. 6.2). Although the moths also nectar on the flowers, pollination does not occur during nectaring (Pellmyr and Thompson 1992). The larvae feed on a small percentage of the hundreds of developing seeds within each flower. Hence, pollination by *Greya* moths imposes a cost to plants for the benefits they provide. In some habitats, the flowers are also visited commonly by copollinators, especially bombyliid flies, and some of these are as effective at pollination as *G. politella* (Pellmyr and Thompson 1996). Because these copollinators do not oviposit into the flowers, pollination by these species imposes a lower cost to the plant. Where the copollinators are abundant, the pollination benefits provided by *Greya* can be outweighed by the costs imposed by larval consumption. Thus there is a strong potential for development of a geographic selection mosaic in this interaction, driven by the relative abundance of *Greya* and copollinators in different habitats.

Across twelve habitats in the northern U.S. Rocky Mountains and adjacent regions, the effect of *G. politella* on pollination varies from mutualism to commensalism to antagonism (Thompson and Cunningham 2002) (fig. 6.3). In some habitats, flowers that receive *Greya* eggs have a higher chance of developing seeds than flowers without eggs, creating the potential for mutualistic hotspots. In other habitats, the probability of developing seeds does not differ between flowers with eggs and flowers without eggs, creating potential coldspots. In yet other habitats, flowers without eggs are selectively aborted in some years, creating potential antagonistic hot spots.

Moreover, the mutualistic, commensalistic, and antagonistic populations are distributed as a mosaic across the region (fig. 6.3). Within this highly mountainous region, populations that appear to be neighbors on a map can actually occur at elevations hundreds of meters apart. Nevertheless, there is no simple geographic pattern to the distribution of the potential coldspots and the two different forms of hotspot. These results are for the outcome of the interaction at the flower and population levels and therefore provide only indirect evidence for mutualistic selection on individual plants in some environments and antagonistic selection in other environments. Instead, the results show that the ecological conditions favoring habitat differences in coevolutionary selection are in place. Some *Lithophragma* populations rely upon *Greya* for pollination, whereas others do not.

SUSTAINED GEOGRAPHIC SELECTION MOSAICS

If current selection on local coevolutionary hotspots is maintained over time, the result is a geographic mosaic of coevolved traits. Research over the past decade has provided several well-studied examples of such mosaics, and these include a diverse array of taxa and forms of interaction: garter snakes and newts, conifers and crossbills, native herbs and fungal pathogens, and invasive plants and insects. These studies illustrate clearly how coevolutionary hotspots can produce strong regional patterns in coevolved traits and are discussed in this and later chapters, beginning here with the elegant studies of coevolution between garter snakes and newts in western North America (fig. 6.4).

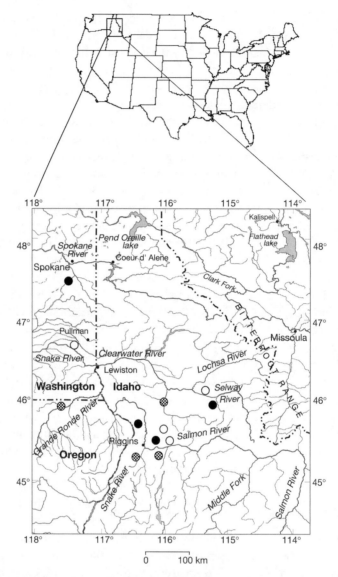

Fig. 6.3 Distribution of a sample of populations of the plant *Lithophragma parviflorum* that rely upon the moth *Greya politella* for pollination at least in some years (black circles), populations that selectively abort flowers with eggs in some years (checked circles), and populations in which the moths have no effect on the probability of a capsule's developing seeds in any year in which it has been studied (open circles). All populations shown have been studied for at least three years, and some populations have been studied for up to seven years. After Thompson and Cunningham 2002.

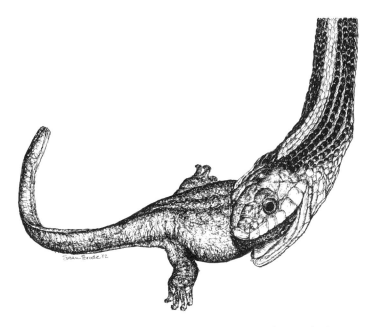

Fig. 6.4   A garter snake (*Thamnophis sirtalis*) eating a newt (*Taricha granulosa*).
Drawing by Susan Brodie.

Newts in the genus *Taricha* and garter snakes in the genus *Thamnophis* are abundant along the North American Pacific coast in southern British Columbia, Washington, Oregon, and California. On some days during the newts' breeding season, it is common to see dozens of individuals during even a short walk along trails. Several species of *Taricha* and *Thamnophis* occur along the Pacific coast, but the most widespread are *Taricha granulosa* and *Thamnophis sirtalis,* and these two species show strong evidence of co-evolution as prey and predator.

All newts in the genus *Taricha* contain in their skin a potent neurotoxin called tetrodotoxin (TTX) (Brodie 1968; Hanifin et al. 1999), which, when ingested by a predator, blocks sodium channels in nerve and muscle tissue (Goldin 2001). The blocked channels are unable to transmit action potentials, and the tissues are therefore paralyzed. After ingesting the toxin, a predator usually dies from respiratory failure (Geffeney et al. 2002). TTX is a highly effective toxin found in species in at least four phyla, including arthropods, mollusks, platyhelminthes, and vertebrates (Brodie, Ridenhour, and Brodie 2002). *Taricha granulosa* is more toxic than other *Taricha* species, and some *T. granulosa* populations harbor very high levels of TTX, enough in some individuals to kill twenty-five thousand white mice (Brodie, Hensel, and Johnson 1974).

The common garter snake, *Thamnophis sirtalis,* is the only potential predator of newts that is resistant to TTX. In fact, it is the only predator known to have high levels of resistance to prey harboring TTX (Brodie and Brodie 1999b). The genus *Thamnophis* has a higher baseline resistance to TTX than related snake genera (Motychak, Brodie, and Brodie 1999), but only some *T. sirtalis* populations along the west coast of North America have resistance elevated to levels that function as effective counterdefenses to TTX. Resistance level is controlled genetically and has a high broad-sense heritability (Brodie and Brodie 1990, 1999b; Ridenhour, Brodie, and Brodie 1999).

TTX resistance provides these garter snakes with an abundant prey species, but it imposes a strong physiological cost. More resistant snakes crawl at slower speeds than those that are less resistant (Brodie and Brodie 1999a). Moreover, after ingesting a newt with TTX, even resistant snakes are incapacitated for anywhere from half an hour to seven hours (Williams, Brodie, and Brodie 2003). This cost of resistance is highly replicable in laboratory racetrack experiments that have tested maternal families of snakes with known levels of TTX resistance. Slower crawling speeds and periods of incapacitation could have potential consequences for fitness in resistant snakes, either by reducing their ability to capture other prey or by making them vulnerable to their own predators. If this physiological cost does, in fact, reflect a fitness cost, then very high levels of resistance are likely to be favored only in populations in which the newts exhibit very high levels of TTX.

Newt toxicity levels and snake resistance levels are closely matched geographically for populations in which the levels have been assessed in both species (fig. 6.5). Moreover, overall levels of newt toxicity and snake resistance vary greatly among populations (Brodie and Brodie 1999a; Hanifin et al. 1999). Populations of the garter snakes outside the range of the newts have low levels of TTX resistance, and levels of TTX resistance in populations within the range of the newts vary by more than an order of magnitude. Among newt populations, those on Texada Island in British Columbia lack TTX, those on the Olympic Peninsula of Washington State have low levels (< 0.2 mg TTX/g skin), and those from west-central Oregon have TTX levels almost an order of magnitude higher (approximately 1.8 mg TTX/g skin).

Extensive testing of resistance levels of garter snake populations along the west coast of North America has suggested two regions that may be long-term coevolutionary hotspots (Brodie, Ridenhour, and Brodie 2002). One hotspot is centered on west-central Oregon, and the other is around San Francisco

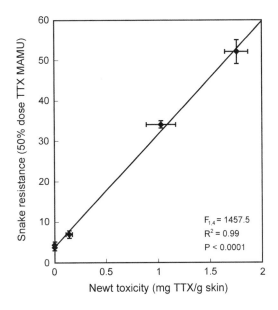

Fig. 6.5   Geographic matching of levels of tetrodotoxin in *Taricha* newts and resistance to tetrodotoxin (TTX) in garter snakes (*Thamnophis sirtalis*) based upon four sympatric pairs of newt and snake populations in far western North America. Combining all populations tested so far, maternal families of snakes vary in their resistance level from 1.1 to 1279 MAMU. A MAMU is a mass-adjusted mouse unit, and a mouse unit, which equals 0.01429 μg TTX, is the amount of TTX that would kill one gram of mouse within ten minutes. MAMU values standardize the mouse units across garter snake populations such that a MAMU is one mouse unit of TTX per gram of snake. After Brodie, Ridenhour, and Brodie 2002.

Bay. These two regions include snake populations with very high levels of TTX resistance, and current results suggest that the TTX resistance levels decrease clinally from these possible hotspots (fig. 6.6). This geographic pattern may result from gene flow among hotspot and coldspot populations or from some environmental gradient controlling overall investment in defense and counterdefense in these species.

Elevated levels of resistance in these two hotspots have arisen independently, as indicated by phylogenetic reconstructions based upon analysis of three mitochondrial DNA genes (Geffeney et al. 2002; Janzen et al. 2002). Populations of *T. sirtalis* from far western North America are divided into three clades: a California clade, a northwestern clade, and an intermountain clade (Janzen et al. 2002). These clades correspond to similar phylogeographic patterns found in other taxa (Calsbeek, Thompson, and Richardson 2003). Comparable data are not yet available for newt populations through-

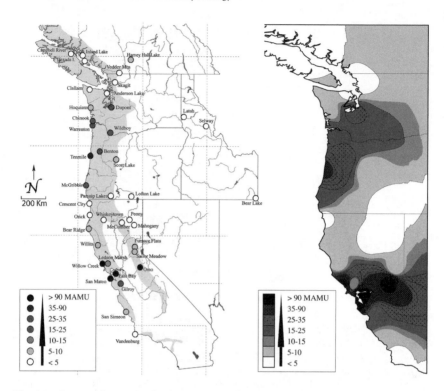

Fig. 6.6   Geographic mosaic of resistance to tetrodotoxin in garter snakes (*Thamnophis sir-talis*) in western North America. Resistance is shown as the mean ability of snakes to handle newts at particular levels of toxicity measured in MAMU units, as described in the caption of fig. 6.5. The map on the left shows actual sampling sites and the mean level of resistance found in snakes from those sites. The geographic distribution of *Taricha* newts is shown in gray. The map on the right shows the results of a model that interpolates the sample points to show isoclines in snake resistance across western North America. After Brodie, Ridenhour, and Brodie 2002.

out the west coast of North America. The results for the garter snakes, however, coupled with the results showing matched levels of toxicity and resistance in a subset of populations (fig. 6.4), provide a clear prediction for future research on geographic patterns of tetrodotoxin levels in newt populations.

The suite of selection pressures favoring not only close matching but also coevolutionary escalation of traits in some regions is not yet known. The current major working hypothesis is that highly resistant but slow snakes may themselves be more susceptible to predation from their own predators. If so, then escalation to very high levels of TTX in the newts and TTX resistance in the snakes may occur only within regions in which predators of the snakes are uncommon and alternative food for the snakes is limited (Brodie and Brodie 1999a, 1999b). The presence of other *Taricha* species in some parts of far western North America could also contribute to the geographic pattern.

If the distribution of coevolutionary hotspots in this interaction is governed by yet other interspecific interactions (i.e., predation on the snakes), then the coevolutionary structure is similar overall to that of *Greya* moths and *Lithophragma,* even though the details of the interactions are completely different. In both cases, the presence of a hotspot requires the regional absence of other interactions that constrain or swamp pairwise coevolutionary selection—predators on snakes in the one case and copollinators in the other. Similarly, in the well-studied coevolutionary interaction between crossbills and lodgepole pines (see chapter 7), coevolution between this pair of species is restricted to regions of the northern Rockies where red squirrels are not present (Benkman 1999; Benkman, Holimon, and Smith 2001). Where red squirrels are present, selection on pines imposed by the squirrels swamps selection imposed by the crossbills.

Coevolutionary hotspots may therefore commonly be driven by geographic variation in the community context of interactions. Coevolutionary selection on pairs or small groups of antagonists or mutualists may be strong and consistent in only some communities, where the effects of other interactions are mitigated. In yet other interactions, strong coevolutionary selection may depend on the presence rather than the absence of other interactions. For example, some defense mutualisms between ants and plants should be strongest in communities in which herbivory or competition imposed by neighboring plants is greatest. Given the tremendous variation in community composition across the earth's complex landscapes, most widespread coevolving interactions probably exhibit coevolutionary hotspots of varying intensity as well as coldspots.

Formalizing the Dynamics of the Geographic Mosaic

MODELS WITHOUT GEOGRAPHIC SELECTION MOSAICS

A diverse group of mathematical models has now evaluated different components of the geographic mosaic of coevolution. All the models have simplified genetic structure, geographic structure, and population dynamics, and there is much still to do in evaluating which results are robust and which rely upon the structure of particular models. Collectively, however, the models have begun to characterize how the geographic structure of species and interactions shapes coevolution across landscapes.

Some models have used a simple matching-allele structure to explore the dynamics of the geographic mosaic of coevolution (box 6.2). In the simplest matching-allele models, allele Y in species 1 matches allele Z in species 2; and allele y in species 1 matches allele z in species 2. If the interaction is antagonistic, as in interactions between parasites and hosts, natural selection favors parasites that match the complementary allele in their hosts (e.g., Y-Z), but it favors hosts that are mismatched to the parasites (Y-z). That is, natural selection favors enemies that are adapted to their victims, and it favors victims that harbor alleles to which the enemies are not adapted. The result at the local level is often fluctuating allele frequencies, in which negative frequency-dependent selection favors hosts of the rarer genotype. Antagonistic interactions between parasites and hosts or predators and prey may therefore sometimes favor allele frequencies in the interacting species, as natural selection favors rare host genotypes (May and Anderson 1983; Seger 1988, 1992). With few interacting alleles and no spatial structure, the models often show increasing swings in allele frequency over time, leading eventually to extinction of one or both populations as alleles are lost or fixed. Sustained fluctuating polymorphisms have been possible in some genetic models, mostly when time lags in evolutionary response have been added or values of variables are maintained within certain limits (Antonovics and Thrall 1994; Morand, Manning, and Woolhouse 1996; Agrawal and Lively 2003).

If, however, the interaction is mutualistic between a pair of species in matching-allele models, then natural selection favors matching of alleles within local populations of both species through positive frequency-dependent selection (Nuismer, Thompson, and Gomulkiewicz 1999; Gomulkiewicz et al. 2000). Genotypes of one species that most closely match those of the other species are favored, and those matched alleles increase to fixation within the local interacting populations. Polymorphisms in both species are therefore transient. Box 6.2 shows a simple model using a matching-allele structure that has been evaluated both for local coevolutionary dynamics in the absence of gene flow and for coevolution between two sets of coevolving populations linked by gene flow.

Such models seem biologically oversimplified, but they can provide important insights into the range of local nonequilibrium dynamics possible between interacting species. The simplest way of incorporating geographic structure into these and other models is to assume that all environments are

Box 6.2

A simple matching-alleles model of the geographic mosaic of coevolution. The local coevolution version of this model was explored by Seger (1988) and Gavrilets and Hastings (1998), and the geographic mosaic version was explored by Nuismer, Thompson, and Gomulkiewicz (1999, 2000, 2003), Gomulkiewicz, Nuismer, and Thompson (2003), Gomulkiewicz et al. (2000), and Thompson, Nuismer, and Gomulkiewicz (2002).

Assumptions

The interaction occurs between a pair of haploid species with discrete generations. The coevolutionary interactions are determined by a single locus with two alleles. Gene flow occurs at rate m and is the same for both species, occurring each generation before natural selection.

Definitions

· Species 1 has alleles Y and y with frequencies of $p_{1,i}$ and $(1 - p_{1,i})$ respectively.
· Species 2 has alleles Z and z with frequencies of $p_{2,i}$ and $(1 - p_{2,i})$ respectively.

In antagonistic trophic interactions, individuals of species 1 always receive a fitness gain by matching corresponding alleles with individuals of species 2 (e.g., Y-Z but not Y-z), but individuals of species 2 receives a fitness loss in antagonistic interactions by matching alleles with individuals of species 1.

In mutualistic interactions, individuals of both species receive a fitness gain by matching corresponding alleles in the other species (i.e., Y-Z or y-z).

In competitive interactions, individuals of both species receive a fitness gain by mismatching corresponding alleles in the other species (i.e., Y-z or y-Z).

Fitnesses and Recursions within Coevolutionary Hotspots

Fitnesses within a single geographic location i are:

$$W_{Y,i} = 1 + c_i p_{2,i},$$
$$W_{y,i} = 1 + c_i(1 - p_{2,i}),$$
$$W_{Z,i} = 1 + b_i p_{1,i},$$
$$W_{z,i} = 1 + b_i(1 - p_{1,i}),$$

where

$W_{x,i}$ is the fitness of genotype x at location i

(Box 6.2 continued on page 114)

(Box 6.2 continued)

c_i is a parameter that specifies the fitness sensitivity of species 1 to changes in the frequency of its matching allele in species 2 at location i

b_i is a parameter that specifies the fitness sensitivity of species 2 to changes in the frequency of its matching alleles in species 1 at location i

If mutation and random genetic drift are weak relative to natural selection, then

$$p^*_{1,i} = \frac{p_{1,i}W_{Y,i}}{p_{1,i}W_{Y,i} + (1 - p_{1,i})W_{y,i}}$$

$$p^*_{2,i} = \frac{p_{2,i}W_{Z,i}}{p_{2,i}W_{Z,i} + (1 - p_{2,i})W_{z,i}}$$

where $p^*_{k,i}$ is the allele frequency after selection for species k at position i after selection.

The rate of change in allele frequencies is determined by the strength of reciprocal selection (i.e., the magnitudes of parameters c_i and b_i).

Geographic Selection Mosaics

Geographic selection mosaics are represented by differences among locations in the relationships between c_i and b_i.

For location i the interaction is antagonistic—in the sense that selection favors matching alleles in one species but mismatching in the other—if $c_i > 0$, $b_i < 0$ or $c_i < 0$, $b_i > 0$.

The interaction is mutualistic—in the sense that selection favors matching alleles in both species—if c_i and b_i are both positive.

The interaction is competitive if c_i and b_i are both negative.

The interaction is commensalistic if either $b_i = 0$ or $c_i = 0$. This condition creates a coevolutionary coldspot, as would the condition in which $b_i = c_i = 0$.

Gene Flow

Assume that the movement of genes among populations is constant through time, symmetric in space, independent of local fitness, and occurs after selection. If in every generation each population is composed of a proportion of migrants $M_k(i,j)$, then

$$p'_{k,i} = \sum_{j=1}^{n} M_k(i,j) p^*_{k,j},$$

where n is the total number of populations and $p^*_{k,j}$ is the allele frequency of species k at location j after selection.

$M_k(i,j)$ can assume any form but is structured here in a Gaussian form so that the proportion of population i composed of migrants from population j decreases with distance from population j and increases with the migration variance of species σ^2_k

$$M_k(i,j) = \frac{exp\left[-\frac{1}{2}\left(\frac{j-1}{\sigma_k}\right)^2\right]}{\sum\limits_{j=1}^{n} exp\left[-\frac{1}{2}\left(\frac{j-1}{\sigma_k}\right)^2\right]}$$

equal so that the models lack geographic selection mosaics and coevolutionary hotspots. These are, in effect, null spatial models for the geographic mosaic theory, incorporating only nonequal movement of individuals across landscapes or gene flow among populations. Such models suggest that even local spatial structure can stabilize some interactions or at least increase their persistence over longer periods of time (Leonard 1998).

Metapopulation dynamics are the next step in moving beyond local coevolutionary dynamics to broader geographic dynamics (Thrall and Burdon 2002). Most metapopulation models that have been developed in recent years assume a set of local populations connected by gene flow but lack a geographic selection mosaic or coevolutionary hotspots. The metapopulation models of interactions between parasites and hosts developed by Gandon and colleagues (Gandon et al. 1996; Gandon 2002a; Gandon and Michalakis 2002), for example, create patterns of local adaptation much different from those observed in models of locally coevolving species. The models explore different spatial and genetic structures. Their initial metapopulation model used a landscape of N populations arranged in a two-dimensional torus to eliminate the different dynamics that arise at the edges of geographic ranges (Gandon et al. 1996). All populations have equal carrying capacity, and a small percentage of individuals from each population disperses during each generation to the surrounding four populations. As the parasites become locally adapted, natural selection favors resistant genotypes in the hosts. Within these models, local host resistance to sympatric parasites depends upon the frequencies of the host and parasite types, the number of populations where the hosts and parasite coexist, the rate of gene flow among populations, and the mutation rate of the host and parasite.

The model uses a simple haploid version of a matching-allele model with multiple alleles at a single locus within each species. Parasites are able to attack all host genotypes except those that have a specific R gene making them resistant to that particular parasite genotype. Each parasite type is able to attack $(n - 1)$ host types, where n is the number of host and parasite types. If all host and parasite types are present within a local population, then the probability of resistance in the host is $1/n$ (Gandon et al. 1996). Consequently, as the number of coevolving alleles increases, hosts become susceptible to an increasing number of parasites.

If the distribution of host and parasite types is initially distributed randomly among the populations, the coevolutionary dynamics in this kind of model depend upon the initial conditions and the subsequent mutation rates and gene flow rates. In the absence of new mutations, gene flow has strong effects on the local pattern of host resistance. If gene flow occurs only in the host, then low levels of gene movement result in high host resistance and high rates of parasite extinction. High levels of gene flow in the host result in an overall lower level of local resistance, because local host adaptation is decreased as susceptible alleles are continually added to the host population. If gene flow occurs only in the parasites, then the probability of resistance in the host populations increases until it reaches an asymptote when all parasite types are locally present. Under this condition, all the host populations are favored for resistance against all the parasite types, but the parasites show adaptation to the frequency of types in the local host population. If gene flow occurs among host populations and among parasite populations, then the probability of resistance converges on $1/n$. When gene flow is equal in the two species, local adaptation occurs in neither species. A high mutation rate in these models acts in ways similar to gene flow by increasing the local number of types of parasites and decreasing the rate of parasite extinction.

Simulation results of the model suggest that current levels of host resistance and patterns of local adaptation depend upon the number of alleles involved in the interaction, mutation rates, the magnitude of gene flow among populations, and the relative levels of gene flow among host and parasite populations. Adjusting any of these factors changes some of the dynamics and the specific conditions under which local adaptation occurs, but not the overall qualitative results (Gandon 2002a; Gandon and Michalakis 2002). Sustained coevolution between a parasite and host under constant selection becomes more likely in a metapopulation than in a local population. Metapopulation structure and broader geographic structure stabilize the interaction in two ways: through a demographic effect that allows regional persistence of the interacting populations despite local instability, as has been suggested in a number of models of metapopulation dynamics, and a genetic effect that stabilizes the coevolutionary dynamics by allowing regional persistence of the polymorphisms. From these results, it appears that geographic differences in patterns of gene flow among metapopulations will result in strong geographic differences in the coevolutionary dynamics of interacting parasites and hosts.

ADDING GEOGRAPHIC SELECTION MOSAICS AMONG COEVOLUTIONARY HOTSPOTS

Allowing fitnesses of genotypes to vary among coevolutionary hotspots creates additional dynamics and patterns of coadaptation across landscapes. One approach has been to develop explicit genetic models based upon geographic variation in coevolutionary selection. Again, the simplest case is a pair of species, living in two different locations, that coevolve through matching alleles (box 6.2). Although the species coevolve at both locations, the forms of fitness interaction differ between the locations. Under these conditions, the dynamics of local coevolution depend upon the form of coevolutionary selection in each location, the relative intensity of coevolutionary selection in the two locations, and the pattern of gene flow between populations of the two interacting species (Nuismer, Thompson, and Gomulkiewicz 1999). The model therefore incorporates major components of the geographic mosaic, although it does not include extinction and recolonization of populations or coevolutionary coldspots. The most extreme version of the model is one in which selection on the interaction is antagonistic in one environment but mutualistic in another, as occurs in the interaction between *Lithophragma* plants and *Greya* moths (Thompson and Cunningham 2002; Gomulkiewicz et al. 2000).

Under these conditions, allele frequencies are selected to oscillate in both species in the environment in which the interaction is antagonistic, as the parasite or predator evolves to match the host or prey genotype and the host or predator evolves to mismatch the parasite or predator genotype (fig. 6.7A). In the absence of gene flow, the oscillations occur and often increase in amplitude over time and never stabilize. By comparison, isolated mutualistic interactions rapidly become fixed for one set of alleles in both species through selection to match alleles (fig. 6.7A). If the two sets of interaction are linked by gene flow, very different dynamics occur (Nuismer, Thompson, and Go-mulkiewicz 1999; Gomulkiewicz et al. 2000). If selection on the mutualistic interaction is stronger ("hotter") than selection on the antagonistic interaction, then equal gene flow in both species between the two environments can result in evolution toward fixation of alleles even in the antagonistic environment (fig. 6.7B, C, and D).

These results suggest that an evolutionary ecologist studying coevolution could document antagonistic selection within one of the local environments, predict selection toward fluctuating polymorphisms, but simply be wrong because the overall coevolutionary trajectory across the two environments is toward fixation of the alleles. If, however, selection is stronger in the antagonistic environment than in the mutualistic environment, then allele frequencies would exhibit fluctuating polymorphisms in both environments, driven by negative frequency-dependent selection in the antagonistic environment. These results caution against any broad interpretation of coevolutionary dynamics from studies of coevolutionary selection on a particular interaction within a single local community.

If an interaction is distributed in these models across a group of populations, rather than just two, then the geographic structure and dynamics of coadaptation show a clinal pattern. The coevolving interaction can produce many different degrees of local adaptation across the cline (Nuismer, Thompson, and Gomulkiewicz 2000, 2003). Such clines in coadaptation are strong evidence that a coevolving interaction is geographically structured. Even if selection is homogeneous across the landscape, a long-term transient cline can develop if the allele frequencies differed initially among the populations prior to development of gene flow (fig. 6.8). Over time, the interactions would reach an equilibrium that is the same across all populations. In some cases, however, the time to equilibrium is so long that it is unlikely to be reached before the pattern of selection changes in at least some populations along the cline. In the simulations shown in figure 6.8, for example, the transient dynamics persist for four thousand generations, which is longer for some species than the number of generations since the end of the Pleistocene.

If we now add a geographic selection mosaic among populations, then even more persistent dynamic clines can result. Examples of such a mosaic include geographic differences in parasite virulence or host defense. An extreme example would be an interaction that is antagonistic in one region but mutualistic in another (fig. 6.9). One possible outcome of this kind of structure is the development of persistent antagonistic oscillations in the antagonistic populations farthest removed from the mutualistic populations, and

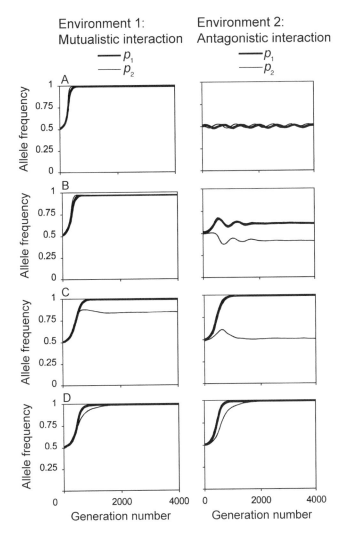

Fig. 6.7 Coevolutionary dynamics between a pair of species interacting in two environments. Within this matching-alleles model, each haploid species has two alleles but only one is shown; p_1 is the frequency of one of the alleles in the parasite population, and p_2 is the frequency of one of the alleles in the host population. In one environment the interaction is mutualistic, but in the other it is antagonistic. In the example shown, coevolutionary selection is stronger in the environment where the interaction is mutualistic than in the environment where it is antagonistic. (A) No gene flow between populations across the two environments. (B) Populations of both species are connected by gene flow across the environments at the rate of 0.002. (C) Gene flow at the rate of 0.015. (D) Gene flow at the rate of 0.05. After Nuismer, Thompson, and Gomulkiewicz 1999.

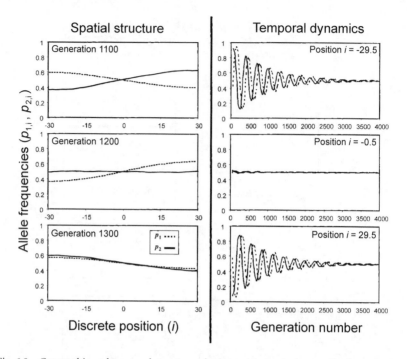

Fig. 6.8 Geographic and temporal structure of a dynamic coevolutionary cline in a parasite-host or predator-prey interaction. The three panels on the left show the spatial pattern of allele frequencies among populations at three points in time. The three panels on the right show the change in allele frequencies over time at three geographical positions. The model has a structure similar to that used in fig. 6.7 but generalized for gene flow across multiple populations. After Nuismer, Thompson, and Gomulkiewicz 2000.

fixation of matched alleles only in the mutualistic populations farthest removed from the antagonistic populations. Intermediate populations show different degrees of ongoing oscillation over time, and they reach different means in allele frequency. These persistent dynamic clines appear to be most likely to develop wherever reciprocal selection is relatively strong and gene flow is relatively low (Nuismer, Thompson, and Gomulkiewicz 2000). As gene flow increases relative to selection, equilibrium clines can develop.

ADDING COEVOLUTIONARY COLDSPOTS

Using the same overall model structure as Nuismer, Thompson, and Gomulkiewicz (1999, 2000, 2003), Gomulkiewicz et al. (2000), and Gomulkiewicz, Nuismer, and Thompson (2003) have shown that coevolutionary selection need not be ubiquitous for coevolution to have important effects on

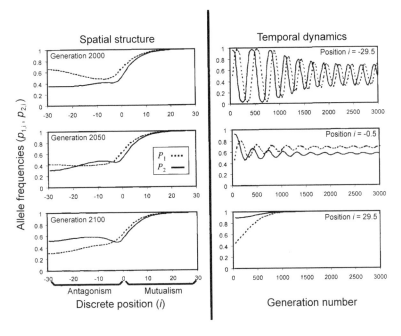

Fig. 6.9   Dynamic coevolutionary clines across a landscape in which an interaction between a symbiont and host is antagonistic in one region and mutualistic in another region. Natural selection is strong in both regions, and gene flow is low but persistent. The three panels on the left show the distribution of allele frequencies among populations at three points in time. The three panels on the right show the change in allele frequencies over time at three positions. See the caption for fig. 6.8 for further explanation of the model structure. After Nuismer, Thompson, and Gomulkiewicz 2000.

the overall evolutionary dynamics of an interaction. Strong coevolutionary hotspots can have large influences on the overall dynamics and trajectory of an interaction, if the populations are connected by gene flow. An extreme example is a coevolving interaction in which a parasite has a detrimental effect on host fitness only in some populations. These populations are coevolutionary hotspots, because they impose reciprocal selection on the interacting species. In other regions, however, the parasite is actually a commensalistic symbiont. Selection acts on the symbiont to become adapted to the host, but the symbiont imposes no detrimental effect on host fitness and therefore no reciprocal selection (i.e., coevolutionary coldspots). In the absence of gene flow between neighboring populations in this model, the interaction within the hotspot produces rapidly increasing oscillations in allele frequencies in both species until the species fix one allele and lose the other. The parallel

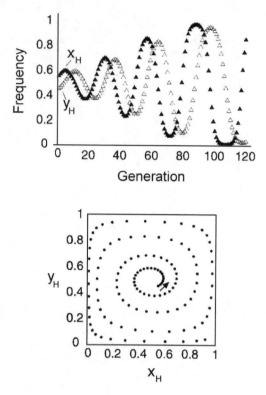

Fig. 6.10   Increasing oscillations in allele frequencies in a parasite (open triangles) and host (filled triangles) in a coevolutionary hotspot, shown on the top as a temporal graph and on the bottom as a phase-plane diagram. In the phase-plane diagram each point shows the current allele frequency of the host (x_H) and the parasite (y_H), beginning at the center, where both frequencies are 0.5. After Gomulkiewicz et al. 2000.

increasing oscillations in both species can also be readily seen in a phase-plane diagram. Unlike a temporal graph, which plots the allele frequencies of the two species separately, a phase-plane diagram plots the two frequencies at any moment in time as a single point on a bivariate graph (fig. 6.10).

The initial polymorphism is lost even faster in a region where the interaction is not antagonistic (i.e., a coevolutionary coldspot) in the absence of gene flow (fig. 6.11A). If the hotspot and coldspot are linked by gene flow (fig. 6.11B, C), polymorphism is maintained across both regions for a longer period of time. Depending upon the question being asked, these novel dynamics could be interpreted as the effect of the hotspot on dynamics in the

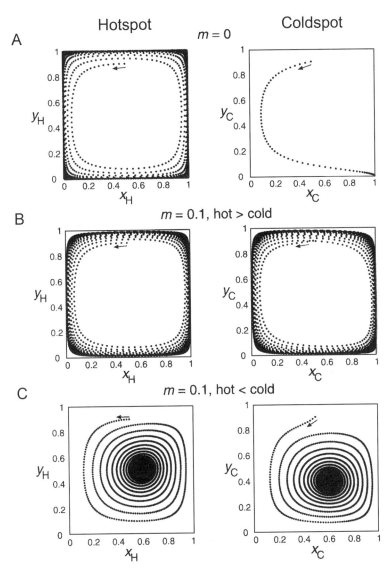

Fig. 6.11 Dynamics of allele frequencies in a pair of species interacting in a coevolutionary hotspot and coldspot. The symbols x_H, y_H, x_C, and y_C refer to the allele frequency of the host (x) and parasite (y) within the hotspot (H) and coldspot (C), respectively. (A) No gene flow between the hotspot and coldspot. (B) and (C) Symmetric gene flow (m) with different relative strengths of selection acting in the hotspot and coldspot. After Gomulkiewicz et al. 2000.

A. Hotspot at the edge of a species range

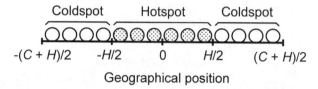

B. Hotspot in the center of a species range

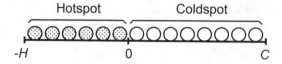

Geographical position

Fig. 6.12   Two extremes in the possible distribution of coevolutionary hotspots and cold-spots along a linear group of populations connected by gene flow. The relative total length of the coevolutionary hotspot (H) and coldspot(s) (C) is the same in (A) and (B). After Nuismer, Thompson, and Gomulkiewicz 2003.

coldspot or the effect of the coldspot on the local coevolutionary outcome. Regardless of the perspective, linked hotspots and coldspots create novel dynamics, which, under some conditions, maintain local polymorphisms over more generations than in the absence of gene flow.

If the coevolutionary hotspots and coldspots are distributed along a linear series of populations linked by gene flow (fig. 6.12), then clines can result that depend on where the hotspots occur along the continuum. When the hotspot is within the geographic center of the distribution of populations, gene flow occurs from both sides. Under this geographic configuration, high levels of gene flow can swamp a small hotspot and prevent formation of clines in coevolving traits across the landscape (Nuismer, Thompson, and Gomulkiewicz 2003). When the hotspot is at the periphery, gene flow from the coldspot is from one side only, increasing the range of conditions that lead to clines in coevolving traits across landscapes.

These genetic models do not include explicit effects of fluctuating population numbers on coevolutionary dynamics, although the Gomulkiewicz et al. (2000) models incorporate different forms of population regulation. One approach to incorporating population dynamics has been to formulate the models using the familiar structure of Lotka-Volterra population models and then evaluate coevolution across interacting populations that differ geographically in productivity (i.e., birth rates) (Hochberg and van Baalen 1998; Hochberg et al. 2000; Sasaki, Kawaguchi, and Yoshimori 2002). These demographic models have explored how coevolutionary hotspots may develop and become distributed across landscapes as local populations of predators and prey or of parasites and hosts differ in productivity. These models have also evaluated how antagonistic or commensalistic interactions could become mutualistic at the edges of geographic ranges as interactions convert local demographic sinks into populations with positive growth rates.

The demographic models begin with the empirical observation that some environments support higher reproductive rates than others (Hochberg and van Baalen 1998). The models then create a gradient in predator and prey productivity from the center to the edge of the geographic ranges of both species. Within the models, productivity is highest at the center of species ranges, decreasing gradually outward to the edge, where the populations form demographic sinks. The analyses suggest that coevolutionary selection for investment in the interaction (i.e., the level of defense in the prey and counterdefense in the predator) will be highest at the center of the range, where prey productivity is highest (Hochberg and van Baalen 1998). Because central populations produce more offspring than peripheral populations, gene flow in the models is mostly from the center to the edge. Consequently, the addition of gene flow results in peripheral populations that show more coevolutionary investment than would be predicted from local coevolutionary selection alone. These models therefore suggest that demographic gradients from the centers to edges of geographic ranges can create coevolutionary gradients of hotspots to coldspots.

LIMITATIONS OF THE MODELS

Additional approaches to modeling the geographic mosaic of coevolution continue to develop, which will help refine our understanding of how geographic selection mosaics, coevolutionary hotspots, and trait remixing collectively shape the long-term trajectories of coevolution across landscapes (e.g., Case and Taper 2000; Sasaki, Kawaguchi, and Yoshimori 2002; Thrall and Burdon 2002). Some future modeling needs are evident. The current geographic models assume that geographic selection mosaics and coevolutionary hotspots are static. We already know that local temporal dynamics in selection can contribute importantly to coevolutionary change. Time lags in local coevolutionary responses (Lively 1999), historical events such as which genotypes colonize an area first (Parker 1999a), and temporally varying selection (Nuismer, Gomulkiewicz, and Morgan 2003) can all shape local co-evolutionary dynamics and therefore potentially change the geographic dynamics. The challenge is to separate the effects of space and time. The main effect of temporally varying selection is to maintain local polymorphisms and, in some interactions, create temporal fluctuations in local coadaptation and maladaptation. Local temporally varying selection alone, however, does not explain most of what we need to know about the coevolutionary dynamics of real species: how the geographic structure found in most species shapes coevolution, how coevolution produces complex geographic patterns of trait distribution, and how interactions persist for thousands or millions of years despite the fact that local populations are often transient. Only by combining local temporal dynamics with geographic dynamics can we understand how real species coevolve.

None of the current models of coevolution has therefore come to grips with complex geographic mosaics. Imagine interactions with strong geographic selection mosaics, coevolutionary hotspots scattered across a matrix of coevolutionary coldspots, differing rates of gene flow among populations, local population fluctuations that change the intensity and direction of selection and the rate of genetic drift, and occasional local extinction of interactions. This is probably the common condition of many coevolving interactions.

Sorting out the geographic scale at which most coevolutionary dynamics take place will demand not only more studies of the same interaction across multiple communities but also analyses of geographic structures that are more complex than those found in the current models. Coevolutionary dynamics may be driven in part by the spatial scale of synchrony in allelic fluctuations (Nuismer, Thompson, and Gomulkiewicz 2000). Fluctuations among neighboring communities may be in closer synchrony than distant populations. These general effects of spatial autocorrelation on coevolutionary dynamics are unknown for natural populations. Fortunately, methods for analyzing spatial autocorrelation in ecological and evolutionary data continue to improve (Koenig 1999), providing a basis for developing techniques applicable to coevolutionary studies.

IMPLICATIONS OF THE MODELS: LONG-TERM PERSISTENCE, DEMOGRAPHY, SPECIES' RANGES, AND MALADAPTATION

Despite their limited complexity, the current models indicate that geographic selection mosaics, coevolutionary hotspots, and trait remixing can all contribute importantly to the long-term coevolution of species. Geographically structured models often allow global persistence of interactions over longer periods of time than models lacking spatial structure. Metapopulation structure and gene flow continually shift combinations of genes and genotypes across landscapes as local populations become demographic sources or sinks. As coevolution proceeds, local and regional adaptation or extinction becomes only part of the raw material that shapes the pattern of long-term persistence in coevolving taxa. Future models that link demographic and genetic approaches over multiple geographic scales will further refine our understanding of how local, regional, and broader biogeographic processes shape the long-term persistence of coevolving taxa.

Current models also suggest that the geographic mosaic of coevolution may shape the geographic ranges of interacting species. That is, not only may geographic structure affect coevolution, but the coevolutionary process itself may reshape the geographic distribution of species. The potential for such feedback is evident in models of coevolution that allow gene flow from the center to the periphery of species ranges (Hochberg and van Baalen 1998; Case and Taper 2000; Hochberg et al. 2000; Nuismer and Kirkpatrick 2003). Combining demographic structure with explicit quantitative genetic structure in a geographic model of coevolution can produce conditions in which gene flow among host populations limits the geographic range of a parasite (Nuismer and Kirkpatrick 2003). Increasing gene flow among host populations can lead to decreases in parasite density during coevolution and subsequent contraction of parasite ranges. As in single-species models and competitive models of the effects of gene flow on geographic range (Kirkpatrick and Barton 1997; Case and Taper 2000; Ronce and Kirkpatrick 2001), this result depends upon geographic differences in the optimum gene combinations favored by stabilizing selection. If it holds within natural populations, then gene flow between the interior and edges of geographic ranges could create dynamic boundaries that fluctuate over time as coevolutionary hotspots or coldspots change across landscapes. The coldspots would include both functional coldspots (e.g., areas of commensalism) and structural coldspots in which peripheral host populations lack parasites.

The models also suggest that geographic selection mosaics, coevolutionary hotspots, and gene flow may commonly create some degree of local maladaptation in coevolving interactions (Lively 1999; Gandon 2002a; Thompson, Nuismer, and Gomulkiewicz 2002; Gomulkiewicz, Nuismer, and Thompson 2003; Nuismer and Kirkpatrick 2003; Nuismer, Thompson, and Gomulkiewicz 2003). When Gould and Lewontin (1979) accused many biologists of viewing organisms as perfectly adapted to their environments in a

Panglossian world, they were in effect reacting to a research approach that interpreted the traits of species as a direct result of current selection in local environments. The adaptations of organisms are indeed often remarkably good, but the past few decades of research have shown us just how jury-rigged are the processes of adaptation and coadaptation. The recent models of the geographic mosaic of coevolution have indicated that populations may often be at least somewhat locally maladapted in their interactions with one another. Local adaptation may be thwarted to varying degrees by the distribution of coevolutionary hotspots and by patterns of gene flow. Random genetic drift during periods of low population numbers is also likely to contribute to the geographic mix of adaptation and maladaptation. The stronger the selection mosaic across geographic landscapes and the higher the gene flow, the greater the likelihood that connected populations will show local maladaptation.

Local maladaptation driven by the coevolutionary process can be evaluated in several ways, and each way suggests something different about the coevolutionary process or the patterns that result (Crespi 2000; Gandon and Michalakis 2002; Thompson, Nuismer, and Gomulkiewicz 2002). The degree of local phenotypic matching between species in the means of coevolving traits provides one measure. The means, however, are interpretable only when evaluated against some standard, such as a predicted optimum value (e.g., from enzyme kinetics). Alternatively, performance (e.g., host resistance, prey escape success, parasite infectivity, or predator handling time) can be measured on sympatric and allopatric populations of the other species to allow an assessment of differential adaptation. The geographic structure of maladaptation could then include an evaluation of either the performance gradient across landscapes or the local performance of a population relative to its average performance in all other populations (Gandon and Michalakis 2002). Yet another approach is to evaluate the deviation between the current best-adapted phenotype in a population and the current average phenotype. This approach provides a measure of the current evolutionary lag in a population under directional selection (Gandon and Michalakis 2002).

Even in the absence of the geographic mosaic, coevolved traits will not always be well matched within local populations by any of these measures. The temporal dynamics of negative frequency-dependent selection in interactions between parasites and hosts, for example, will create transient, oscillating patterns of relative local adaptation and maladaptation over time (Dybdahl and Lively 1998; Kaltz and Shykoff 1998; Lively 1999; Nuismer, Gomulkiewicz, and Morgan 2003). This is maladaptation only in the structural sense produced by the temporal dynamics of the coevolutionary process. Each evolutionary shift in one partner changes the position of the adaptive peaks for the other partner. Local populations must track these constantly changing coevolutionary adaptive peaks. Because the frequencies of genotypes cannot change instantaneously within populations, these coevolutionary responses will show time lags, as has been observed within coevolving interactions (Dybdahl and Lively 1998). Point samples of the coevolutionary structure of local populations are therefore likely to show at least some degree of mismatching of coevolved traits relative to the current local optimum, unless selection is highly intense and the evolutionary response time of both or all species is very fast. Such fast response may occur in some species, as indicated by the rapid evolutionary changes that have been observed in some local interspecific interactions (Dybdahl and Lively 1998; Burdon and Thrall 1999; Zangerl and Berenbaum 2003). More likely, any interaction studied across multiple populations will show a range in the degree of matching and mismatching of locally coevolved traits.

The geographic mosaic, however, can further modify the local and regional patterns of coevolutionary mismatches (fig. 6.13). Depending upon the level of gene flow among populations, local populations within coevolutionary hotspots could often be far removed from what would otherwise be a local adaptive peak for each of the interacting populations. If groups of connected populations exhibit coevolutionary hotspots amid coldspots, then clines in adaptation can occur, with most intermediate populations along the cline exhibiting small to moderate maladaptation (Nuismer, Thompson, and Gomulkiewicz 2003). If, however, the interaction affects both the gene frequencies and the densities of the interacting populations, this geographic effect may potentially be mitigated or modified in other ways (Nuismer and Kirkpatrick 2003).

Coevolutionary models showing the interaction of gene flow, relative population sizes, and mutation show the wide range of the degree of adaptation that can result even in the absence of geographic selection mosaics (fig. 6.14). These models show a high degree of local adaptation in either parasites or hosts under only a subset of the range of possible conditions. Because all populations fluctuate in size and in number of immigrants over time, local populations will continually shift in the degree of local adaptation shaped by coevolution.

Empirical studies of species interactions have increasingly shown evidence of local maladaptation in some populations of interacting species. Examples include the lack of local matching in chemical defenses and counter-

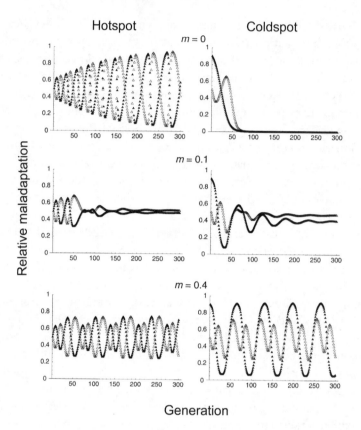

Fig. 6.13   Changes over time in local maladaptation in two local populations of a coevolving parasite (open triangles) and its host (filled triangles). One pair of populations is in a coevolutionary hotspot, whereas the other is in a coevolutionary coldspot. The model is a simple matching-alleles model with two alleles in each haploid species. Only one allele is shown for each species. In the top pair of panels, there is no gene flow between the hotspot and coldspot populations. The bottoms of two sets of panels show the effects of different degrees of gene flow (m) on the local temporal dynamics of maladaptation. Maladaptation in this model is assessed relative to the local adaptive peak for that population at that moment in time. After Thompson, Nuismer, and Gomulkiewicz 2002, based upon the model structure of Gomulkiewicz et al. 2000.

defenses in some populations of wild parsnip and parsnip webworm (Berenbaum and Zangerl 1998; Zangerl and Berenbaum 2003), lower infectivity of a population of the parasitic plant *Cuscuta europea* on its sympatric host population than on allopatric populations (Koskela, Salonen, and Mutikainen 2000), and antipredator defenses in *Ambystoma barbouri* salamanders in some streams (Storfer and Sih 1998). Some instances of local maladaptation

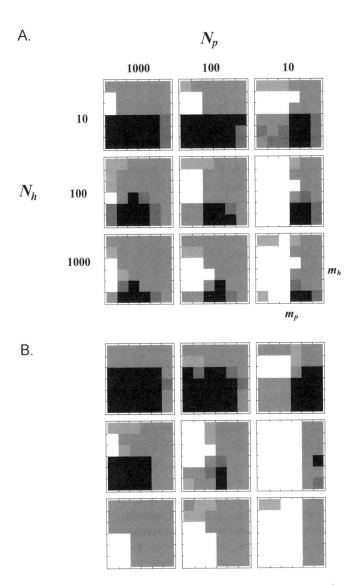

Fig. 6.14 Degree of local adaptation (or maladaptation) in host or parasite populations under different levels of mutation, relative population sizes, and gene flow. The model assumes two loci with four alleles in each haploid species and a matching allele coevolutionary structure. Adaptation is measured here as local performance of a population relative to its average performance in all other populations. (A) The top set of nine panels shows patterns of adaptation in the absence of mutation. (B) The bottom set of nine panels allows mutation on each locus at a rate of 10^{-4}. Gene flow varies within each panel from 0.0001 on the left to 1.0 on the right for the parasite (m_p) and from 0.0001 on the bottom to 1.0 on the top for the host (m_h). Population size varies among panels both in the host (N_h) and in the parasite (N_p). Black boxes indicate high parasite adaptation (or low host resistance) and white indicates the other extreme. Shades of gray indicate intermediate levels of adaptation. After Gandon and Michalakis 2002.

may be only apparent maladaptation, resulting from the local temporal dynamics of fluctuating polymorphisms. Others, however, appear to be due to gene flow or conflicting local selection pressures, as in the case of the parsnip webworm in regions where it has two sympatric host species rather than one (Zangerl and Berenbaum 2003). The result is a mosaic of traits and current selection across landscapes in these interactions.

Yet other interactions, however, show a combination of low regional differentiation in traits important to the interaction and evidence of local maladaptation that may result either from metapopulation dynamics, extensive gene flow, and/or low genetic variation for coevolving traits in one of the species. A combination of those factors appears to shape, for example, the interaction between the plant *Silene latifolia* and the fungal pathogen *Microbotryum violaceaum* in Europe (Kaltz et al. 1999; Kaltz and Shykoff 2002). Which of these factors contributes most to local maladaptation in this particular interaction is unknown, but the result holds for multiple traits across more than a dozen European populations.

Darwin understood that the jury-rigged nature of adaptation provided some of the best evidence for evolution. In the same way, a geographic mix of relatively coadapted and maladapted populations that differ in their jury-rigged coevolving traits provides some of the strongest evidence for the geographic mosaic of coevolution. Instances of local mismatches in coevolved traits are not a failure of the coevolutionary process, and differences among populations in coevolved traits are not evidence of unimportant coevolutionary meanderings. Instead, local differences in traits and degrees of maladaptation associated with those traits are an inevitable part of the geographic mosaic of coevolution. The geographic interplay between coadaptation and maladaptation drives ongoing coevolution and the diversification of species, and deviations from local coevolutionary optima drive ongoing coevolutionary selection. In fact, one way of thinking about antagonistic coevolution is as a process favoring traits that make the other species more maladapted (Thompson, Nuismer, and Gomulkiewicz 2002). Even in mutualistic interactions, the evolution of cheater genotypes can generate locally maladapted populations. In the absence of ongoing selection to mitigate the effects of cheaters, mutualisms could erode through a ratchet of maladaptation as the fitness benefits of the interaction erode over time through the accumulation of cheaters.

Local mismatching of traits therefore may be an important driving force in ongoing coevolutionary dynamics, because it drives ongoing evolutionary change in populations not residing on an adaptive peak. The geographic mosaic drives ongoing coevolution as the geographic mix of local adaptation and maladaptation changes over time and across landscapes. Mismatches in adaptation and counteradaptation within local communities are therefore not, in themselves, evidence against coevolution. If the geographic mosaic theory of coevolution holds, then mismatches should be evident within some local communities, although there should be evidence of reciprocal change when many populations of the interacting species are evaluated together.

The Geographic Mosaic and Adaptive Landscapes

The geographic mosaic theory of coevolution suggests a worldview of co-evolutionary dynamics closer to Wright's shifting balance theory of the structure of evolutionary change than to Fisher's theory of large population sizes (Thompson 1994). As Wade and Goodnight (1998) have argued, Wright viewed the essential problem of evolution to be the origin of adaptive novelty in a constantly changing world. In his view, the major evolutionary processes driving change are a combination of local natural selection involving epistasis and pleiotropy, random genetic drift, gene flow, and interdemic selection. These processes occur in an ecological context that often includes small, subdivided populations and alleles whose effects vary among environments. In contrast, Fisher emphasized the continuing refinement of adaptation in large panmictic populations within stable or slowly changing environments. Within that view, mutation and selection are the dominant evolutionary processes and act on the additive effects of alleles.

The current geographic mosaic models of coevolution have not analyzed the effects of epistasis, random genetic drift, or interdemic selection. As in Wright's view, however, most of these geographically structured models emphasize dependence of the selective value of particular alleles on both the environment and the genotypes in which they occur. This is bound to be so in coevolutionary biology, because the selective value of an allele depends not only on the genotype of the individual in which it occurs but also upon the population-genetic structure of the other species and the environment in which the interaction occurs. Moreover, as in Wright's view, the geographic mosaic models indicate that deviations from panmixis can have important effects on the rates and trajectories of evolutionary change. In addition, the models suggest that geographically structured coevolution may partially generate the continual change in environment that was a cornerstone of Wright's perspective.

The geographic mosaic theory of coevolution, however, differs in emphasis from Wright's particular view of the shifting balance of evolution in important respects. Wright was concerned with the problem of how to shift from one adaptive peak to another. In the geographic mosaic of coevolution, the problem is how to keep species coevolving over millennia as coevolutionary peaks continue to shift in fitness space across real geographic landscapes that include geographic selection mosaics and coevolutionary hotspots and coldspots.

The constantly shifting geographic mosaic of coevolution involves too many variables to predict the long-term trajectories of particular coevolving traits in interacting species. But that should not be the objective of coevolutionary analyses anyway. Rather than knowing whether alleles Y or y will be maintained in a species through coevolution, we need to know more broadly how coevolving polymorphisms are maintained within and among populations and at what geographic scales. We need to understand the large-scale ongoing ecological and genetic processes that hold interactions together over long periods of time through the geographic mosaic of coevolution. We also need to understand how often local maladaptation results from the geographic mosaic of coevolution and how the geographic pattern of mismatched traits shapes subsequent coevolution. That is, the goal of coevolutionary analyses must be to understand the structure and dynamics of the process itself and its effects on the overall structure of species, species networks, and biodiversity.

Conclusions

Coevolutionary biology is inherently a hierarchically structured science, built upon the phylogeographic and phylogenetic structure of interactions as they are continually reshaped across constantly changing landscapes. It has taken decades to develop a science of coevolutionary change that goes beyond either local coevolutionary dynamics at the one extreme or typological descriptions of coevolution at the other extreme. Empirical and mathematical studies of coevolving interactions over the past decade have indicated that geographic selection mosaics, coevolutionary hotspots, and trait remixing produce regional patterns in coevolving interactions. Coevolved traits differ among populations, and few of these coevolved traits are likely ever to spread across all populations. Moreover, the varying community context of an interaction across landscapes generates coevolutionary hotspots and coldspots, and the geographic mosaic of coevolution creates regions that vary in their degree of local adaptation. Coevolving traits are well matched in some localities and mismatched in others. The result is a highly dynamic ecological process that continues to reshape the seven major classes of coevolutionary change across landscapes. We will turn to those dynamics in part 2 of the book, after exploring how the geographic mosaic of coevolution may shape the diversification of taxa.

7 Coevolutionary Diversification

The geographic mosaic of coevolution may contribute to speciation and the diversification of phylogenetic lineages. As geographic selection mosaics create coevolved populations that differ in ecological outcome and traits from other populations, some of these divergent populations may become new species. Diversifying coevolution may produce hybrid inferiority or promote isolation among populations more quickly than occurs during adaptation of populations to different physical environments, because the genetic feedback of coevolution can drive and sustain rapid evolutionary change. Coevolution therefore has the potential to produce not only rapid genetic changes within populations but also rapid diversification among populations (Thompson 1987a, 1994).

Diversifying coevolution can be defined as the process by which populations diverge from each other through the geographic mosaic of coevolution. This diversification is the next level in the hierarchical structure of the coevolutionary process, beginning with local coadaptation and continuing through the dynamics of the geographic mosaic to the eventual formation of new species and lineages. It is a specific form of ecological speciation (Schluter 1996a, 2000a), driven by coevolution. Through coevolutionary selection, a local population diverges in traits in ways that decrease the likelihood of successfully mating with other populations of the same species. Postmating isolating mechanisms may arise through the juxtaposition of coevolutionary hotspots and coldspots or the juxtaposition of hotpots in which local populations are coevolving in different ways. Premating isolating mechanisms may develop through reinforcement, as natural selection disfavors individuals that mate with individuals from the other population. These are components of the standard model of speciation mediated by selection, but with one important difference. Diversifying coevolution implies that geographic differences in coevolution drive the process of species formation, either directly or indirectly.

Most of the dynamics of diversifying coevolution will be masked by any phylogenetic approach that focuses solely on patterns of divergence at the levels of species and higher taxa. The patterns seen in the analysis of cladograms show only the highly filtered results of coevolution and little of the transient coevolutionary dynamics underlying the patterns. Consequently, any purely phylogenetic theory of coevolution that ignores the geographic mosaic of coevolution cannot evaluate the coevolutionary processes that contribute to the diversification of lineages. Combined with ecological, geographic, and genetic studies, however, phylogenetic approaches are central to understanding the long-term consequences of diversifying coevolution.

This chapter explores diversifying coevolution as a process that results from the geographic mosaic of coevolution, and it links that process to larger-scale phylogenetic patterns found in diversifying lineages. By necessity, it focuses on the potential role of coevolution in speciation and the diversification of lineages rather than on any demonstrated role. Little of speciation theory has been devoted to the problem of how reciprocal selection may shape the speciation process. It is only now that studies are being designed to evaluate how coevolution may directly shape hybrid inferiority or promote isolation among populations. The available studies, however, point the way to future work by suggesting that specialization resulting from coevolution may contribute to population divergence and, hence, to speciation.

From the Geographic Mosaic to Speciation

DIVERSIFYING COEVOLUTION IN NATURAL POPULATIONS

Much of the history of the diversification of life is a history of divergence in specialization to other species. Major phylogenetic lineages are collections of species that differ in the set of other species they depend upon for survival or reproduction. Moreover, many species can successfully attack, defend themselves against, compete with, or mutualistically benefit from only a subset of the populations of those other species. The geographic selection mosaic becomes a mosaic of specialized populations.

Even a small genetic difference among individuals can sometimes have large effects on specialization to other species and thereby potentially on diversification. Within the Sierra Nevada of California, two species of monkey flower, *Mimulus lewisii* and *M. cardinalis,* differ in floral color and overlap in distribution. The pink-flowered *M. lewisii* is pollinated by bumblebees; the red-flowered *M. cardinalis,* by hummingbirds. Substituting the yellow upper (YUP) allele in near-isogenic lines of one species with the allele for the other species results in large differences in patterns of floral visitation in the field. Visits by hummingbirds to *M. lewisii* with the substituted allele can increase sixty-eight-fold, and visits by bees to *M. cardinalis* can increase seventy-fourfold (Bradshaw and Schemske 2003).

When the geographic differences in specialization occur in both species, the potential arises for rapid speciation through diversifying coevolution. Coevolution between red crossbills (*Loxia spp.*) and conifers in North America has become the clearest and best-studied example of how the geographic mosaic of coevolution may contribute to speciation (Benkman 1999; Benkman, Holimon, and Smith 2001; Parchman and Benkman 2002; Benkman 2003). Through diversifying coevolution, some populations of crossbills and conifers have diverged significantly from other populations since the end of the Pleistocene. The process is mediated partially by red squirrels (*Tamiasciurus hudsonicus*), which are a third crucial coevolutionary partner in these interactions (Smith 1970; Benkman, Holimon, and Smith 2001). In fact, analysis of coevolution between red squirrels and conifers in the Rocky Mountains was one of the pioneering studies in evolutionary ecological approaches to coevolutionary biology in the 1960s and 1970s (Smith 1968, 1970; Elliott 1974; Smith 1981). The overall coevolutionary structure of these interactions, however, has become interpretable only in recent years, through an elegant series of studies that has evaluated the structure of geographic selection mosaics and coevolutionary hotspots across the Rocky Mountains and, more recently, in eastern North America.

Crossbills are holarctic in distribution and specialize on extracting seeds from the partially closed cones of conifers. All crossbills have a characteristic bill that crosses at the tip, which allows the birds to pry apart the scales of conifer cones to reach the seeds (fig. 7.1). Red crossbills are a complex of North American species and incipient species specialized to different conifers. There are about nine recognized call types (Benkman 1989, 1993; Groth 1993; Benkman 1999), at least six of which correspond to birds specialized on different conifers (Benkman 1989, 1993, 1999). Specialization on different conifer species therefore appears to be one of the factors driving divergence among crossbill populations. These species or incipient species include, for example, specialists on lodgepole pine (*P. contorta*), ponderosa

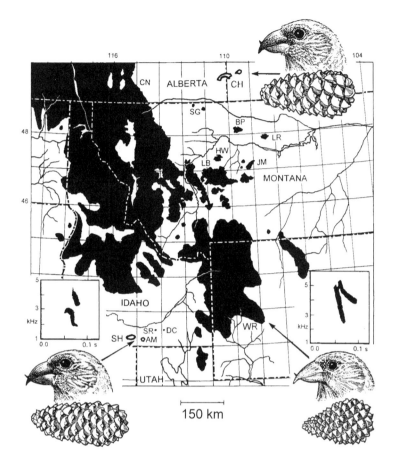

Fig. 7.1   The distribution of lodgepole pine (*Pinus contorta*) in the Rocky Mountains of the northern United States and southern Canada (black shading). Populations that have undergone intense study are indicated by letters and arrows. The crossbill and lodgepole pine cone shown in the lower right are representative of populations in regions with red squirrels. The other two crossbills and cones are representative of populations found in the isolated mountain ranges lacking red squirrels (*Tamiasciurus*). Sonograms of call types are shown for South Hill (SH) birds in comparison with birds from the less isolated parts of the Rocky Mountains. From Benkman 1999.

pine (*P. ponderosa*), Douglas-fir (*Pseudotsuga menziesii*), and western hemlock (*Tsuga heterophylla*). The specialists differ not only in calls, but also in bill size and shape, body size, and geographic ranges.

Within the central and northern Rocky Mountains, crossbills that are lodgepole pine specialists often move nomadically, searching regionally for good cone crops. The birds, however, impose little or no selection on lodge-pole pines within these mountains, because their effects are swamped by selection imposed by red squirrels (Benkman, Holimon, and Smith 2001). The squirrels are abundant, and they harvest and cache large numbers of cones in early fall, depleting the cones available to crossbills. The cone caches are the major winter food for the squirrels, and the caches are inaccessible to crossbills. The crossbills feed on the subset of cones remaining on the trees.

Rocky Mountain lodgepole pines have therefore coevolved primarily with red squirrels, and their cones have a set of traits that serve as defenses against harvesting by the squirrels (Smith 1970; Benkman, Holimon, and Smith 2001). Throughout the Rocky Mountains, from the Yukon to Colorado, the cones are short and wide at the base (fig. 7.1), making it difficult for the squirrels to bite the cones off the branch. Squirrels tend to avoid cones that are particularly wide at the base (Smith 1970; Elliott 1974). Each of these cones contains few seeds, and the ratio of cone mass to seed mass is nearly one hundred to one (Benkman 1999). As a result, the reward per harvested cone is low. To reach the seeds, a squirrel bites off scales at their base, beginning with the proximal end of the cone and moving distally. Most seeds are located under the distal scales.

Bill morphology of crossbills within the Rocky Mountains has evolved to cope with cone defenses against squirrels, as shown in experiments comparing seed harvesting rates by these birds on cones from their normal lodgepole populations with harvesting rates on cones from other populations (Benkman, Holimon, and Smith 2001). The crossbills, however, do not appear to exert significant selection on cone morphology among Rocky Mountain lodgepole pine populations. The Rocky Mountains are therefore a coevolutionary coldspot for the interaction between the crossbills and the lodgepole pines, but a coevolutionary hotspot for the interaction between the squirrels and the pines.

In some regions peripheral to the Rocky Mountains, red squirrels are absent. In these isolated areas of southern Idaho, southern Alberta, and north-central Montana, the crossbills are resident, and, as the only major predator of lodgepole cones, they impose strong selection on lodgepole cone morphology (Benkman et al. 2003). Cones within these regions are narrower and have thicker scales in the distal ends, where most of the seeds are located (fig. 7.2). This suite of cone traits makes seed harvest less efficient for crossbills, and the resident crossbills have evolved larger and more strongly decurved bills, thereby increasing their efficiency at harvesting these cones. Crossbills

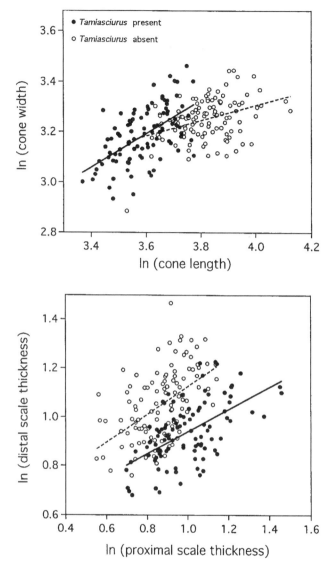

Fig. 7.2   Relationships between cone traits in lodgepole pines subject to natural selection by squirrels (filled circles) or by red crossbills (open circles). Selection by crossbills favors longer, narrow cones (top panel) and greater distal scale thickness (bottom panel) relative to proximal scale thickness. After Benkman, Holimon, and Smith 2001.

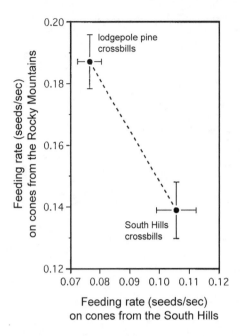

Fig. 7.3   Feeding rate (mean ± standard error) of red crossbills on sympatric and allopatric lodgepole pine cones. The tested crossbills from the Rocky Mountains (called here lodgepole pine crossbills) feed on cones adapted to red squirrels, whereas the tested crossbills from the South Hills feed on cones adapted to crossbills. After Benkman, Holimon, and Smith 2001.

in the Rocky Mountains and in the peripheral regions show evidence of local adaptation to cone morphology through divergent bill morphology (fig. 7.3).

Cone traits and bill traits have coevolved in similar ways in southern Idaho and southern Alberta populations where squirrels are absent. They are presumably independent coevolutionary events, although phylogeographic analyses are not yet available. These independent coevolutionary hotspots are therefore similar to the geographically independent origins of hotspots found in the interaction between *Taricha* newts and garter snakes along the west coast of North America (see chapter 6).

One mountain range in Montana does not fit the pattern found among other peripheral mountain ranges. In the Little Rocky Mountains, where squirrels are absent, distal scale thickness has increased as predicted, but cone size has not (Siepielski and Benkman 2004). Cone morphology in this isolated range is shaped not only by crossbill-mediated selection, but also by lodgepole pine borer moths (*Eucosma recissoriana*). Moths reach densities here ten times higher than in other ranges lacking squirrels and impose strong selection on cone morphology. This isolated mountain range therefore adds yet another dimension to the geographic selection mosaic on these coevolving interactions.

Peripheral crossbill and lodgepole pine populations appear to have diverged from populations in the main mountain chain over the past ten thousand years, since the end of the Pleistocene isolated these populations from others in the Rocky Mountains. Resident crossbill populations in southern Idaho and southern Alberta were able to evolve adaptations to the pines and impose strong selection on the pine populations. During this period of isolation, the birds have developed novel behaviors (resident rather than nomadic movements), novel morphologies, and novel songs. Diversifying coevolution has produced what may now be at least incipient species of crossbill in these peripheral populations. There is no evidence that the lodgepole pines have undergone comparable differentiation to suggest speciation. But that in itself is instructive for our understanding of the coevolutionary process. Diversifying coevolution need not create symmetric geographic patterns of speciation in coevolving taxa. The genetic and geographic structures of coevolving species will rarely be identical, because the genetic mechanisms of speciation will often differ between the interacting species.

Remarkably, similar diversifying coevolution appears to have occurred in populations of crossbills and black spruce (*Picea mariana*) in Newfoundland (Parchman and Benkman 2002). Black spruce is distributed across the northern forests of North America from Alaska to Newfoundland and extends south to the Great Lakes (fig. 7.4), and a subspecies of white-winged crossbill (*Loxia leucoptera leucoptera*) is specialized for feeding on these partially closed cones (Benkman 1987). Red squirrels also feed on black spruce and, as in the case of lodgepole pines, shape the evolution of cone morphology in regions where they are present. The squirrels, however, have been absent from Newfoundland during the past nine thousand years, since the retreat of the glaciers. In their absence, black spruce coevolved with a unique form of red crossbill (*L. curvirostra percna*), called the Newfoundland crossbill (fig. 7.4).

Cones of black spruce in Newfoundland differ from those on the nearby mainland in some of the same ways in which peripheral lodgepole pines differ from those within the Rocky Mountains. The cones on Newfoundland are larger, contain more seeds, and have a greater ratio of seed mass to cone mass. The pattern of shifts in cone traits in response to the lack of red squirrels and the presence of crossbills is therefore the same in two unrelated conifer species on opposite sides of North America (fig. 7.5).

Newfoundland crossbills had relatively deep bills in comparison to red crossbills that feed on ponderosa pine or lodgepole pine (fig. 7.6). Unfortunately, these divergent crossbills are now presumed to be extinct, probably as a result of the introduction of red squirrels in 1963. Following squirrel intro-

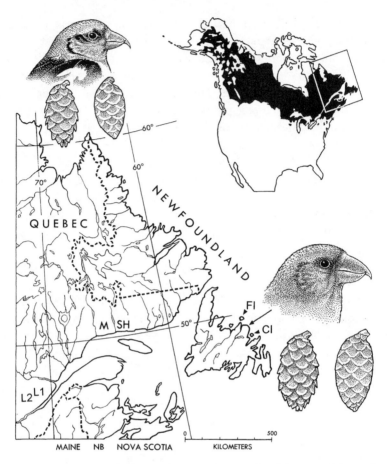

Fig. 7.4   White-winged crossbill (*Loxia leucoptera leucoptera*) and the size and shape of partially open and closed cones of black spruce from the mainland (upper left) as compared with a Newfoundland crossbill (*L. curvirostra percna*) and Newfoundland black spruce cones (lower right). The letters and numbers on the map indicate populations sampled for cone size and shape. The inset in the upper right shows the overall geographic distribution of black spruce in North America. From Parchman and Benkman 2002.

ductions, crossbill numbers dropped precipitously by the mid-1970s and there have been only a few purported sightings since the mid-1980s (Parchman and Benkman 2002).

Experiments with mainland red crossbills have shown that birds with relatively deep, short bills, approximating those found in Newfoundland crossbills, are more efficient at harvesting black spruce cones from Newfoundland

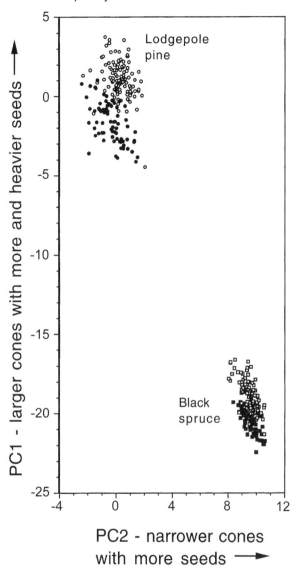

filled symbols - *Tamiasciurus* present
open symbols - *Tamiasciurus* absent

Fig. 7.5   Differences among populations and species in traits of conifer cones subject to selection by red squirrels as compared with cones from regions lacking squirrels, analyzed using principal components analysis. After Parchman and Benkman 2002.

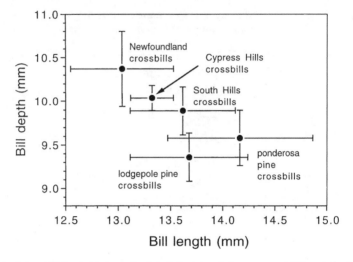

Fig. 7.6   Mean bill depth ($\pm$ 1 standard deviation) in relation to mean bill length for five taxa within the red crossbill species complex. The Cypress Hills and South Hills crossbills feed on lodgepole pines in the absence of red squirrels; the Newfoundland crossbills feed on black spruce in the absence of crossbills. The Newfoundland and Cypress Hills crossbills are now presumed extinct and were measured from museum specimens. After Parchman and Benkman 2002.

than are other crossbills. The combined data on cone morphology, bill morphology, and foraging efficiency suggest that black spruce and crossbills underwent diversifying coevolution on Newfoundland over the past nine thousand years. The rapid decline of the birds following the introduction of red squirrels reinforces the conclusion that these squirrels significantly affect the interaction between crossbills and conifers.

The studies of crossbills and conifers on both sides of North America suggest that diversifying coevolution can produce significant divergence among populations of species that are long lived and wide ranging over relatively short periods of geologic time. If populations of birds and trees are capable of significant regional divergence through coevolution over less, possibly much less, than twenty thousand years, then diversifying coevolution has the potential to be a major force in diversification in most of the earth's species. Relative to birds and conifers, many species are even more genetically differentiated among populations, and most have faster generation times. That is why the results for crossbills and conifers, together with the results for garter snakes and newts (see chapter 6), are so important for our understanding of coevolutionary dynamics. They suggest that the geographic mosaic of co-evolution is potentially as important in long-lived, wide-ranging taxa as it is in short-lived taxa.

The next step in evaluating diversifying coevolution will be to understand which of the increasing number of studies on the geographic mosaic of co-evolution are likely to be incipient cases of speciation through diversifying coevolution. We will need a new generation of mathematical models of co-evolution to help pinpoint the general conditions favoring speciation through diversifying coevolution, given the fast rate of divergence possible through coevolutionary selection.

DIVERSIFYING COEVOLUTION IN THE LABORATORY

Laboratory studies on rapid coevolution between bacteria and bacteriophages offer an important middle ground between field studies and mathematical models. Although ecologically simplified, the genetics are real. Bacteria and phages have complex genomes that go beyond anything possible in a mathematical model. Careful long-term laboratory studies of *Escherichia coli* have shown that evolution is highly dynamic under laboratory conditions even in the absence of major spatial structure. Rapid ongoing evolution occurs both in isolated bacterial cultures and in cultures interacting with phages (Lenski and Travisano 1994; Bohannan and Lenski 2000a, 2000b; Cooper and Lenski 2000; Remold and Lenski 2001). Recent studies of the bacterium *Pseudomonas fluorescens* and the naturally virulent phage SBW25ϕ2, however, have shown that geographic structure can create populations that diversify rapidly through coevolution (Buckling and Rainey 2002). The trajectory of coevolution in the bacteria and the phages is toward the ability to resist or infect an increasingly wide range of enemy genotypes. Through differential coevolution among allopatric populations, bacterial populations become resistant to different combinations of phage while the phages increase their infectivity to different combinations of bacterial genotypes.

Coevolutionary experiments with *Pseudomonas* have taken advantage of results from earlier studies showing that isogenic populations of this bacterium rapidly diversify into niche specialist types when grown in a spatially structured environment in the absence of phage. The environment is a static glass microcosm containing a nutrient-rich medium for which the bacteria compete. Through mutation, heritable new forms of the bacteria appear, and those forms that compete least with the dominant forms increase in numbers. Although numerous niche specialists evolve in these cultures, they fall into three general groups. Smooth morphotypes, which resemble the ancestral form, inhabit the liquid phase. Wrinkly spreader morphotypes form a biofilm at the air-broth interface. Fuzzy spreader morphotypes colonize the less aerobic bottom of the vials.

In the absence of phages, each microcosm maintains multiple niche types if the medium is replenished by transferring 1 percent of the culture to a fresh microcosm every two days. Under these conditions of alternating food abundance and competition, the morphotypes compete for food but are transferred to new medium before the microenvironment for any morphotype is completely eliminated. Replicate microcosms maintain the same morphotypes at similar frequencies. In the absence of phages, then, the microcosms exhibit little geographic (i.e., among replicate) structure.

In the presence of phages, morphotype diversity rapidly decreases within individual microcosms, but it increases among replicate microcosms (fig. 7.7). Under the conditions of the experiment, the phages spread rapidly throughout each microcosm and impose strong directional selection on the bacterial culture. Within two subsequent transfers to new medium, the bacteria within each microcosm evolve from being entirely sensitive to the phages to being almost completely resistant. Throughout six more transfers to new medium, the frequency of resistance in all microcosms remains above 95 percent. Mutations that confer resistance, however, appear by chance in the different morphotypes. The first morphotype to develop a resistant mutant is favored within that culture. Consequently, microcosms rapidly diverge in the morphotype that dominates.

From Speciation to Large-Scale Diversification: Escape-and-Radiate Coevolution

The results for crossbills and pines and for *Pseudomonas* and phage suggest ways in which diversifying coevolution may contribute to the formation of new species as populations diverge across landscapes. Although speciation theory is advancing through new analytical approaches (Gavrilets 2003), we lack specific models on how coevolution alters the genetics, rates, and geographic requirements for speciation. We are also still in the early stages of developing specific hypotheses of how the geographic mosaic of coevolution translates over the long term into even broader patterns of diversification of lineages.

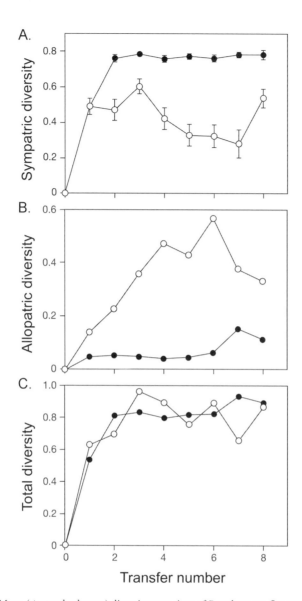

Fig. 7.7   Mean (± standard error) diversity over time of *Pseudomonas fluorescens* morphotypes evolving in isolation (filled symbols) and with phage (open symbols) in spatially structured microcosms. The three panels show (A) diversity within microcosms (sympatry),
(B) diversity among microcosms (allopatry), and (C) total diversity within and among all
microcosms. Sympatric diversity was calculated using a measure that determines the probability that two randomly chosen morphotypes are different, given the proportions of the different morphotypes in the sample. Total diversity was calculated in a similar way. Allopatric
diversity was calculated as a variance among populations. After Buckling and Rainey 2002.

The simplest pattern is parallel divergence of interacting lineages, which is often called parallel cladogenesis or cospeciation. As interacting populations diverge geographically, they may form new species in parallel and thereby create matched cladograms. The most ancestral species in one lineage interacts with the most ancestral species in the other lineage, and each subsequent speciation event in one lineage is matched in the other lineage.

Everything we have learned about the geographic mosaic of coevolution suggests that cospeciation should be a rare outcome. The geographic ranges of interacting species continually change, bringing previously allopatric populations of closely related species into secondary contact. A local habitat that previously held one species of genus A and one of genus B may now have two species of A and three of B. These instances of secondary contact can occur in multiple regions and involve different combinations of species from two or more interacting lineages. There is, therefore, plenty of opportunity over millions of years for populations to try out new partners as geographic selection mosaics continue to change across landscapes over time. True parallel cladogenesis at the species-for-species level is therefore likely to be common only for symbiotic mutualistic or commensalistic relationships between hosts and their symbionts that are obligately transmitted directly from mothers to offspring (see chapter 12).

In 1964 Paul Ehrlich and Peter Raven put forward the first hypothesis invoking true reciprocal selection during clade diversification and predicting something much more interesting than parallel cladogenesis. Their hypothesis focused on how reciprocal selection could produce higher-level patterns of diversification within phylogenetically conserved interacting lineages. The example they chose, butterflies and plants, was grounded in the strong phylogenetic conservatism of butterflies in their choice of host plants. Most butterfly lineages have diversified primarily through specialization on different host species within a single plant family, as have many other phytophagous insect lineages. These higher-level patterns of diversification correspond broadly to differences within and among plant families in their profile of chemical defenses (Ehrlich and Raven 1964; Berenbaum 1995; Becerra 1997; Farrell and Mitter 1998; Janz and Nylin 1998; Cornell and Hawkins 2003).

The formal version of the Ehrlich-Raven hypothesis, now generally called escape-and-radiate coevolution (Thompson 1989), posits that reciprocal diversification between parasites and hosts results from a combination of reciprocal adaptation, geographic differentiation and speciation, and periods of noninteraction in the diversifying lineages.

1. The process begins with a new mutation that spreads throughout a host population, allowing mutant individuals to avoid attack by parasites (or other enemies).

2. The mutant population enters a new adaptive zone, allowing it to spread and diversify across landscapes in the absence of attack by enemies. Population diversification produces a starburst of speciation in the mutant host clade primarily during the time period in which the interaction is not occurring.

3. Subsequently, a mutant parasite population overcomes the new host defenses, resulting in a starburst of parasite speciation as parasite populations specialize on different species within the newly diversified mutant host clade.

Any test of escape-and-radiate coevolution must evaluate the major predictions of the hypothesis (Thompson 1999e).

Prediction 1. Novel defenses and counterdefenses occur *among* clades within lineages. A host clade resulting from the escape-and-radiate process should be evident as a starburst of speciation associated with evolution of a new defense. A parasite clade resulting from this process should be evident as a subsequent starburst of speciation associated with evolution of a new counterdefense that allows attack of that new host clade. Overall, then, coevolutionary diversification of hosts and parasites should be evident as reciprocal starbursts of speciation associated with major new defenses and counterdefenses.

Prediction 2. The parasite clade has not colonized each new species within a host clade in a systematic fashion, beginning with the most ancestral host species and proceeding in lockstep through the more derived species. Because diversification of the new host clade occurred in the absence of the interaction, there should be no consistent ancestral-descendant pattern of escalation of defenses within each starburst of speciation in the host lineage. In fact, consistent parallel cospeciation is direct evidence against escape-and-radiate coevolution.

Overall, evidence for the hypothesis can come only from a combined demonstration of lack of pattern within starbursts of host speciation and novelty of defense among starbursts of speciation. There are, however, two caveats to the predicted patterns. The new mutant parasite lineage may colonize the mutant host lineage before the host lineage has undergone extensive speciation. That is, initial diversification of the host clade could occur during the absence of the parasite, but some later diversification of the host clade may occur after the mutant parasite has colonized it. This situation may appear as a pattern in which the new parasite lineage does not colonize the basal host species in an ancestral-descendant sequence, but it does colonize the more derived host species (i.e., those that appeared after colonization by the parasite) more or less in sequence. This is further complicated by the possibility that, as each new parasite species develops within the new lineage, some of these parasites may colonize more-basal rather than more-derived species within the host lineage. Wherever coevolving interactions are distributed as geographic mosaics across landscapes, even multiple colonizations of the diversifying host lineages may occur.

The second caveat is that Ehrlich and Raven were not concerned exclusively with repeated starbursts of speciation between two lineages. The focus on reciprocal starbursts, which I have outlined here, is the most strictly coevolutionary version, now formalized as escape-and-radiate coevolution. Ehrlich and Raven's initial view, however, allowed for even broader general patterns of escape and radiation of lineages in which subsequent insect colonizers might not even be part of the insect group from which the plants initially escaped. Nevertheless, neither the more strictly coevolutionary view nor the more general view predicts parallel cladogenesis at the species level. Nothing about the hypothesis predicts that a new mutant parasite will colonize first the most ancestral host in the mutant host lineage and then radiate in lockstep up through the more derived host species. Because the new host lineage underwent diversification during the period in which there was no interaction, there is no reason to expect that there will be any escalating pattern of defenses that makes more-derived hosts more resistant than more-ancestral hosts *within* the new host clade. A mutant parasite may therefore be just as likely to colonize a derived species as it would an ancestral species within the mutant host lineage. The geographic mosaic of coevolution will assure a complex pattern of host shifts within each of the host starbursts, with some occasional host shifts onto more distant host taxa. Any pattern of escalation of defenses and counterdefenses consistent with escape-and-radiate coevolution will occur only at higher taxonomic levels. The patterns at these levels would reflect repeated escapes and species radiations in hosts and subsequent colonizations and speciation events in parasites.

Evaluation of escape-and-radiate coevolution therefore requires a comparative analysis of patterns of defense and counterdefense within and among clades. We are only now in a position to begin evaluating just how diversifying coevolution through mechanisms such as escape-and-radiate co-evolution shapes these kinds of larger patterns in the diversification of life. There are increasingly good data on the traits under selection in some inter-actions, powerful molecular and statistical methods for reconstruction of phylogenies of interacting taxa, and increasing numbers of fossils that can anchor the timing and pulse of diversification.

Nevertheless, it will still take more time before we have fully robust analy-ses of any coevolving lineages. Obtaining reliable phylogenies from both (or all) interacting lineages has turned out to be time consuming and requires collaboration among researchers who are expert in very different taxa and techniques. Nonetheless, there are increasingly good phylogenies for some lineages that have been the subject of intense study of coevolving inter-actions. These include major insect groups such as prodoxid moths (includ-ing yucca moths and *Greya* moths; Pellmyr et al. 1996; Pellmyr 1999; Pellmyr and Balcázar-Lara 2000), fig wasps (Machado et al. 2001), and beetles (Far-rell 1998; Farrell and Mitter 1998; Becerra and Venable 1999; Farrell 2001) and their host plants, and other major interactions such as those between leaf-cutter ants and their fungal gardens (Mueller, Rehner, and Schultz 1998; Aanen et al. 2002), lycaenid butterflies and ants (Pierce et al. 2002), fungi and algae in lichen associations (Romeike et al. 2002), and aphids and *Buchnera* bacteria (Clark et al. 2000).

Systematics, Parallel Speciation, and Patterns of Codiversification

Running parallel to discussions over coevolution and diversification is a long history of studies of comparative speciation between parasites and hosts, reaching back to the work of parasitologists and comparative systematists early in the twentieth century (Klassen 1992). These studies have focused on evaluation of cladistic patterns in the phylogeny of interacting lineages and have been concerned primarily with comparative patterns of divergence be-tween parasites and hosts. The goal of these analyses has been primarily to evaluate the degree of parallel speciation and host switching and the pattern of biogeographic vicariance in the historical associations of interacting taxa. These are potentially valuable studies for our understanding of the geo-graphic mosaic of coevolution, because the techniques coming from these approaches attempt to link phylogeny and geography beyond the study of in-dividual lineages.

Unfortunately, following publication of Ehrlich and Raven's (1964) paper, some biologists co-opted the word "coevolution" for these analyses. In these studies, coevolution became synonymous with a variety of related phrases, including parallel speciation, phylogenetic tracking, parallel cladogenesis, and cospeciation. Some of the usage came from the observation that reciprocal selection can result in parallel speciation in some mutualistic symbioses (see chapter 13), but some came from a simple co-opting of a convenient word to use as a general synonym for codiversification. Part of the usage, however, appears to have stemmed directly from a misreading of Ehrlich and Raven's original paper. Although Ehrlich and Raven's hypothesis could not, by definition, result in parallel speciation, their paper has sometimes been cited for exactly this prediction. Such citations often ignored the central role of reciprocal evolution that Ehrlich and Raven gave to the process.

This alternative use of the word "coevolution" to mean parallel speciation lives on as a historical artifact, primarily within the field of comparative systematics. That usage continues to create some confusion in discussions on the kinds of evidence for coevolution obtainable from phylogenetic analyses. Increasingly, however, those studying the comparative phylogeny of interacting taxa use "cospeciation" or "parallel cladogenesis" for these studies and restrict the word "coevolution" to the process of true reciprocal evolution between interacting taxa driven by natural selection.

The development of sophisticated phylogenetic and statistical approaches has created the opportunity to evaluate more rigorously the patterns of codiversification found in interacting lineages (Brooks and McLennan 1991, 2002; Page 2003b). Analyses of interspecific interactions in the fossil record are helping date the timing of major events during the diversification of some interactions (Labandeira et al. 1994; Vermeij 1994; Labandeira 1998; Vermeij and Lindberg 2000; Wilf et al. 2000; Wilf et al. 2001; Vermeij 2002b). Still other approaches are evaluating the structure of genetic variation and genetic correlations during diversification and its potential influences on patterns of codiversification (e.g., Futuyma, Keese, and Funk 1995; Knowles et al. 1999; Futuyma 2000). Together these studies are helping to move analyses of codiversification beyond questions about the degree of parallel speciation evident in extant taxa. Because all real phylogenies are partially unresolved, evaluating the probability of competing patterns remains a problem. A new generation of approaches, however, is introducing alternative ways of evaluating patterns of diversification by incorporating parsimony (Ronquist 2003), cost-based methods that build in the likelihood of particular events (Charleston and Perkins 2003), Bayesian statistical methods that include uncertainty in the reconstructed phylogenies (Huelsenbeck, Rannala, and Larget 2003), and hierarchical analyses that compare gene trees at the within-population, between-population, and species levels (Rannala and Michalakis 2003). These competing methods for comparing patterns of codiversification should eventually provide us with robust phylogenetic templates upon which to evaluate the role of reciprocal evolutionary change driven by natural selection in the diversification of lineages.

Coevolution and Codiversification: Developing the Questions and the Tools

Achieving better integration between studies of codiversification and studies of coevolution will require more careful phrasing of questions on how the two are linked. The overall process of diversifying coevolution does not itself result in any specific pattern in the diversification of lineages, because there is no necessary connection between coevolution in general and particular patterns of speciation. Nevertheless, as first suggested by Ehrlich and Raven (1964), particular forms of coevolution may produce identifiable patterns of codiversification. They suggested one kind of process and the pattern that should result, now called escape-and-radiate coevolution. We still require, however, a group of alternative hypotheses to ask whether there are identifiable signatures of codiversification in coevolving interactions resulting from specific modes of interaction, phylogeographic structure, or the genetic architecture of coevolving traits. We can begin to search for those signatures only by incorporating what we know about the geography, genetics, and ecology of interspecific interactions.

USE OF GEOGRAPHIC STRUCTURE

Studies on the geographic structure of extant interactions should provide the starting point for any analysis of the way in which coevolution contributes to codiversification of lineages. These studies can show the genetic and ecological structure of interactions across landscapes and the scale of local coadaptation. They provide an informed basis for evaluating the processes of population differentiation that may create new tips on phylogenetic branches.

For example, the geographic pattern of diversification of crossbills resulting from coevolution with lodgepole pines or black spruce discussed earlier in this chapter suggests a relationship between coevolution, peripheral isolates, and diversification in these interactions. In this case, the presumably derived crossbill species or incipient species are geographically isolated and highly coevolved with local populations of a conifer that shows similar patterns of geographic divergence. Similar studies mapping the geographic structure of other coevolving and diversifying lineages will help us further unravel the role of coevolution in codiversification.

Toward that end, some components of geographic structure may be particularly useful as tools. Pleistocene glaciers covered large expanses of the earth that have been recolonized only in the past ten thousand years. Studies that compare patterns of coevolution and codiversification following glacial retreat can provide a relatively clear geographic template on which to develop interpretations. In some cases, the interactions in their post-Pleistocene environments may be less diversified than in nonglaciated areas, which provides us an opportunity to study the structure of coevolution and diversification over relatively short periods of time in comparison to longer periods following extensive diversification. In the post-Pleistocene environments of the Pacific Northwest, the moth *Greya politella* feeds upon and pollinates only the plant *Lithophragma parviflorum* in most populations. In contrast, some of the phylogenetically older populations in California are involved in more phylogenetically complicated interactions (Thompson 1997). Some localities have two *Greya* species and two *Lithophragma* species, and both moths use both *Lithophragma* species in some areas. Evaluating the north–south gradient in the complexity of this interaction can provide a way of evaluating how interactions diversify and reticulate over various lengths of time.

Hybrid zones are another crucial component of the geographic mosaic of coevolutionary diversification and reticulation. We currently have no mathematical models of how hybrid zones may shape the geographic mosaic of coevolution or the phylogenetic structure of codiversification between coevolving taxa. We know, however, from molecular and field analyses that hybrids among species are common in some taxa (Harrison 1993; Rieseberg 1997; Hewitt 2001; Rieseberg et al. 2003). Reticulation of coevolving interactions is not inevitable through hybrid zones, but at the very least these regions can be coevolutionary hotspots or coldspots that may shape the overall geographic mosaic of coevolution among species. By providing regions of genetic continuity between two host species, hybrid zones may act as genetic bridges for parasites, allowing them to shift onto different hosts in some geographic regions (Floate and Whitham 1993).

The most comprehensive data sets currently available regarding the effects of hybridization on interspecific interactions are for plants and their enemies and mutualists. Over one hundred studies have shown that plant hybrid zones affect the geographic pattern of attack by herbivores and fungi (Strauss 1994; Whitham et al. 1999; Campbell et al. 2002), and some of these

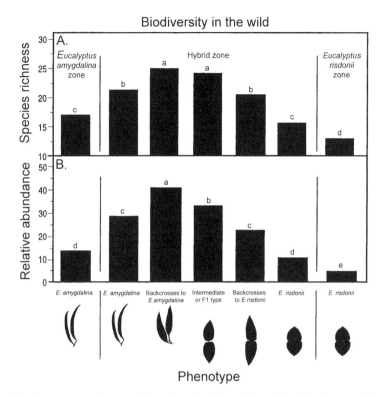

Fig. 7.8 The number of insect and fungal species (A) and the relative abundances of these species (B) on naturally occurring *Eucalyptus* plants within a hybrid zone as compared with parental populations outside the hybrid zone or nonhybrid plants within the zone. Different letters indicate significant differences among groups. The analysis is based upon forty insect and fungal species, with relative abundances calculated by averaging the standardized relative abundance measures for each taxon. After Whitham et al. 1999.

studies show highly elevated levels of herbivory within these zones (e.g., Morrow et al. 1994; Whitham, Morrow, and Potts 1994; Christensen, Whitham, and Keim 1995; Floate, Martinsen, and Whitham 1997). These studies also suggest that plant hybrid zones have strong effects on the overall assemblage structure of species attacking local plant populations (Fritz, Nichols-Orians, and Brunsfeld 1994; Fritz et al. 1996; Fritz 1999; Whitham et al. 1999). Geographic analyses of two *Eucalyptus* species in Australia suggest that hybrid *Eucalyptus* populations harbor greater numbers of insect and fungal species and higher relative abundances of species than parental *Eucalyptus* populations outside these zones (fig. 7.8).

Natural selection on pollinators and plants can also vary across the clines created by hybrid zones (Campbell, Waser, and Melendez-Ackerman 1997), directly shaping diversification or reticulation in plant taxa. Even where pollinators are almost completely specialized to different plant species, occasional movement among sympatric plant species can create hybrid zones (Leebens-Mack, Pellmyr, and Brock 1998). Those events, in turn, could shape the geographic mosaic of coevolution among other species. Some of those effects may be on completely different forms of interspecific interactions. For instance, avian nesting assemblages differ between hybrid zones of cottonwoods *(Populus* spp.) and parental nonhybrid regions in the southwestern United States (Martinsen and Whitham 1994).

USE OF GENES

In some taxa, coevolution and speciation may both be driven by a small number of genes responsible for specialization in defense, counterdefense, or mutualistic outcome, as indicated by the example earlier of flower color in monkey flowers *(Mimulus)* (Bradshaw and Schemske 2003). These studies may provide important tools for evaluating the pace of codiversification linked to coevolution, especially if they can be linked to diversification of entire lineages.

The predatory marine gastropod genus *Conus* may be a group in which diversification and specialization of toxin genes are simultaneously major components of species diversification (Duda and Palumbi 1999). Studies of comparative rates of diversification in these genes and in the counterdefense genes in their prey could provide powerful analyses of how coadaptation and codiversification interact across landscapes and over time. The genus *Conus* is a diverse group of about five hundred predatory snails that has evolved specialization to different prey taxa. All *Conus* species use venom, which stuns the prey by blocking a variety of ion channels and neuronal receptors (Olivera 2002). The venom mixture is deployed through hollow harpoonlike radular teeth that are individually moved into the proboscis (fig. 7.9).

Each conopeptide is the product of a single gene, and each *Conus* species has a unique mixture of 100–200 of these peptides (Olivera 2002). These mixtures are specialized for stunning particular prey taxa, in some cases even a single prey species (Olivera 2002). This remarkable diversification in venoms has occurred over millions of years, as *Conus* has evolved clades with venoms, morphologies, and behaviors specialized to different species. The

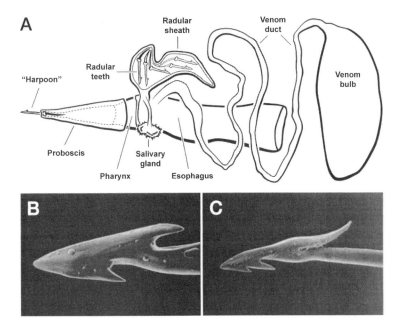

Fig. 7.9   (A) The venom deployment apparatus in the marine gastropod genus *Conus*. Scanning electron micrographs show the harpoonlike radular teeth of (B) *C. purpurascens* and (C) *C. obscurus*. From Olivera 2002.

fossil record suggests that the genus originated in the Eocene, and the molecular phylogenetic data suggest that the genus has diversified into clades specialized on prey taxa ranging from polychaetes to gastropods to fish (Duda, Kohn, and Palumbi 2001). As suggested by Olivera (2002), the *Conus* peptide genes are molecular readouts of the biotic interaction of each species.

Two distantly related *Conus* species that have been studied in detail for the genetics of conotoxins illustrate the high rate of divergence in these genes relative to other genes. The toxin regions of the genome have diverged at an average rate of 1.7–4.8 percent per million years, based upon nonsynonymous substitutions. This rate is much higher than nonsynonymous rates generally reported for mammals or *Drosophila*. The rates are also much higher than rates of substitution of other genes that have been studied within *Conus* (Duda and Palumbi 1999). These genes therefore provide a tremendous opportunity for future studies of the comparative rates of coadaptation with prey and diversification within and among clades over time.

USES OF MODES OF INTERACTION

Different modes of interaction impose very different selection pressures on coevolving taxa, which may potentially result in different rates and patterns of speciation (Price 1980; Thompson 1982, 1994). Comparisons of closely related taxa that differ in their mode of interaction with other species therefore provide powerful tools for studying the role of coevolution in diversification.

Figs are a highly diverse group, with an estimated 750 species (Weiblen and Bush 2002). They include some of the most important species in tropical forests and have been called keystone mutualists for some communities (Gilbert 1980). All figs are pollinated by wasps in the family Agaonidae, but not all fig wasps are mutualistic. Although the antagonists are in the same family as the mutualists, they are in different subfamilies. Pollination by fig wasps is widespread across continents and highly diversified throughout tropical and subtropical environments worldwide (Herre et al. 1996). Figs produce a highly unusual and specialized enclosed inflorescence with a tiny opening at the top through which a pollinating fig wasp must squeeze in order to gain access to the flowers. Flowers line the internal cavity of the fig, and, once inside, pollen-carrying females lay their eggs into a subset of the floral ovaries. The wasp larvae feed on the developing seeds, pupate, and mate as adults within the fig cavity. Pollen-carrying females then leave the fig in which they developed and fly to other receptive figs to lay their eggs.

Figs are attacked by a diverse group of insects that take advantage of this mutualism and lay their eggs in the developing fruits without contributing to pollination (Herre 1996). Among these is a genus of agaonid wasps from the subfamily Sycophaginae whose species pierce the outside of the ovary wall with their ovipositors and induce the fig ovaries to develop into large galls (Weiblen and Bush 2002) (fig. 7.10). These *Apocryptophagus* wasps are specialized only to figs in one monophyletic subgenus, *Sycomorus*, which is pollinated by one genus of mutualistic moths, *Ceratosolen*, in the subfamily Agaonidae (Weiblen 2000, 2001). This set of interactions among the figs, mutualists, and antagonists is therefore among three highly specialized monophyletic groups.

Since the mutualistic and antagonistic wasps use the same fig species, it is possible to evaluate directly how the mode of interaction shapes patterns of diversity. Using sixteen species in the subgenus *Sycomorus* from Melanesia and three additional species from outside the region, Weiblen and Bush (2002) compared patterns of codiversification in these taxa. The patterns were based upon a ribosomal phylogeny of the plants and mitochondrial

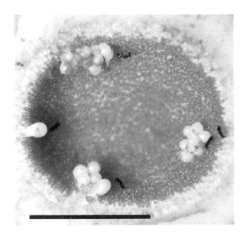

Fig. 7.10   Cross-section of a fig showing the interior. Female *Ceratosolen*
wasps are pollinating and laying eggs
into normal-sized ovaries. The large
flowers are galls induced by the *Apocryptophagus* wasps. Scale bar: 1 cm.
From Weiblen and Bush 2002.

phylogenies of the wasps. They used TREEMAP software (Page 2003b) to
generate the most parsimonious trees, which maximized cospeciation and
minimized losses and duplications of associations. According to this method
of phylogenetic reconstruction, the mutualistic fig wasps showed more evidence of codiversification with figs than did the parasites (Weiblen and Bush
2002). Like all methods of phylogenetic reconstruction, TREEMAP includes
assumptions that are somewhat unrealistic. The algorithms belong to a class
of methods that assumes the trees for parasites and hosts are known without
error (Page 2003a). Nonetheless, these methods, which continue to improve,
provide working hypotheses for future research. In this particular example,
future analyses will require evaluation of newer results, which show previously unsuspected cryptic speciation in fig wasps (Molbo et al. 2003). The
overall analysis, however, points to the kinds of comparative evaluations of
codiversification that should be possible with other coevolving interactions.

Patterns of codiversification between plants and their pollinators and
floral parasites may not be representative of the relative effects of mutualistic
coevolution as compared with antagonistic coevolution on speciation. The
effects of pollinators that are also floral parasites, such as fig wasps and yucca
moths, may be uncommon as modes of interaction. The mutualistic insects
directly determine patterns of gene flow in their plant hosts. As a result,
wherever speciation in the pollinator is accompanied by divergence in the use
of host plants, plant speciation can result as a direct consequence. Some other
forms of interaction also directly shape patterns of gene movement among
individuals within species. Examples include symbionts that affect the sex ratio of hosts or the reproductive compatibility among host individuals (see

chapter 10). Consequently, a productive area of future research will be to evaluate patterns of codiversification in coevolving interactions that affect patterns of sexual reproduction in comparison with coevolving interactions that do not.

Conclusions

The simplest possible relationship between coevolution and diversification is cospeciation, but it may be the least common. The complex geographic structure of most interspecific interactions creates selection mosaics, coevolutionary hotspots, and opportunities for trait remixing that are bound to break down any simple patterns of strict cospeciation in most interactions. Although diversifying coevolution can produce rapid divergence among populations, the opportunities for speciation vary across landscapes. The exception may be vertically transmitted mutualistic symbioses, where sustained parallel speciation can sometimes occur during coevolution.

By definition, some coevolutionary processes, such as escape-and-radiate coevolution between parasites and hosts, predict patterns of diversification that preclude cospeciation. We still require, however, a range of alternative hypotheses that link particular modes of coevolution and geographic structure with particular patterns of diversification. The emerging protocol for coevolutionary research calls for studies that integrate phylogenetic, ecological, geographic, genetic, developmental, and statistical approaches. That integration should result in more robust evaluations of how coevolution shapes diversification across clades and landscapes.

8 Analyzing the Geographic Mosaic of Coevolution

The hierarchical structure of coevolution requires approaches to research that incorporate local, geographic, and phylogenetic analyses. The results discussed in the chapters of part 1 suggest that, without those combined perspectives, it is impossible to understand the full dynamics of the coevolutionary process. In this chapter I summarize the forms of evidence that indicate an interaction is coevolving. I also put forward a general four-step protocol for coevolutionary research (modified from Thompson 1997). My purpose is not to propose specific methods but rather to argue for the need for complementary studies that address the hierarchical structure of coevolution as we continue to evaluate how the coevolutionary process shapes the organization of biodiversity.

Forms of Evidence for Coevolution

In 1980 Daniel Janzen wrote a short paper called "When Is It Coevolution?" as an attempt to stop the tendency of some researchers to assume that every trait involved in an interspecific interaction is a direct result of coevolution (Janzen 1980). In the subsequent decades, the science of coevolutionary biology has matured, and we now understand much better the forms of evidence needed to address how and where reciprocal evolutionary change shapes interactions. Researchers in different subdisciplines and those studying different kinds of interaction have focused on different forms of evidence. Studies of character displacement in species competing for limited resources, for example, have focused strongly on a set of six primarily ecological criteria for demonstrating character displacement (Schluter 2000a, 2000b). These criteria are discussed in chapter 15 and are incorporated into the more hierarchically structured list proposed below. Studies of potentially coevolving parasites and hosts have often relied more strongly on analyses of the genetic structure and dynamics of interactions or on the phylogenetic distribution of traits that may be under coevolutionary selection. Studies of mutualism have often focused on the ratio of fitness costs to fitness benefits resulting from an interaction or on interpretation of how particular traits contribute to the mutualism.

Together, the sets of criteria and approaches used in different studies suggest eleven major forms of evidence, divided into four groups, that can be marshaled in evaluating whether coevolution has shaped an interaction. Few interactions are likely to be analyzed for all these forms of evidence, and the point of coevolutionary studies should not be to march through all eleven. Instead, the list is intended as a guideline for mixing and matching evidence as we attempt to evaluate whether an interaction has been shaped by reciprocal selection. Some of the forms of evidence overlap somewhat but differ in methods of assessment. The list emphasizes the hierarchical structure of the evidence and the role of the geographic mosaic of coevolution. Each form of evidence contributes to our understanding not only of whether coevolution is involved in the evolution of an interaction, but also of the processes involved in shaping coevolutionary dynamics. For example, coevolution may be important in shaping patterns of adaptation but not patterns of speciation in some interactions.

Local Level

1. Individuals participating in a local interspecific interaction differ in Darwinian fitness from individuals excluded from the interaction.

2. Differences among individuals in the ecological outcome of a local interaction depend upon the traits of the interacting individuals.

3. Traits influencing differences in ecological outcome affect the fitnesses of the locally interacting individuals and are heritable.

Geographic Level

4. The geographic pattern of divergence in the traits of interacting species cannot be explained solely by phylogeographic patterns identified using neutral molecular markers. That is, geographic patterns are not simply due, for example, to parallel phylogeographic divergence as an obligate commensalistic species adapts to a host species.

5. Selective forces other than the interaction cannot fully explain the pattern of geographic covariation in these traits (e.g., adaptation strictly to a gradient in physical environmental conditions).

6. Traits important to the interaction are less exaggerated in the di-

rection predicted by reciprocal selection within presumed coevolutionary coldspots than within coevolutionary hotspots.

7. Although coevolving traits may be mismatched in some localities, they are well matched in other localities.

8. Varying degrees of mismatching of coevolving traits across landscapes can be explained by asynchronous fluctuating selection on those traits among populations, presence of additional species imposing selection on the interaction in some localities, or gene flow from other populations within a metapopulation or across a cline.

Species and Phylogenetic-Lineage Levels

9. Traits important to the ecological outcome and fitness of interacting individuals are unique or exaggerated in those species relative to closely related ancestors.

10. Speciation within interacting lineages results, directly or indirectly, from selection on coevolving traits.

Multispecific and Multilineage Networks

11. Additional species converging on the interaction do so through evolutionary convergence on the same traits as other species in the interaction (e.g., same detoxification mechanisms or same floral shapes) or through the evolution of traits that are complementary to or divergent from other species within the interaction.

For example, garter snakes (*Thamophis sirtalis*) and newts (*Taricha granulosa*) (see chapter 6) show multiple forms of evidence of coevolution. The interaction can affect the fitness of individuals. The outcome depends upon the level of toxicity in the newts and the level of resistance to tetrodotoxin (TTX) in the snakes. The degree of TTX resistance is heritable. TTX levels in the newts and TTX resistance in the snakes vary geographically in ways consistent with a geographic mosaic of coevolution but not with phylogeographic codivergence of snakes and newts across western North America. Selective forces other than coevolution are unlikely explanations for the observed geographic patterns of levels of defense and counterdefense. The levels of toxin and TTX resistance are exaggerated in two suggested coevolutionary hotspots, and TTX resistance in the snakes is lower outside the geographic range of the newts. Trait mismatches have not been identified, but extensive studies of geography of newt toxicity to match similar studies on TTX resistance are underway (E. D. Brodie III, personal communication). Tetrodotoxin occurs in all members of *Taricha*, but levels vary among species.

Thamnophis sirtalis shows a much greater and higher range of TTX resistance than other garter snakes.

Other examples from part 1 (e.g., red crossbills and pines; wild parsnips and parsnip webworms) and additional examples from part 2 all provide multiple forms of evidence for coevolution. These examples will provide the best model systems for future research on the coevolutionary process.

The Practicalities of Obtaining Evidence for Coevolution

Obtaining the different forms of evidence for coevolution requires multiple observational and experimental approaches that bridge subdisciplines. At a practical level, an analysis that provides some data on all these forms of evidence would involve four steps.

Step 1: Use ecological observations and experiments to identify (1) pairs or groups of species that have the potential to exert reciprocal effects on fitness, (2) the potential sources of differential selection on the interaction among communities, and (3) the possible conditions favoring coevolution. Coevolution is an inherently ecological process resulting from reciprocal selection imposed on interacting species within and among complex biological communities. Ecological studies are therefore the core of any coevolutionary analysis. Studies of the local ecological structure of an interaction are the starting point for coevolutionary analysis, and these kinds of studies have been the mainstay of research in evolutionary ecology for almost forty years. They are the crucial testing grounds for evaluating which of the millions of possible interactions worldwide have the greatest potential for providing informative results about the coevolutionary process.

This first step should no longer be biased toward interactions that are most likely to exhibit tight pairwise coevolution throughout the geographic range of the interaction. Rather, it should involve carefully choosing interactions that are amenable to robust analyses of how reciprocal selection can shape interactions within and among natural communities. We need to know the multispecific context of coevolving interactions, and we need to understand how groups, not just pairs, of species coevolve. This requires choosing interacting species in which it is possible to evaluate the potential multispecific structure of coevolutionary selection. Nothing can substitute for a good understanding of natural history in making these choices.

For example, the interaction between the pollinating floral parasitic moth *Greya politella* and its host plants discussed in earlier chapters became useful for studying the geographic mosaic of coevolution for three reasons. First, the interaction is widespread and common across western North America, making it relative easy to study the interaction in multiple environments. Second, the outcome of this pairwise interaction can be assessed within the full multispecific structure of natural communities, because the moths leave a record of their visit to each flower (i.e., eggs). Hence, it is possible to assess the effect of moth visits on pollination without artificially preventing all the other interactions such as visits by copollinators. Third—and this leads to step 2 in the process of coevolutionary analyses—knowing that *Greya* moths are closely related to yucca moths (and being aware of their well-known coevolutionary relationships with yuccas) provided a broader context right from the start for thinking about ways of studying the coevolutionary structure of this interaction (Thompson 1986b, 1987b; Davis, Pellmyr, and Thompson 1992).

Step 2: Establish robust phylogenies for the interacting species to determine the overall patterns of diversification and trait evolution that are potentially driven by coevolution. Phylogenetic analyses provide the foundation for understanding the origins and diversification of traits that are likely the focus of co-evolution in an interaction. Each species is a collection of hundreds, if not thousands, of identifiable morphological, physiological, biochemical, behavioral, and life-history traits that may affect the ecological outcome of an interaction. Phylogenetic analyses provide a way of honing the search for those traits most likely to be coevolved. These analyses can suggest the pattern of modification of traits shared with ancestors, the origin of traits "unique" to an interaction, the relative evolutionary malleability of coevolving traits, multiple origins of interactions, and the relationship between coevolving traits and patterns of lineage diversification (Thompson 1999d). Phylogenetic analyses therefore can help constrain the universe of possible explanations for a potentially coevolved trait to the small subset that is highly probable given the phylogenetic context of the interacting species.

Even if phylogenetic analyses show that 99 percent of the variation in traits involved in an interaction is accounted for at the genus level or higher, that would not by itself constitute evidence that ongoing coevolution is unimportant. The remaining 1 percent of traits may be the intense focus of current natural selection on those species. It is not the number or percentage of coevolving traits that matters, but rather the intensity of selection pressure on those traits. Arguing that coevolution is unimportant because only a few traits are coevolved would be like arguing that evolution in general is unimportant because species are polymorphic at only a small percentage of their loci or because only a small percentage of DNA codes for proteins.

Phylogenetic analyses of coevolving interactions also make it possible to evaluate how traits have been modified by ongoing coevolving interactions. For example, phylogenetic studies have shown that the unique pollen-holding tentacles of yucca moths are highly modified components of the characteristic lepidopteran proboscis (Pellmyr and Krenn 2002), and they have shown that active pollination of figs by fig wasps may have originated more than once (Cook et al. 2004). These kinds of analysis can help establish which traits are highly malleable and which are highly constrained during the diversification of lineages (e.g., Futuyma, Keese, and Scheffer 1993; Futuyma, Keese, and Funk 1995; Pellmyr et al. 1996; Armbruster et al. 1997; Thompson 1998b; Herrera et al. 2002; Herrera 2002; Pellmyr and Krenn 2002). Such studies can, in turn, provide a solid basis for undertaking experiments in evolutionary developmental genetics ("evo-devo" studies) to evaluate why some potentially coevolving traits are more malleable than others (Brakefield, French, and Zwaan 2003). Few coevolving interactions have even begun to be evaluated in this way to determine why interactions diversify in particular ways over space and time.

A central focus of all these studies should remain on how specialization evolves across populations, species, and lineages during codiversification (Thompson 1994; Armbruster and Baldwin 1998; Termonia et al. 2001). Again using the yucca moths as an example, these species are highly host specific to the yuccas with which they have coevolved. Nevertheless, extensive studies of host specificity throughout the family Prodoxidae have shown that extreme host specificity is characteristic throughout this moth family (Davis, Pellmyr, and Thompson 1992; Pellmyr, Leebens-Mack, and Thompson 1998). Hence, extreme host specificity in yucca moths and their close relatives the *Greya* moths is probably not a direct result of coevolutionary selection. Instead, it is an ancestral condition that may have increased the opportunity for coevolutionary interactions.

Step 3: Use molecular studies to evaluate the phylogeographic structure of the interaction and the potential for gene flow and trait remixing among populations. These studies provide the context for interpreting large-scale patterns of genetic differentiation among the interacting species. They provide a context for understanding the sizes of "tiles" within the geographic mosaic. Geographic differentiation in a particular interaction may arise from differential coevolutionary selection among regions or from a shared history of phylogeographic differentiation, with many co-occurring species driven by climate and geography (Brooks and McLennan 2002). The increasing availability of comparative phylogeographic analyses for entire regions (see chapter 2) makes it possible to evaluate what is shared with other taxa in the phylogeographic structure of a particular interaction and what is unique to that interaction. Some major phylogeographic breaks in the interaction between *Greya* moths and their host plants, for example, correspond to similar breaks found in other plant and animal species (Thompson and Calsbeek 2004). Coevolution between the moths and the plants is therefore an unlikely cause of these particular major phylogeographic breaks. Other geographic differences in traits and ecological outcomes, however, do not correspond to phylogeographic breaks shared with many other taxa. These regions of differentiation specific to a particular interaction can be useful for analyses of how the interaction itself may act directly to shape population differentiation.

Phylogeographic studies therefore provide a crucial template for teasing apart the effects of gene flow, hybridization, random genetic drift, and natural selection in creating the geographic mosaic of coevolution. If traits involved in an interaction show strong regional patterns of phenotypic differentiation but little or no molecular phylogeographic structure, the likelihood is increased that selection rather than random genetic drift has shaped these regional differences. Nevertheless, exactly how to link studies of comparative phylogeography with studies of coevolution remains a major challenge in coevolutionary research.

Step 4. Evaluate the actual geographic selection mosaic, the distribution of coevolutionary hotspots, and the pattern of trait remixing across populations. Steps 1, 2, and 3 provide the background needed for evaluating the geographic scale at which to undertake analyses for step 4. By knowing traits that are the focus of coevolution, the ecological conditions that favor different ecological outcomes and coevolutionary hotspots, and the structure of gene flow, it becomes possible to study the coevolutionary process much more efficiently. These studies provide the context for understanding the genetics, ecological structure, and overall dynamics of coadaptation in an interaction. Moreover, they provide the basis for understanding how new species may form along phylogenetic branches.

The fourth step is easier for some interactions than for others, but studies on the geographic mosaic of coevolution over the past decade have shown that it is possible to directly study these components of the coevolutionary process (e.g., Benkman, Holimon, and Smith 2001; Brodie, Ridenhour, and

Brodie 2002; Burdon, Thrall, and Lawrence 2002; Thompson and Cunning-ham 2002; Zangerl and Berenbaum 2003). The studies described throughout part 1 of this book show the variety of possible approaches to confronting the reality that we cannot study every population of every species in every way. Some components of interactions are more amenable to study than others, but each study contributes to our understanding of the role of the geographic mosaic in holding together some coevolved interactions over long periods of time.

A major part of step 4 involves analysis of geographic variation in traits and ecological outcomes. Procedures for evaluating geographic covariation in traits in coevolving interactions are developing quickly. Some current approaches use logistic regression to evaluate the structure of reciprocal selection (Benkman 2003) or the degree of matching and mismatching of coevolved traits across landscapes (Zangerl and Berenbaum 2003). These procedures have proven to be useful for evaluating the geographic structure of selection on particular sets of coevolving traits.

Brodie and Ridenhour (2004) have proposed a direct extension of Lande and Arnold's (1983) covariance approach for estimating selection within natural populations. The approach evaluates conditions under which fitness of one species depends upon the phenotype of a second species. This covariance approach allows simultaneous assessment of direct and indirect selection on multiple quantitative traits. The approach assumes that individuals interact at random, which is more relevant for some interactions than for others. Its power lies in its ability to allow exploration of the interaction fitness surfaces for various combinations of traits in the interacting species. For example, consider the geographically variable interaction between toxic newts (*Taricha*) and garter snakes in western North America. The covariance approach makes it possible to evaluate how the combined effects of the level of newt toxicity, snake skeletal muscle resistance, and snake speed determine snake fitness in populations where average newt toxicity is high compared with populations where average toxicity is low (fig. 8.1). The illustrated example is a hypothetical data set, but it shows how the structure of selection on phenotypes in coevolving interactions may differ among environments.

Two kinds of experimental analysis can help further define the geographic structure and scale of the coevolutionary adaptive landscape. Challenging local populations with sympatric and carefully chosen allopatric populations of the other species can determine how deviation from local means in coevolving traits affects the fitness of individuals (e.g., Benkman 1999; Thrall, Burdon, and Bever 2002). Allopatric comparisons must be chosen in a way

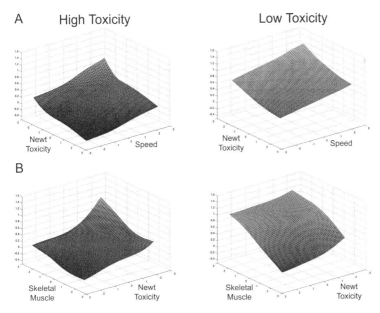

Fig. 8.1   Hypothetical interaction fitness surfaces for interactions between toxic newts and garter snakes in a region where average newt toxicity is high as compared with regions where it is low. The panels under each average level of toxicity (i.e., high toxicity; low toxicity) show how snake fitness (vertical axis) varies depending upon the phenotypes of the interacting individuals (i.e., newt toxicity and either speed or skeletal muscle resistance in the snakes in response to that level of toxicity). From Brodie and Ridenhour 2004.

that makes it possible to capture the geographic scale of local adaptation and coadaptation in the interaction. In addition, artificial selection on coevolving traits can provide direct measures of the rate at which these traits can coevolve and the degree to which evolution of these traits is coupled with evolution of other traits. Such experimental approaches have become increasingly important to our overall understanding of the population and geographic structure of all forms of adaptation (Brakefield 2003; Conner 2003; Pigliucci 2003).

Finally, empirical studies of trait remixing can help in evaluating how the geographic mosaic of coevolution contributes to the dynamics of local mismatching of coevolving traits. The mathematical models have shown that gene flow and metapopulation dynamics can have powerful influences on coevolutionary dynamics. Nevertheless, studies of trait remixing are the uncharted waters of coevolutionary research. There are currently no direct experimental studies on how different levels of gene flow shape the geographic mosaic of coevolution. Through the use of molecular markers, such studies will contribute to future work on the geographic scale of coadaptation and coevolutionary dynamics.

In research on actual coevolving species, evaluation will never, of course, proceed as a simple ordered set of steps. Individual studies and collaborations among research groups generate new results for each of the steps over time. As the developing protocols for analysis of coevolutionary selection and trait matching are applied to each study, they can show which populations and geographic regions will provide the most useful prospects for future research on the scale and process of coevolutionary change. I have outlined here a general set of steps for coevolutionary research only to suggest how different subdisciplines and approaches can contribute collectively to an overall evaluation of the coevolutionary process.

Together, these components of coevolutionary analysis can help us understand how geographic selection mosaics shape traits across landscapes, how local mismatches in traits develop, and why few coevolved traits spread across all populations. These analyses are also the means for evaluating how the geographic mosaic of coevolution shapes the seven major trajectories of local coevolutionary change. I turn to those trajectories in part 2 of the book.

Part 2

Specific Hypotheses
on the Classes of
Coevolutionary Dynamics

9 Antagonists I

The Geographic Mosaic
of Coevolving Polymorphisms

Part 2 explores how the geographic mosaic shapes the seven major classes of local coevolutionary dynamics. The specific hypotheses discussed in these chapters are extrapolations and syntheses of observations and experiments from the past decade, which I present with the objective of suggesting fruitful topics for future coevolutionary modeling and empirical research. Each hypothesis is developed with one or more empirical examples. These examples provide some support for the hypotheses, but they also highlight gaps in what we currently know.

This chapter and the following two focus on the geographic mosaic of antagonistic trophic coevolution between parasites and hosts, grazers and victims, and predators and prey (see table 9.1). These modes of antagonistic interaction represent a continuum from symbiotic relationships to those among free-living species and differ in the opportunities they create for specialization, defense, and counterdefense (Price 1980; Thompson 1982, 1994). A similar continuum influences coevolution in mutualisms, and the hypotheses on mutualism in chapters 12–14 will, in part, parallel those on antagonism in these three chapters by using the mode of interaction as a major determinant of how interacting species coevolve. Throughout these chapters I use the word "symbiont" in its original sense, which is any organism that lives intimately throughout its lifetime or a major life-history stage on or within a host individual. This usage spans parasites, commensals, and mutualistic symbionts.

I begin in this chapter by evaluating two ways in which the geographic mosaic of coevolution contributes to coevolving polymorphisms in antagonistic trophic interactions: oscillating polymorphisms and optimal allelic diversification.

Table 9.1   Ways by which the geographic mosaic shapes coevolution in antagonistic trophic interactions and the kinds of evidence currently available in published studies

| | Current evidence suggests | | |
	Geographic selection mosaics	Coevolutionary hotspots	Trait remixing
Oscillating polymorphisms	yes		yes
Optimal allelic diversification	yes		
Sexual reproduction	yes	yes	
Coevolutionary alternation	yes	yes	

A Plethora of Polymorphisms and a Role for the Geographic Mosaic of Coevolution

Coevolving polymorphisms are one of the seven major classes of coevolutionary dynamics observed in mathematical models of local coevolution. Most species have more than one allelic form at many genetic loci. This simple fact has lead to the development of the fields of molecular ecology, molecular systematics, phylogeography, molecular forensic science, molecular medicine, molecular epidemiology, and comparative genomics, and some subfields of applied mathematics, statistics, and computational science. Many polymorphisms appear to be selectively neutral, but others have major phenotypic effects.

These phenotypically expressed polymorphisms may be maintained by multiple forms of interaction between species (e.g., sex-limited selection on Batesian mimics) and also by sexual selection and intraspecific competition (Sinervo and Zamudio 2001). Coevolution with parasites, however, appears to be a particularly important means by which polymorphisms are maintained within species, and this chapter concentrates on these interactions. Gene-for-gene coevolution between plants and pathogens is driven by fluctuating polymorphisms, both in natural populations and in agricultural breeding programs (Thompson and Burdon 1992; Simms 1996; Burdon 1997; Crute, Holub, and Burdon 1997; Thrall and Burdon 2002). Similarly, coevolution between the vertebrate immune system and parasites is based upon tremendous allelic diversification (MHC Sequencing Consortium 1999; Wegner, Reusch, and Kalbe 2003). The more we learn about coevolutionary dynamics, the more evident it is that ongoing coevolution is a major cause of the origin and maintenance of genetic diversity.

Coevolution and the observed commonness of genetic polymorphisms are therefore tightly linked, and studies of coevolving polymorphisms are becoming one of the most important interfaces between evolutionary biology and human society. Genomic approaches to medicine are moving toward the long-term goal of treatment based upon polymorphisms found within hosts, parasites, and vectors, and the clinical outcomes that result from different combinations of host and parasite alleles as expressed in different environments (Hoffman et al. 2002). Moreover, studies of the structure and dynamics of polymorphisms are becoming increasingly important in efforts to develop effective control against major diseases such as malaria. These studies have increased our need to understand how the geographic mosaic of coevolution shapes interspecific interactions.

The development of DNA technologies has shown that genetic polymorphisms are distributed throughout the genomes of most species. Many polymorphisms are probably transient, appearing through mutation and then quickly lost through natural selection or random genetic drift. Many others are probably maintained at least partly by variable selection imposed by heterogeneous physical environments (Zhivotovsky and Feldman 1993; Mitton 1997). For example, even small differences along light, temperature, and moisture gradients can favor different alleles in some plant populations (Bazzaz 1996). Similarly, because enzymes differ in their temperature optima, in some animal species different alleles can be favored even over short distances under different thermal regimes (Watt et al. 2003).

The proportion of polymorphic genes actually under selection at any moment within a population is unknown. Moreover, given the current state of our knowledge, it is impossible to make any reliable guess about the proportion of polymorphic genes maintained by coevolutionary selection. We can, however, now say with some confidence that coevolution contributes to the long-term maintenance of polymorphisms. The process may often involve the geographic mosaic of coevolution, because few populations harbor more than a small subset of the polymorphisms that occur within species, and many polymorphisms are often distributed nonrandomly across space. In addition, some polymorphisms have been maintained within species for millions of years, which is far beyond the length of time during which most local populations persist. Hence, the working hypothesis that best captures our current understanding of coevolving polymorphisms is as follows:

The hypothesis of the geographic mosaic of oscillating polymorphisms.

Coevolving polymorphisms are shaped by natural selection on parasites to adapt to the most common local host genotypes and by selection on hosts to recognize and resist attack by parasites. Selection on hosts favors individuals with genes that allow detection of multiple parasite species or genotypes, but it also favors rare genotypes to which the local parasites are not adapted. Local persistence of a small number of coevolving polymorphisms is maintained by multiple forms of selection, but maintenance of high allelic diversity of coevolving polymorphisms within species across millennia requires geographic selection mosaics, coevolutionary hotspots, and trait remixing.

In principle, polymorphisms can be maintained within host species through natural selection in four major ways: heterozygote advantage, local temporally varying selection, local frequency-dependent selection, and the geographic mosaic of coevolution. Heterozygotes may exhibit higher resistance against enemies than homozygotes. Such heterozygote advantage (i.e., overdominant selection) has long been proposed as a hypothesis for the maintenance of sickle-cell polymorphisms within human populations that live where malaria is endemic. Individuals who are heterozygous for the trait have lower fitness in the absence of malaria, but higher fitness in malarial regions because sickle cell protects them from *Plasmodium* infection. In other cases, heterozygote advantage could result from an increased ability to recognize a broader range of sympatric pathogens. If a locus functions primarily to create proteins that detect the presence of pathogens and that locus is codominant, then heterozygotes may be able to detect two kinds of parasites or pathogens, whereas homozygotes could detect only one (Hughes and Nei 1992). Maintenance of resistance polymorphisms through this mechanism becomes less likely at the level of local populations if the heterozygotes and homozygotes differ in fitness (Lewontin, Ginsburg, and Tuljapurkar 1978; Hedrick 2002). In general, although heterozygote advantage could contribute to the long-term maintenance of some polymorphisms, it cannot explain polymorphisms at loci lacking overdominance or the maintenance of very high levels of allelic diversity found at many loci within species.

Temporally varying selection can contribute to the maintenance of polymorphisms by favoring different alleles in different generations, as shown in recent models (Hedrick 2002; Nuismer, Gomulkiewicz, and Morgan 2003). Two situations have been explored in detail. In one set of models, each allele confers resistance against different pathogens (Hedrick 2002). The two pathogens differ in their proportions over time, resulting in temporally varying selection on the host population. Under some conditions, resistance polymorphisms can be maintained by such varying selection within a local population. Heterozygosity is highest when pathogens vary considerably in presence among generations (i.e., low autocorrelation). In the other set of models, an interaction varies among generations from antagonism to mutualism, and polymorphism is maintained through their extreme shifts in selection among generations (Nuismer, Gomulkiewicz, and Morgan 2003). Whether local polymorphisms and high allelic diversity could be maintained over millennia in less extreme situations through temporally varying selection alone is unknown.

Frequency-dependent selection contributes to the maintenance of coevolving polymorphisms by favoring individuals with rare alleles, as outlined in chapter 5. Under this form of selection, multiple alleles cycle within local populations, with each allele favored when rare but disfavored when common in at least one of the interacting species. As applied to coevolution between parasites and hosts, a rare host genotype increases in frequency within a local population until it becomes disfavored by selection imposed by parasites that are simultaneously under selection to attack common host genotypes. Under these conditions, the local dynamics of coevolving polymorphisms arise from oscillating selection driven by negative frequency dependence. This coevolutionary process produces time-lagged cycles in allelic frequencies in both species, which maintains coevolving polymorphisms over at least some period of time. The extent to which allelic cycles of the two species are in phase or out of phase depends upon the relative rates at which the two species evolve. High mutation rates, high recombination rates, and fast generation times can all decrease the phase difference between the two species (Kaltz and Shykoff 1998).

The genetic architecture of coevolution can influence the dynamics of locally coevolving polymorphisms. Even coevolution controlled by only a few genes can have very different genetic architectures among species. For example, although a matching-alleles structure is the simplest form of coevolutionary interaction, many plants and pathogen coevolve at least in part through a gene-for-gene interaction. In gene-for-gene interactions, each resistance gene in a host population matches and recognizes a corresponding avirulence gene in the pathogen population. In the simplest case, a pathogen is able to attack a host only if the host lacks the corresponding resistance gene or the pathogen carries a different allele from the one recognized by the host resistance allele (table 9.2).

Matching-alleles models and gene-for-gene models represent the two ends

Table 9.2   Matching allele and gene-for-gene (GFG) models of coevolution

	Pathogen genotype			
Host genotype	*AB*	*Ab*	*aB*	*ab*
Matching-alleles model				
AB	I	...	...	...
Ab	...	I	...	...
aB	...	...	I	...
ab	...	...	...	I
	vv	*vV*	*Vv*	*VV*
GFG model				
rr	I	I	I	I
rR	I	...	I	...
Rr	I	I	...	...
RR	I	...	...	...

Source: After Otto and Michalakis 1998 and Lively 1999.
Notes: I = infection by a parasite or successful attack by a predator. In the GFG model, R = resistant, r = susceptible, v = virulent, and V = avirulent.

of a continuum in simple genetic architectures, and there has been considerable discussion of how to interpret the fit of empirical results along this continuum (Frank 1994b; Parker 1994; Agrawal and Lively 2002, 2003). Allele frequencies often fluctuate faster and with greater amplitudes in matching-allele models than in gene-for-gene models (Parker 1994; Agrawal and Lively 2002). The genetic architecture of coevolving polymorphisms could even vary over time as the number of polymorphic loci changes, or the architecture could involve a multistep process with both matching-gene and gene-for gene components—or "specific" and "nonspecific" components of resistance—adding further complexity to the dynamics of fluctuating polymorphisms (Frank 2000; Agrawal and Lively 2003).

The potential for polymorphisms to be maintained by frequency-dependent selection has long been known (Haldane 1949; Hamilton 1980), and models that attempt to capture the allelic complexity of actual populations and allow long-term persistence of that complexity are still being explored. In fact, the first model of coevolution (Mode 1958) was created to explore the coevolutionary dynamics of polymorphism in plants and pathogens. In the subsequent decades, dozens of models have explored coevolutionary dynamics involving frequency-dependent selection. The models differ in their tendency to create stable coevolutionary cycles, but a broad range of models permit local coevolutionary cycles for multiple generations (e.g., May and Anderson 1983; Marrow, Law, and Cannings 1992; Seger 1992; Dieckmann, Marrow, and Law 1995; van der Laan and Hogeweg 1995; Abrams and Matsuda 1996; Gavrilets 1997; Abrams 2000a; Bergelson, Dwyer, and Emerson 2001; Agrawal and Lively 2003).

Most of the models have assumed that a small number of major genes govern the coevolutionary dynamics. Some recent models, however, explore the dynamics of polygenic inheritance, and these models, too, can produce coevolutionary cycles similar to those found in major-gene models, at least under some conditions (Gavrilets 1997). Many quantitative genetic approaches to evaluating fitness and selection are possible in coevolutionary models (e.g., Taper and Case 1992; Frank 1994c; Abrams and Matsuda 1996), and much work still remains to be done on how the genetic architecture of coevolution shapes the local maintenance and dynamics of polymorphisms. Nevertheless, it appears that multiple forms of genetic architecture can maintain fluctuating polymorphisms for many generations.

Heterozygote advantage, temporally varying selection, and frequency-dependent selection capture only part of the actual structure of coevolving polymorphisms within species. The presence and abundance of enemies varies among populations, and the selective effects of these enemies also vary, depending upon the expression of genes in different environments and the community context in which the interaction occurs. Even if there is no sustained difference among populations in the presence of particular enemies, the interaction of all the mechanisms that favor polymorphisms will almost inevitably result in a geographic mosaic of traits and selection pressures. Populations will cycle in gene frequencies out of phase with other populations. Consequently, allele frequencies of resistance and counterresistance genes will vary broadly across both space and time. An allele driven to extinction within a local population may persist in other populations. Through the geographic mosaic of coevolution, polymorphisms may be maintained at the regional and species levels, even when they are transient at the local level.

Appreciation of the potential importance of the geographic mosaic of coevolution in maintaining a high diversity of coevolving polymorphisms within species is increasing. The new models outlined in chapter 6 show that geographic selection mosaics, coevolutionary hotspots, and trait remixing can increase the chance of maintaining polymorphisms between parasites and hosts over long periods of time. Even small differences in coevolutionary selection among populations connected by gene flow can help maintain polymorphisms. These results suggest that maintenance of coevolving poly-morphisms across long periods of time may be explained most simply by the commonness of geographic structure in interacting species.

In fact, some results now show that polymorphisms driven by inter-specific interactions may have been maintained by selection not only for thousands but for millions of years as local populations have come and gone across landscapes. The *Rpm1* gene in *Arabidopsis thaliana,* for example, con-fers resistance against *Pseudomonas* by allowing individuals to detect the presence of pathogens carrying *AvrRpm 1* or *AvrB* genes (Stahl et al. 1999). *Rpm1* is one of more than one hundred *R*-genes distributed throughout the *Arabidopsis* genome (Tian et al. 2003). It codes for a peripheral plasma mem-brane protein, which allows individuals to detect the presence of pathogens carrying those particular complementary genes (Boyes, Nam, and Dangl 1998). Homologues of *Rpm1* occur in related plant genera such as *Brassica* and *Arabis,* suggesting that the gene has persisted during at least part of the diversification of the family Brassicaceae. Its distribution, however, is not ubiquitous within species. One analysis of twenty-six populations of *Ara-bidopsis thaliana* collected from North America, Europe, Asia, and Africa showed much variation within and across continents in the presence of *Rpm1* (Stahl et al. 1999).

Coalescence analysis has shown that the geographic pattern of presence of *Rpm1* is not due simply to historical subdivision of the species into regions with different alleles. Instead, ongoing natural selection appears to maintain the polymorphism. The pattern of divergence across these regions is statisti-cally consistent with maintenance of the polymorphism by natural selection but not with divergence through neutral substitutions of base pairs. DNA se-quencing of the *Rpm1* locus and flanking regions in these twenty-six popula-tions shows a strong peak of polymorphism at the *Rpm1* site in comparison with the flanking regions. In addition, the age of this polymorphism within *A. thaliana* suggests an ongoing role for natural selection. The coalescence analyses of molecular divergence at *Rpm1* suggest that the polymorphism is at least 9.8 million years old (Stahl et al. 1999).

Although highly effective as a defense mechanism, *Rpm1* imposes a fitness cost on individuals that harbor it. Such costs of resistance have long been an important part of the theory of why defense genes do not always become rapidly fixed within populations, but only in recent years have robust experiments begun to evaluate the structure and level of those costs (Antonovics and Thrall 1994; Bergelson and Purrington 1996; Mitchell-Olds and Bradley 1996; Gemmill and Read 1998; Bohannan, Travisano, and Lenski 1999; Kraaijeveld and Godfray 1999; Strauss et al. 2002; Hare, Elle, and van Dam 2003; Marak, Biere, and Van Damme 2003). The creation of transgenic lines that differ only in the *Rpm1* allele has allowed especially precise tests of the effect of harboring one particular allele rather than another. That technique has shown that maintenance of *Rpm1* requires ongoing selection imposed by enemies.

Demonstrating this effect has required a combination of molecular and ecological studies. *Rpm1* is known to be a single locus. Tian et al. (2003) inserted the *Rpm1* gene, along with its promoter and terminator, into a susceptible ecotype. Further crosses produced four independent pairs of homozygous lines that differed only in the presence of the *Rpm1* gene. The lines were then transplanted into a field plot and allowed to grow to maturity in the absence of detectable disease. Bacterial isolates were collected from the plants throughout the experiment, but none were pathogenic, and none carried *AvrRpm1* or *AvrB*, which are the bacterial genes recognized by *Rpm1*. At senescence, plants with the *Rpm1* gene produced, on average, nine percent fewer seeds than their paired counterparts lacking the gene (fig. 9.1).

The molecular and biochemical mechanisms that produce this cost are unlikely to be due to the costs of *Rpm1* synthesis itself, because the constitutive levels are very low (Tian et al. 2003). Instead, the costs may be due either to direct induction of some plant defense pathways in *Rpm1* plants even in the absence of pathogen attack, or to indirect induction of plant defense pathways through interaction between *Rpm1* and other resistance (R) genes (Tian et al. 2003). The high costs that harboring this gene imposes may not be typical of resistance genes. Many plants harbor multiple resistance genes, implying that the costs of all such genes cannot be high and additive or multiplicative (Burdon and Thrall 2003). If the cost of *Rpm1* is indeed unusually high, the results are even more impressive. Despite these high costs, natural selection has maintained this gene over millions of years, because it provides an effective defense against pathogens.

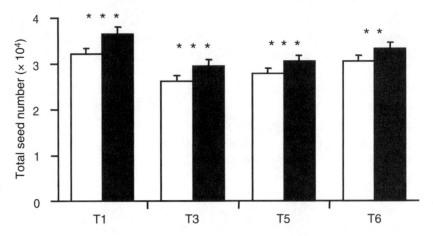

Fig. 9.1　Total number of seeds produced by four pairs of homozygous lines (T1–T6) of the plant *Arabidopsis thaliana* that differ within each line only in the presence (white bars) or absence (black bars) of the pathogen resistance gene *Rpm1*. The asterisk (*) indicates significant differences between treatments within each homozygous line, and the number of asterisks indicates the level of statistical significance. After Tian et al. 2003.

The Dynamics of Gene-for-Gene Coevolution within and among Natural Populations

Although difficult enough in itself, it has been easier to obtain data on the costs of resistance in hosts than it has been to obtain data on the actual dynamics of coevolving polymorphisms within natural populations. The detailed and careful studies of coevolution between Australian wild flax (*Linum marginale*) and flax rust (*Melampsora lini*) led by Jeremy Burdon are unquestionably the most comprehensive and important data available on the dynamics of coevolved genetic polymorphisms across natural landscapes (e.g., Jarosz and Burdon 1991; Burdon 1994; Burdon and Thompson 1995; Burdon and Thrall 2000; Burdon, Thrall, and Lawrence 2002; Thrall, Burdon, and Bever 2002). The relative frequency of genetic morphs of Australian wild flax that provide protection against different genetic morphs of flax rust has fluctuated broadly within and among local populations across years (Burdon and Thompson 1995; Burdon and Thrall 2000). Two decades of research on multiple local populations have demonstrated conclusively that polymorphisms associated with defense and counterdefense can evolve rapidly within and among host and pathogen populations. These studies have become the exemplar of how coevolving polymorphisms are maintained within species.

Australian wild flax is an herbaceous perennial endemic to southern Australia. It is widely distributed, ranging from lowlands under sparse woodland canopies to subalpine meadows. It occurs throughout its range as local populations that include hundreds or thousands of individuals. The mating system varies among populations, and subalpine populations especially show high levels of inbreeding (Burdon, Thrall, and Brown 1999).

Melampsora lini, the causative agent of flax rust, is an autoecious pathogen that occurs on multiple continents but in Australia is naturally restricted to *Linum marginale.* The pathogen attacks living tissue and produces lesions that are readily visible on plants in the field. Except for occasional attack by a seed-feeding insect (J. J. Burdon, personal communication), it is the only major symptomatic pathogen or parasite of its host. The rust significantly affects survival and reproduction in wild flax, and population declines of up to 80 percent have followed epidemics (Jarosz and Burdon 1992). In some areas, rust populations appear to be capable of both sexual and asexual reproduction, while in other regions, populations show strong linkage disequilibrium, suggesting that reproduction in those populations is asexual (Burdon and Roberts 1995).

The most detailed studies of coevolutionary dynamics between Australian wild flax and flax rust have been on populations distributed across the Kiandra Plain near Mount Kosciusko, in southeastern Australia (fig. 9.2). The plain is an extensive, open subalpine region of grassland and meandering streams. *Linum marginale* grows in discrete local populations across the plain and westward down the mountains to the western plains. Each deme has up to several hundred individuals, and demes separated by only a few hundred meters can differ greatly in their genetic composition (Thrall, Burdon, and Young 2001).

The open habitat of the Kiandra Plain makes it possible to evaluate attack on all individuals within a deme. Yearly studies of marked populations since the mid-1980s have shown that both plant and rust populations undergo large changes in numbers from year to year. Both species show evidence of a metapopulation structure, and the pathogen populations are particularly susceptible to local extinctions—followed by recolonizations—across multiple spatial scales (Burdon and Thrall 2000).

Wild flax and flax rust coevolve through a gene-for-gene structure. Within natural populations, host individuals vary in the resistance genes they carry, and pathogens vary in their ability to overcome different resistance genes. Virulence in gene-for-gene interactions is generally defined as the ability of different pathogen genotypes to infect host genotypes. Highly virulent

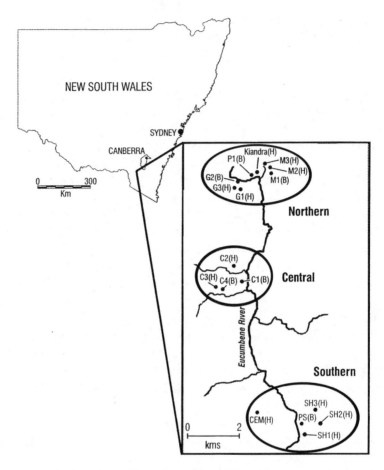

Fig. 9.2   Distribution of populations of Australian wild flax, *Linum marginale,* chosen for intensive study by Burdon and colleagues within the Kiandra metapopulation. Other populations occur within the mapped area but are not shown. After Burdon and Thrall 2000.

genotypes are those capable of attacking many host genotypes. (This definition of virulence, widely used in plant pathology, differs from that used in some other subdisciplines of epidemiology and evolutionary ecology, where it commonly refers to the negative effects of parasites and pathogens on host fitness.)

Through gene-for-gene coevolution, local populations can accumulate polymorphisms across multiple loci. One subalpine population of Australian flax studied intensively on the Kiandra Plain between 1981 and 1991 in-

Table 9.3   Resistance phenotypes found within the Kiandra population of Australian wild flax (*Linum marginale*) in New South Wales, Australia

Resistance phenotype	Pathogen isolate								
	N44	P13	K18	MDI	N5	GIN	A54	HSC	WA
I	S	R	R	R	R	R	S	R	R
II	R	R	R	R	R	R	S	R	R
III	R	R	S	S	R	R	S	R	R
IV	R	R	S	R	R	R	S	R	R
V	R	R	R	S	R	R	S	R	R
VI	S	S	S	R	R	R	S	R	R
VII	S	S	R	R	R	R	S	R	R
VIII	S	R	S	S	R	R	S	R	R
IX	S	S	S	S	R	R	S	R	R
X	R	R	S	R	S	R	S	R	R
XI	R	S	S	S	R	R	S	R	R
XII	S	S	R	S	R	R	S	R	R
XIII	S	R	R	S	R	R	S	R	R
XIV	R	S	S	R	R	R	S	R	R
XV	R	S	R	R	R	R	S	R	R
XVI	R	R	R	R	S	R	S	R	R
XVII	S	R	S	R	R	R	S	R	R
XVIII	R	S	S	R	S	R	S	R	R

Source: From Burdon and Thompson 1995.
Notes: Resistance was determined by challenging plants with known isolates of the rust *Melampsora lini*. R = resistant to that particular pathogen isolate. S = susceptible to that isolate.

cluded eighteen resistance phenotypes, as assessed using inoculations of nine pathotypes of *M. lini* (table 9.3). These local frequencies of resistance phenotypes can change rapidly within a deme (fig. 9.3). The two most common resistance phenotypes within the Kiandra population in 1986 each accounted for 33–44 percent of all plants, dominating the population. Following an epidemic in 1989 after the appearance of a new pathotype (A54), the frequency of these phenotypes plummeted, making them among the least common phenotypes during the subsequent several years (Burdon and Thompson 1995).

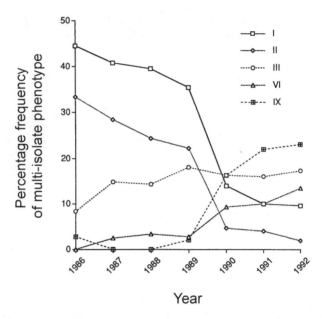

Fig. 9.3   Changes over time in the frequency of resistance phenotypes of Australian wild flax (*Linum marginale*) within the Kiandra population of *Linum marginale*. The resistance phenotypes given in Roman numerals correspond to the phenotypes in table 9.3. After Burdon and Thompson 1995.

There is a strong spatial mosaic of resistance and virulence across the Kiandra Plain. Local host populations differ in their average resistance, and local pathogen populations differ in the number of host genotypes they can infect. Moreover, local plant and fungal populations differ in the specific resistance and avirulence genes they harbor. If the plant and pathogen populations are very locally coadapted within the plain, they should show a distance effect in patterns of coadaptation. Evidence of coadaptation should decrease as individuals of one species are challenged by individuals from populations at ever-greater distances away. There are two ways of conducting this experiment (Ebert, Zschokke-Rohringer, and Carius 1998; Gandon et al. 1998; Kaltz and Shykoff 1998; Kaltz et al. 1999; Lively 1999; Lively and Dybdahl 2000; Mutikainen et al. 2000), and both kinds of experiments have been performed on these species. One is to test a local pathogen population with its sympatric host and with allopatric hosts at different distances away. The other is to test a local host population with its sympatric pathogen and with allopatric pathogens at different distances away. Congruence between these tests provides strong evidence for coadaptation at a very local scale. Lack of congruence in these two tests suggests that regional coevolutionary dynamics prevent simple relationships between local coadaptation and distance.

Within the Kiandra Plain, pathogens are generally better able to attack plants within their local population than plants from other populations, but distant pathogens within the region are no worse than local pathogens at attacking a host population (Thrall, Burdon, and Bever 2002). The lack of congruence in these tests suggests that these coevolving polymorphisms are coadapted at the metapopulation scale, because this is the scale at which the regional pools of resistance and virulence genes are cycling. Recent models have suggested that the overall pattern of coadaptation within such metapopulations depends upon differences between hosts and pathogens in rates of gene flow, generation times, mutation rates, the strength of selection on resistance and virulence genes, and the spatial distribution of the populations (Gandon 2002a; Gandon and Michalakis 2002).

The ability of rust isolates to infect multiple host genotypes varies broadly among flax populations. These differences appear to be driven by geographic differences in the genetic diversity of resistance within flax populations. Host populations with a high diversity of resistance genes tend to harbor pathogens capable of attacking a broad range of host genotypes (Thrall and Burdon 2003). In these gene-for-gene interactions, this correlation suggests that high resistance diversity in hosts favors a high diversity of genes for overcoming resistance. The correlation, however, could only develop if the ability to attack multiple host genotypes creates a fitness cost in the pathogen. Otherwise, natural selection should always favor pathogen genotypes able to attack a broad range of host genotypes. Such a cost occurs in *Melampsora,* although it is a highly variable cost. Pathogen genotypes differ greatly in their spore production on hosts, but no pathogen isolates capable of attacking multiple host genotypes are able to attain the high levels of spore production found in some more-specialized isolates (fig. 9.4).

Additional geographic structure is evident in this interaction over continental scales. Pathogen isolates from Western Australia are capable of attacking a wider range of host resistance lines from Western Australia than from other parts of Australia (Burdon, Thrall, and Lawrence 2002). Pathogen isolates from eastern Australia and Tasmania also differ in their ability to attack plants from various populations throughout this region. At the overall continental scale, distant pathogen isolates are generally less able to attack local

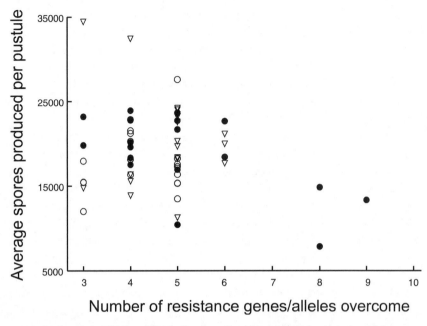

Fig. 9.4   The relationship between the number of resistance genes that an isolate of the rust *Melampsora lini* is able to overcome when attacking its host, Australian wild flax (*Linum marginale*), and the average number of spores produced per pustule by that isolate. Symbols identify isolates from the two most susceptible flax populations (open circles), the two intermediate populations (open triangles), and the two most resistant populations (filled circles). After Thrall and Burdon 2003.

hosts than are local isolates (Burdon, Thrall, and Lawrence 2002). The gene-for-gene interaction between Australian wild flax and flax rust therefore shows evidence of significant geographic structure at all geographic scales, ranging from metapopulations through biogeographic regions to the entire continent. No other coevolving interaction has been studied from both sides of the interaction with such breadth and scale of approach.

Gene-for-gene polymorphisms similar to those found in Australian flax and flax rust are known from a wide range of plant species, and several cases of gene-for-gene interaction occur between plants and animals, including insects or nematodes (Thompson and Burdon 1992; Crute, Holub, and Burdon 1997; Mauricio et al. 2003). Much research in agricultural plant breeding uses the paradigm of gene-for-gene interactions with pathogens (Crute, Holub, and Burdon 1997). Experiments on *Drosophila melanogaster* and the parasitoid *Leptopilina boulardi* suggest that this coevolving relationship is governed at least in part by interactions among major genes that may function in a gene-for-gene-like manner (Dupas, Carton, and Poirié 2003). Few other interactions, however, have been evaluated within natural populations in ways that would allow assessment of whether the interaction coevolves as a gene-for-gene relationship. It is therefore currently impossible to assess the extent to which gene-for-gene relationships occur in antagonistic interactions between parasites and hosts, and almost nothing is known about the potential for gene-for-gene coevolution in interactions in which both participants are animals. Coevolutionary biologists studying interactions between invertebrates and their parasites commonly assume a matching-alleles structure, but some theoretical work has shown that some interactions could coevolve through a two-step process involving both matching alleles and gene-for-gene interactions (Agrawal and Lively 2003).

Many coevolving relationships between parasites and hosts undoubtedly involve complex interactions among genes rather than relatively simple gene-for-gene interactions or matching-allele interactions. For example, breeding experiments with willows have shown that variance in resistance to thirteen insect species and one mite species involves additive, dominance, and epistatic components for each herbivore species (Fritz et al. 2003). Complex patterns of coevolution driven by a combination of major genes, modifier genes, and genetic interactions are likely to be common in coevolving polymorphisms. As we learn more about the coevolutionary process, we will be able to refine our understanding of how the genetic architecture of coevolving traits shapes coevolutionary dynamics, and how coevolution itself shapes the genetic architecture of interacting species.

The Hypothesis of Multispecific Coevolution through Optimal Allelic Diversification

One of the biggest challenges in the study of the genetic architecture of coevolution is to understand how natural selection shapes the level of allelic diversity at the loci that undergo coevolutionary selection. Defense and counterdefense are multistep processes, and selection on the level of allelic diversity may differ among the steps. Matching-allele interactions and gene-for-gene interactions provide two genetic mechanisms for confronting the enemy-recognition step of interactions between parasites and hosts. These forms of genetic interaction may also be involved in other steps of the process of parasite attack and host defense. Evolution of the vertebrate immune system has created an even more complex, but highly flexible, way of handling enemy recognition and resistance. Within natural host populations, all these kinds of enemy-recognition systems are surely under relentless selection to recognize multiple parasites.

In highly industrialized Western cultures, in which public-health efforts have worked wonders to reduce the threat of infectious disease, it is easy to think of parasites and diseases as something unusual. Yet parasitism and disease remain pervasive within biological communities (e.g., Roy et al. 2000), and most well-studied species are known to be attacked by multiple parasite species. Even now, infectious diseases cause more than a quarter of human deaths worldwide each year. The World Health Organization's annual reports catalog the grim numbers. The WHO report for 2003 estimated that 26.2 percent of human deaths in 2002 resulted from infectious and parasitic diseases, including respiratory diseases. The biggest killers, each responsible for over a million deaths, were respiratory infections (almost four million people), HIV/AIDS (almost three million), diarrheal diseases (almost two million), tuberculosis (over one and half million), malaria (over a million), and childhood diseases such as measles and pertussis (over a million). The spread of AIDS, epidemics of hemorrhagic fever, and books with titles such as *The Coming Plague* (Garrett 1994) and *Plague Time* (Ewald 2000) remind us of the ubiquity and continued evolution of parasites. A review in 2001 reported 1415 species of infectious organism known to be pathogenic to humans, and new human diseases continue to be discovered (Taylor, Latham, and Woolhouse 2001). Hepatitis C and E viruses, guaranito virus (Venezuelan hemorrhagic fever), sabia virus (Brazilian hemorrhagic fever), the cholera bacterium *Vibrio cholerae* O139, and human herpes viruses 6 and 8 were all first identified within the past thirty years (Taylor, Latham, and Woolhouse 2001). There is no reason to believe that humans are in any way special in either the large number of diseases that afflict us or the geographic differences in prevalence found among our diseases.

Faced with an ever-changing diversity of parasites, species have evolved multiple genetic mechanisms to recognize attack by parasites. These enemy-recognition genes are often finely tuned to identify, attack, and then set in motion other defenses that act over days, months, or years. Populations, however, are unlikely to simply accumulate an ever-increasing number of alleles to recognize different parasite. Instead, the current working hypothesis can be stated as follows:

The hypothesis of optimal allelic diversification in enemy-recognition systems.

Coevolution between hosts and multiple parasites proceeds through selection on hosts to accumulate alleles that allow recognition of parasite attack. Simultaneously, selection acts on each parasite species to be sufficiently different from other parasites (e.g., in its outer membrane proteins) that individuals escape detection by host alleles evolved to detect other parasite species. Because parasite assemblages differ geographically, host populations will differ in the recognition alleles they accumulate, and parasite populations will differ in how they diverge from co-occurring parasites to escape host detection. Host-recognition alleles will diversify within local populations to encompass the range of detectably different local parasites and may reach an upper limit, beyond which alleles may interfere with each other in parasite detection.

The hypothesis is based upon theoretical studies suggesting that interference among recognition alleles may increase as the number of alleles increases, and upon empirical studies suggesting that vertebrate hosts with moderate diversity of major histocompatibility complex alleles have potentially higher fitness (e.g., Nowak, Tarczy-Hornoch, and Austyn 1992; De Boer and Perelson 1993; Wegner, Reusch, and Kalbe 2003; Kurtz et al. 2004). The coevolutionary process of optimal allelic diversification simply requires the evolution of genes that function in a gene-for-gene or a self/nonself way with particular parasites or groups. Once in place, the process can function either in defense against a single parasite or against many parasites. The hypothesis is therefore a multispecific extension of the arguments on coevolving polymorphisms in pairwise interactions between hosts and parasites. Even in the absence of multiple enemies, coevolution involving enemy-recognition genes is likely to favor accumulation and diversification of multiple alleles over evolutionary time. The empirical results for gene-for-gene coevolution have shown that multiple polymorphisms for *R*-genes may be maintained within a plant population through interactions with a single pathogen species, especially in geographically structured interactions. The extension to multispecific coevolution is that natural selection should favor local combinations of parasite species that are sufficiently different with respect to their gene products that any single enemy-recognition gene in the host population cannot recognize the full range of local parasite species.

As parasite-recognition alleles accumulate and diversify within each local host population, they should reach the local optimum or upper limit in allelic diversity determined by multiple genetic and population processes, including local parasite diversity, interference among gene products from different alleles, stochastic loss of alleles during population fluctuations, and varying selection pressures over time. Similarly, an optimal upper limit in allelic diversity should occur in parasites through a combination of selection for allelic novelty, stochastic loss of novel but infrequent alleles, and competitive interactions among parasite strains or species. That optimum, however, may differ greatly among populations, and stochastic population processes may prevent local populations from reaching the optimum. Recent mathematical models have suggested that short-term nonspecific immunity in hosts may also play a role in determining the number of alleles. If infection of a host by one parasite strain leads to short-term nonspecific immunity against other strains, a density-dependent constraint is placed on the overall incidence of infection. That constraint limits the effective population size of the parasite and can thereby prevent allelic accumulation (Ferguson, Galvani, and Bush 2003).

Pandemics can contribute to the optimum or upper limit in host allelic diversification, because they can temporarily lead to some homogenization in parasite alleles across populations. Influenza A pandemics, for example, can lead to the widespread replacement of one subtype of influenza by another. Consequently, pandemics can shape species-wide allelic diversity in hosts, mitigating complete allelic divergence among local host populations. In the absence of repeated pandemics, natural selection will favor divergence among populations in allelic diversity in hosts and parasites, thereby creating a geographic mosaic of coevolved alleles.

The immune system of vertebrates may be one of the most sophisticated mechanisms ever developed through multispecific coevolution by allelic diversification. Both components, innate immunity and adaptive immunity, exhibit high levels of polymorphism (Frank 2002; Parham 2003). Innate immunity, the first line of defense, occurs in invertebrates as well as vertebrates (Lazzaro, Sceurman, and Clark 2004). In vertebrates, innate immunity functions through natural-killer (NK) cells, which include a variety of receptors that are coded by different genes. Individuals differ greatly in the number and types of genes that code for these receptors (Parham 2003), and loci may interact with one another in ways that affect the progression of disease (Martin et al. 2002). Although NK cells are highly effective defenses, viruses have evolved multiple mechanisms to elude activation of these cells, thereby limiting the efficacy of this first line of defense (Scalzo 2002).

The ability of vertebrates to defend themselves against rapidly evolving pathogens is greatly enhanced by the remarkable flexibility of the second line of defense: the customizable adaptive immunity that develops in response to specific infections, largely through the major histocompatibility complex (MHC). There is still debate over whether the MHC system initially evolved as a defense system or as a kin-recognition system, but there seems no question that a major part of its current function is defense against parasites. MHC Class I and II molecules bind to pathogen or parasite peptides, which triggers a T-cell response. Class I molecules bind to intracellular foreign peptides, causing the infected cells to die. Class II molecules bind to extracellular foreign peptides, stimulating antibody production and production of macrophages in peptides (Zinkernagel and Doherty 1974; Guillet et al. 1986; Meyer and Thomson 2001).

The allelic diversity is astonishing. The MHC in humans, for example, is a region on chromosome 6 called the human leukocyte system (HLA). The region includes over 220 genes, some of which are highly polymorphic. The HLA Class II DRB1 gene alone has over 350 known alleles, and new alleles continue to be discovered. Across the HLA region as a whole, more than 1000 HLA Class I alleles and more than 660 HLA Class II alleles have been identified so far among human populations (Robinson et al. 2003). The sheer number of loci and alleles generates a staggeringly high potential for allelic diversity within and among populations.

The genetic differences in HLA found among human populations suggest that the immune system is not a diffuse response to coevolution with multiple parasites and pathogens. Individuals and populations differ in the allelic structure of their immune systems. Resistance to malaria, leprosy, tuberculosis, and HIV alike has been linked to differences in HLA (Hill 1998), and the prevalence of these diseases varies geographically. Moreover, prominent human parasites such as *Helicobacter pylori* and *Mycobacterium tuberculosis* show strong geographic subdivision in their genetic structure (Kremer et al. 1999; Falush et al. 2003). Hence, human populations are likely to have differed historically in HLA alleles at least partly as a result of differences in exposure to various combinations of parasite species and genotypes.

The immune response mediated by HLA (and, more generally, MHC) genes is induced by the presence of particular parasite genotypes. Having all available antibodies ready at all times would be metabolically expensive. Hence, induction of defenses is part of the process of multispecific coevolution through optimal allelic diversification. One effective counterdefense for parasites is to mask the cues used by hosts in mobilizing antibodies. Some parasites achieve that masking through rapid evolution of their outer membrane protein. In the few species of Rickettsiae that have been tested, the surface protein sequences evolve faster in parasitic species than in mutualistic species (Jiggins, Hurst, and Yang 2002). Some parasites such as *Plasmodium* and trypanosome have pushed this counterdefense further by evolving the ability to change antigenic forms following infections. *Plasmodium falciparum* is partially able to escape detection by the immune system by changing the antigenic properties of attacked erythrocytes (Evans and Wellems 2002).

This cat-and-mouse coevolutionary game should result in a strong match between local HMC allelic diversity and parasite diversity. Although simple in concept, testing for such geographic matching will require more than a simple count of parasite species. Allelic diversity within parasite species and the combined abilities of different parasite species to change antigenic forms must also be part of any full evaluation. Nevertheless, correlations between local MHC diversity and parasite species diversity provide a first level of analysis.

Few interactions between vertebrates and parasites have been analyzed in this way. Studies of threespine stickleback fish (*Gasterosteus aculeatus*), however, provide some evidence for a coevolving geographic mosaic in MHC and parasite diversity. The sticklebacks occur in a wide range of aquatic environments, including estuaries, lakes, and rivers, and the number of parasites they harbor varies across environments. One study in northern Germany found fifteen macroparasite species among eight stickleback populations (Wegner, Reusch, and Kalbe 2003). Both the number of parasites and their relative abundance varies among the populations. The number of MHC class IIB alleles also varies among environments, ranging from a mean of about twelve alleles in the populations in rivers to fifteen in the populations in estuary habitats (fig. 9.5). Among these eight populations, MHC allelic diversity is greatest in the habitats with highest parasite diversity, as estimated by Shannon's information index. This index combines numbers and relative abundances of species and is highest when the number of species is large and population abundances are similar among species. As a partial control on overall genetic diversity, the study also estimated microsatellite diversity among these populations. The relationship between microsatellite diversity and MHC diversity provides a test of whether MHC diversity simply reflects overall differences in genomic diversity among these populations. The partial correlation between MHC diversity and parasite diversity is stronger than that between MHC diversity and microsatellite diversity, suggesting that se-

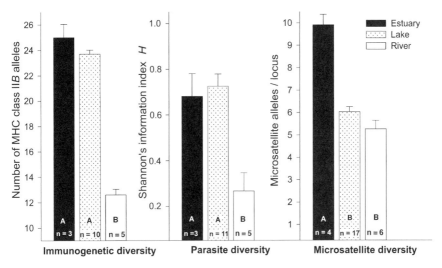

Fig. 9.5   Number and diversity (Shannon's information index) of MHC class IIB alleles in threespine sticklebacks (*Gasterosteus aculeatus*) among three sampled habitats in northern Germany. The bars show the means ($\pm$ 1 standard error) for each habitat, where n = the number of samples of up to twenty-five fish within each habitat type. The Shannon index is highest when the number of species is large and population abundances are similar among species. Also shown is the number of alleles per locus at seven polymorphic loci as a control on overall differences in genetic diversity among populations. The letters inside each bar indicate no significant difference (same letter) or significant difference (different letter) from neighboring bars. After Wegner, Reusch, and Kalbe 2003.

lection mediated by parasite diversity may at least partially explain geographic differences in MHC diversity in this species.

There is therefore nothing "diffuse" about the hypothesis of multispecific coevolution through optimal allelic diversification. By this view, allelic diversity in parasite-recognition genes becomes a direct function of the number and genetic diversity of parasites that attack local host populations and the geographic structure of gene flow and metapopulation dynamics among the populations. In some simulation models, coevolution is able to maintain higher levels of local polymorphism than alternative mechanisms, such as selection that favors MHC heterozygotes, when host populations are large (Borghans, Beltman, and De Boer 2004). The larger problem for coevolutionary biology is to understand how different kinds of parasite assemblages and geographic structures favor different levels of allelic diversity within and among host and parasite populations that are constantly changing in size.

Expanding the analysis to include the geographic mosaic of coevolution therefore shifts the question. Rather than asking how a host species or a local population coevolves with many parasite species, the question becomes how the geographic structure of species shapes the coevolving dynamics of allelic diversity within and among host and parasite populations.

Polymorphisms and Trait Mismatches in Parasites That Differ Geographically in Hosts

If parasites differ geographically in the hosts they use, coevolving polymorphisms may become mismatched in some local populations. The geographic mosaic of defense and counterdefense in wild parsnips (*Pastinaca sativa*) and parsnip webworms (*Depressaria pastinacella*) suggests how the geographic structure of polymorphisms may change under these conditions (fig. 9.6). These rich and varied studies provide the best data available on the local and regional structure of chemical and physiological polymorphisms in coevolving species (Berenbaum and Zangerl 1998; Zangerl and Berenbaum 2003). Wild parsnip is an herbaceous European plant that was naturalized as a common weed in eastern and midwestern North America by the early 1800s. Larvae create webs on flowering umbels and feed on the early stages of the developing seeds. All researchers who have studied this plant species in midwestern North America have noted the severe damage to seed production caused by webworm attack (Riley 1889; Thompson and Price 1977; Hendrix 1979; Berenbaum, Zangerl, and Nitao 1986). Because a parsnip plant flowers only once during its lifetime, the moths have the potential to reduce plant fitness significantly. Larval feeding can reduce seed output by 25–45 percent (Thompson 1978).

In some regions the moths also use a native plant species, cow parsnip (*Heracleum lanatum*), as a host. Wild parsnip and cow parsnip are in closely related genera within the Apiaceae, but they differ in their chemical profiles and in their distributions within North America. The geographic structure of this interaction therefore offers the possibility of exploring how an introduced interaction diversifies as it spreads within a new continent and fits within a wide range of local biological communities.

Throughout the midwestern United States, wild parsnip populations grow as early successional weeds in abandoned fields and along roadsides. Local populations can form dense patches, almost monocultures, of hundreds of individuals. As succession proceeds within an abandoned field, the patches expand in size and density and then begin to dissipate as new, small

Fig. 9.6   Wild parsnip (*Pastinaca sativa*) and a parsnip webworm (*Depressaria pastinacella*). The left panel shows a wild parsnip plant in a successional field in Illinois. The right panel shows a parsnip webworm that has made a web within an umbel at the top of the plant. Photographs by John N. Thompson.

rosettes die from competition and other factors before reaching the threshold size necessary for flowering (Thompson 1978). Local patches are therefore sometimes ephemeral, and metapopulations may shift across the fragmented landscapes of the Midwest. Nevertheless, the interaction between the plants and the moths can remain intact regionally. During the past thirty years the interaction between these species has been studied continuously in Champaign County, Illinois (Thompson and Price 1977; Thompson 1978; Berenbaum, Zangerl, and Nitao 1986; Zangerl and Berenbaum 1990; Berenbaum and Zangerl 1992; Zangerl and Berenbaum 1997; Berenbaum and Zangerl 1998; Zangerl and Berenbaum 2003). These studies have shown that the pattern of attack on the plants varies with plant size, position within patches, and plant chemistry. Plant chemistry, however, has turned out to be the key to understanding the local and geographic dynamics of coevolution in this interaction.

Both species show high heritability of coevolving traits. Wild parsnip plants are defended by furanocoumarins, a group of toxic compounds that vary in amount and composition among individuals. Heritabilities for individual furanocoumarin compounds have been estimated at 0.54–0.62 within natural populations of the plants (Berenbaum, Zangerl, and Nitao 1986). Parsnip webworms use cytochrome P450-mediated metabolism to detoxify these furanocoumarins, and heritability for P450 activity levels tested against some furanocoumarin substrates has been estimated at 0.33–0.45 within one population (Berenbaum and Zangerl 1992). Moreover, both species have been shown to impose phenotypic selection on each other. Local moths survive poorly on plants if their cytochrome P450 monooxygenases do not match their host's chemical profile (Zangerl and Berenbaum 1993, 2003). This interaction therefore has a high potential for rapid and ongoing coevolution.

Four furanocoumarins—bergapten, xanthotoxin, isopimpinellin, and sphondin—dominate the chemical defenses of wild parsnip. The plants and moths can be grouped into clusters, based upon the different proportions and levels of furanocoumarin production in the plants and P450 activity in the larvae. A preliminary analysis showed a good match between the local structure of plant defense and the local structure of insect counterdefense at three of four sampled localities. The one exception was in Urbana, Illinois, where the frequencies of the plant and insect phenotypic clusters were mismatched (fig. 9.7).

Such mismatches are predicted by the geographic mosaic theory of coevolution, but they can also result from slow local fluctuations in gene frequency imposed by frequency-dependent selection. Historical data support the view that frequency-dependent selection contributes to the maintenance of polymorphisms in these interactions. The overall levels of these compounds among populations have remained similar for more than one hundred years, suggesting that they are more likely to be fluctuating over time rather than escalating. Except for sphondin, which occurs at low levels in all these populations, the furanocoumarins show concentrations well within the range found in herbarium specimens dating back to the late 1800s. (Sphondin levels, however, are higher in contemporary populations than in historical populations [M. Berenbaum, personal communication], suggesting that some directional selection may currently be acting on this compound.)

Taken alone, the mismatches could also have two other causes, both of which can be eliminated as likely explanations for the observed patterns. The differences do not seem to be due to any obvious cline in phenotypic frequencies across populations (Berenbaum and Zangerl 1998). It also seems

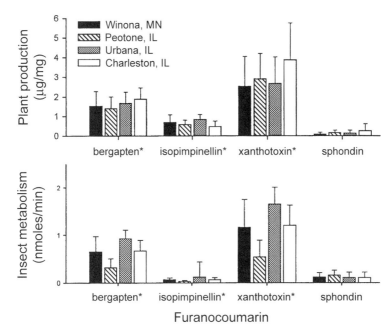

Fig. 9.7   Mean (± SD) production of four furanocoumarins in four populations of wild parsnip in the midwestern United States and the corresponding cytochrome P450 activity in the local population of parsnip webworms. Matching of clusters of plant defense and insect counterdefense is indicated in the figure as similar levels of plant production and insect metabolism for each compound in each population. Furanocoumarins that differed significantly among populations are indicted by an asterisk (*). After Berenbaum and Zangerl 1998.

unlikely that the differences among populations are due solely to chance initial events in which plants with particular furanocoumarins colonized an area, allowing subsequent colonization only by those webworms that already had the right combination of P450 activities. All plant populations have the same furanocoumarins and differ only in their relative frequencies (Berenbaum and Zangerl 1998).

A major follow-up study has suggested that the local mismatches result from differences in current local coevolutionary selection and the nearby presence of cow parsnip in some populations (Zangerl and Berenbaum 2003). Coevolutionary hotspots maintain a tight matching of local chemical profiles, but conflicting selection pressures in other regions create mismatches. In the follow-up study twenty populations of the moths and plants were analyzed, but some of these populations occur in regions where cow parsnip occurs nearby. Twelve of the populations show chemical matching of furanocoumarin profiles in the plants and detoxification profiles in the webworms (fig. 9.8). The other eight populations show significant mismatches. Using logistic regression, the best predictors of local mismatches in this interaction are the nearby presence of cow parsnip and high xanthotoxin production within the local wild parsnip population. Frequency of attack is lower in populations with overall high xanthotoxin levels, but how these levels contribute to mismatches is still under study. The effect of nearby cow parsnip on chemical matching turns out to be not just a matter of conflicting selection on webworms feeding on two hosts. Local populations of wild parsnip growing near cow parsnip have chemical profiles that are intermediate between those found in these plant species where they are growing isolated from each other (Zangerl and Berenbaum 2003). Hence, the nearby presence of cow parsnip may affect selection not only on the moths but also on wild parsnips.

One of the most remarkable aspects of these results is the close phenotypic matching in most populations that are not near cow parsnip populations. Even in a coevolutionary hotspot, snapshots of coevolving polymorphisms should show periods of mismatching as selection switches from favoring one set of alleles to another. In this interaction, however, the combination of high heritabilities and very strong reciprocal selection appears to create coevolutionary hotspots with very fast evolutionary response times. The relative frequencies of the chemical compounds change significantly among years, and populations of the moths show very local adaptation to their host plants. In one set of experiments, larvae transferred to plants just thirty meters from their local population survived less well than larvae transferred as controls within their local host population (Zangerl and Berenbaum 2003).

Overall, these results suggest that the chemical polymorphisms found in these interactions are maintained both locally and geographically through a combination of coevolutionary hotspots and coldspots, mediated in part by the presence of a third species. This kind of effect, in which a third species can significantly shape the geographic mosaic of coevolution, is becoming a common theme as more interactions are analyzed across multiple populations. Other examples include the interactions between crossbills and conifers discussed in chapter 7 and those between *Greya* moths and *Lithophragma* plants, and also possibly those between garter snakes and newts, discussed in chapter 6.

In addition, these studies suggest that invasive species can coevolve rapidly with enemies in new environments and create geographic mosaics.

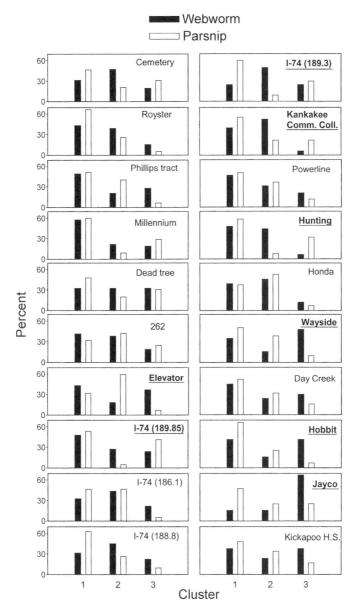

Fig. 9.8 Frequency of wild parsnip and parsnip webworm chemical phenotypes within each of twenty populations in the midwestern United States. The plants and insects were grouped into statistical clusters based upon five plant furanocoumarin production characters and five detoxification characters. Populations with underlined names show significant mismatches in plant and insect chemical profiles. After Zangerl and Berenbaum 2003.

When Riley (1889) first described the interaction between these two species in the midwestern United States, introduced wild parsnip was expanding its geographic range, following the plowing of the Midwest in the nineteenth century. The current geographic mosaic of defense and counterdefense is therefore a result of less than two hundred years of coevolution across these fragmented landscapes.

Conclusions

The tremendous diversity of phenotypically expressed polymorphisms found in most organisms results to at least some extent from the ongoing dynamics of coevolution between parasites and hosts. We now know that coevolving polymorphisms can fluctuate greatly in natural communities within decades. One of the central challenges for coevolutionary biology is therefore to understand how coevolving polymorphisms can remain within species for long periods of time amid rapid ongoing coevolutionary change. Heterozygote advantage, temporally fluctuating selection, and frequency-dependent selection may all help maintain polymorphisms across many generations, but the maintenance of a high diversity of coevolving polymorphisms within species probably relies upon the geographic mosaic of coevolution, as indicated by the detailed results for natural populations now available.

10 Antagonists II

Sexual Reproduction and the Red Queen

The ubiquity, diversity, evolvability, and host specificity of parasites guarantee that multispecific coevolution with parasites is a relentless process for most species. Natural selection often favors rare host genotypes, because local parasites become adapted to the most common genotypes within host populations. At the same time, selection often favors parasites that are best able to attack and spread among the most common host genotypes, but it also favors novel parasite traits that make it more difficult for a host to detect the presence of the parasite or eliminate it. Which genotypes are favored depends on current relative frequencies of alleles in the host and parasite populations.

The dynamics produced by coevolving polymorphisms and optimal allelic diversification (chapter 9) can contribute to the maintenance of rare host and prey genotypes. Sexual reproduction provides a further powerful mechanism to achieve locally rare combinations of genotypes by decreasing linkage disequilibrium among genes. By producing offspring that differ genetically from each other, a sexual female increases the chance that one or more of those offspring will do better than average in warding off parasite attack. In turn, sexual populations may persist longer on average than asexual populations, which will tend to accumulate deleterious mutations over time. With additional caveats and varying emphases, these have become the standard arguments for the evolution and commonness of sexual reproduction.

In this chapter I explore how the geographic mosaic of coevolution shapes selection for sexual reproduction. The chapter focuses on two hypotheses: Red Queen mosaics and Red Queen captured. The first hypothesis focuses on selection favoring sexual reproduction in hosts, whereas the second focuses on the role of parasites in preventing sexual reproduction in their hosts.

Background: Muller's Ratchet, the Red Queen, and Sexual Selection

All else being equal, natural selection should favor individuals that produce offspring genetically identical to themselves. Sexual individuals therefore harbor an inherent fitness cost in the absence of any other selection pressures that can offset that cost. Nevertheless, almost all orders of plants and animals include sexual species. Exclusively asexual major lineages, such as the rotifer class Bdelloidea and the ostracod family Darwinulidae, are noteworthy exceptions (Judson and Normark 1996; Butlin, Schöen, and Martens 1999; Mark Welch and Meselson 2003). In most major lineages, asexual species often appear as terminal nodes on phylogenetic branches, arising repeatedly within lineages but rarely leading to much subsequent diversification (Simon et al. 2003). Most putative ancient asexual clades of animals, plants, and fungi include only a few dozen species, at most (Judson and Normark 1996; Normark, Judson, and Moran 2003).

Coevolution with parasites decreases that inherent genetic advantage to asexual reproduction. If attack by multiple parasites favors rare host genotypes, then natural selection will favor in hosts any genetic mechanism able to produce rare allelic combinations that thwart parasite attack. Such rarity can continue to arise among eukaryotic individuals within local populations in three main ways: new mutations, gene flow among populations, and recombination through sexual reproduction. An asexual population accumulates genetic differences only as fast as new mutations appear locally or arrive through gene flow. If rare genotypes are favored, natural selection on asexual lineages could favor asexual clones with high mutation rates for genes that assist in defense against enemies. Some experiments on *E. coli* have shown that populations subjected to a new environment sometimes produce strains with high mutation rates (Sniegowski, Gerrish, and Lenski 1997). In this particular example, selection on *E. coli* was not driven by parasite-mediated selection, but the outcome of this experiment suggests that selection can favor different mutation rates for some genes in different environments.

Relying upon mutation alone creates two evolutionary problems for asexual organisms. Each eukaryotic clonal lineage harbors only the mutations that have appeared within that lineage. A beneficial mutation that appears in another lineage can never be incorporated into that lineage's genome. The inability of asexual clones to mix and match beneficial mutations that initially arise in different lineages was one of the earliest arguments for the potential evolutionary advantage of sexual reproduction (summarized in Crow and Kimura 1965). The argument relies upon a higher long-term rate of extinction in asexual populations compared with sexual populations, which can more readily combine beneficial mutations.

The more immediate problem is that asexual lineages cannot rid themselves of detrimental mutations. Once a deleterious mutation appears within a clonal lineage, it is there forever or until another mutation eliminates it or masks its effects. As genetic lineages with few mutations are lost by chance within a small population or as those lineages accumulate yet more deleterious mutations, the overall deleterious mutational load increases in the population. This problem, called Muller's ratchet, is particularly acute for small populations and populations subject to high mutation rates (Felsenstein 1974; Bell 1988; Rice 1994; Lynch, Conery, and Burger 1995; Moran 1996; Howard and Lively 2002). Each mutation contributes to the irreversible turn of the ratchet. The genetic structure of asexual taxa is therefore shaped by ongoing local selection among clonal lines, random genetic drift among lines, and movement of clones among populations. Coevolution with parasites could proceed only through differential perpetuation of host clones as they interact with different parasite genotypes across landscapes.

A sexual population solves both problems through the recombination of genes. When individuals mate, their genes recombine in their offspring, creating genetically unique individuals. Beneficial mutations that arise in different parents can end up in a single offspring. In addition, offspring that inherit deleterious mutations or combinations of deleterious mutations are eliminated from the population by natural selection. Hence, such mutations are routinely jettisoned from sexual populations. For a species faced with Muller's ratchet, the advantage of sexual reproduction is that it produces some offspring with relatively low mutational loads, thereby setting the ratchet back. Sexual populations therefore have the potential not only to be much more genetically diverse than asexual populations, but to be more diverse in ways that reduce deleterious mutational load and differentially preserve combinations of beneficial mutations. The ability to mix and match alleles creates the potential for rapid local adaptation of populations across landscapes.

Recombination, however, also breaks up beneficial combinations of genes that were previously favored by natural selection. Multiple hypotheses have been proposed for the ways in which natural selection could favor sexual reproduction and the genetic diversity it produces, despite this problem. Part of the solution is that if selection on beneficial combinations oscillates within a population, such that today's favorable combinations are disfavored by selection in the near future, then the advantage of sexual reproduction is retained (Peters and Lively 1999). The best-supported current view that involves oscillating selection is the Red Queen hypothesis, which argues that such oscillations are maintained by coevolving interactions between parasites and hosts (Lively, Craddock, and Vrijenhoek 1990; Lively 1992; Dybdahl and Lively 1998; Lively and Jokela 2002). More specifically, the hypothesis asserts that local adaptation in parasites results in their disproportionate success on the most common host genotypes, which, in turn, favors sexual host females that produce genetically rare offspring that are more likely to escape parasite attack (Glesener and Tilman 1978; Jaenike 1978; Bremermann 1980; Hamilton 1980; Lloyd 1980). By this view, then, sexual reproduction is an outcome of frequency-dependent selection favoring rare host genotypes that are resistant to parasites specifically because they are locally rare.

It is still unclear, though, which is more important in maintaining sexual reproduction in most species over the long term: thwarting Muller's ratchet through mutational clearance, or racing ahead of enemies through rare combinations of beneficial genes. Both explanations or some variants of them are likely to be important, and their relative importance probably varies among populations over time (Howard and Lively 1998; West, Lively, and Read 1999; Jokela et al. 2003). These processes, of course, may be related. As asexual populations accumulate deleterious mutations, they may rapidly lose their defenses against enemies, regardless of whether their enemies are developing new ways of overcoming the defenses. Mathematical models of the Red Queen and Muller's ratchet have suggested that the two mechanisms working together can favor the maintenance of sexual reproduction when either mechanism is insufficient by itself (Howard and Lively 1994, 1998, 2002).

Additional Influences on the Evolution of Sexual Reproduction

Most research on the potential role of coevolution in the maintenance of sexual reproduction is based on the ubiquitous interactions between parasites and eukaryotic hosts. Before proceeding further with that discussion, however, it is worth pausing to consider the other potential influences on the evolution of sexual reproduction, which will ultimately require more complete integration into our current hypotheses on the coevolutionary dynamics of recombination and sex.

Most organisms harbor transposable elements that have commonly been considered to be parasitic genetic elements. These elements blur the lines between mutational load and coevolution in discussions of parasite-driven sexual reproduction. Transposable elements multiply by making copies of themselves within host genomes. The elements make up a high proportion of some eukaryotic genomes, and species such as humans harbor millions of copies of retrotransposons (Eickbush and Furano 2002). The highly repeated copies can have highly deleterious effects on host genomes, and natural selection should therefore favor hosts that are able to suppress or regulate copy number (Charlesworth, Sniegowski, and Stephan 1994). Regulation, however, is not simple. If the rate of multiplication of transposable elements is faster than the rate of multiplication of the host genome, then transposable elements can spread through host populations even if they have deleterious effects. Host fitness must decline quickly as the number of element copies increases to prevent continued expansion of copy number (Charlesworth and Charlesworth 1983). Sexual reproduction may therefore provide a means to control copy number. Some offspring of sexual females may have a lower than average copy number and an associated higher than average fitness. Sexual reproduction, however, could also favor highly aggressive forms of transposable elements. In asexual clones the fate of the elements is linked to the fate of the host clone, but in sexual species transposable elements can continue to invade new host genomes during meiosis (Wright and Finnegan 2001; Nuzhdin and Petrov 2003).

Transposable elements may therefore be part of the cost of sexual reproduction, acting as sexual parasites in some species. The sporadic occurrence of transposable elements within the ancient asexual lineage of bdelloid rotifers provides evidence for this view (Arkhipova and Meselson 2000). These asexual lineages appear to be remarkably free of these elements. If elimination of transposable elements were the major problem faced by eukaryotes, though, sexual reproduction would be rare rather than overwhelmingly common among species. There is clearly much more to understand about the problem of transposable elements, but it is in some way part of the broader problem of how coevolution with different kinds of parasites shapes the evolution of sexual reproduction.

An additional complication in interpreting selection on sexual reproduction is that, once established, maintenance of sex may not depend solely on the benefits of sexual reproduction. Sexual populations are subject to sexual selection, which can act to favor divergence of traits between the sexes and development of coadapted gene complexes in males and females. Maintenance of sexual reproduction in some species may therefore result from selection imposed both by coevolution with parasites and by sexual selection.

Sexual reproduction may even become fixed within some lineages due to a combination of natural and sexual selection and the coadapted genetic systems that canalize individuals into genetic males and females.

There are at least two specific ways in which parasite-driven sexual reproduction and sexual selection could become intertwined. Hamilton and Zuk (1982) suggested that the exuberant mating displays and garish colors of males in many species are signals to females that these are particularly healthy males with genes that are highly effective against parasites. Under such conditions, sexual selection should favor females that mate with these males rather than more lethargic or less colorful males. Alternatively, females in some species could prefer to mate with males because they share the fewest alleles in common at disease-resistance loci, using cues that may or may not include conspicuousness (Howard and Lively 2003). Either mechanism would link parasite-mediated coevolutionary selection and sexual selection.

The interaction between coevolutionary selection and sexual selection could shape the local persistence of populations. Strong sexual selection driven by evolution of conspicuousness could result in reduced population size due to greater susceptibility to predation or to greater physiological wear and tear on individuals. There is some evidence that dichromatic bird species undergo higher local extinction rates than species in which males and females do not differ greatly in plumage (Doherty et al. 2003). If, instead, sexual selection is driven by choice of mates differing in disease-resistance loci, then reduced local persistence due to sexual selection would be unlikely. In fact, this focus of sexual selection could potentially enhance population stability by increasing the evolvability of a population (Howard and Lively 2003). The important point for future research is that parasite-driven sexual reproduction, sexual selection, and the ongoing geographic dynamics of coevolution may be linked in some species.

A final separate problem is that hypotheses on sexual reproduction and coevolution are primarily about the way in which the Red Queen shapes reproduction in eukaryotes. Much of the earth, however, is populated by prokaryotes that may exchange genes through horizontal gene transfer. These exchanges are now thought to be possible among even highly unrelated species (Jain et al. 2003), and may sometimes occur in eukaryotes as well as prokaryotes (Archibald et al. 2003). The extent to which horizontal gene transfer has shaped patterns of evolutionary change is still unknown, and current views differ widely (Doolittle et al. 2002; Kurland, Canback, and Berg 2003; Raymond and Blankenship 2003). Nevertheless, the coevolved

geographic and phylogenetic mosaics that these transfers produce are becoming a new frontier of research in coevolutionary biology.

The Geographic Mosaic and Parasite-Mediated Sexual Reproduction

The typical coevolving interaction envisioned by the Red Queen view of the
maintenance of sexual reproduction is of a eukaryotic host plagued with
infectious parasites that disproportionately attack the locally most common host genotype. Hypotheses that posit a direct role of ongoing parasite-
mediated selection in maintaining sexual reproduction therefore implicitly
assume that species are collections of genetically differentiated populations in
which local adaptation is possible in parasites and hosts. By this view, sexual
populations are more likely to persist than asexual populations because sexual host individuals are more likely to produce offspring with rare genotypes
able to thwart locally adapted parasites, and because they are most likely to
escape Muller's ratchet (West, Lively, and Read 1999). As patterns of parasite
infection and local adaptation continue to shift across landscapes, maintenance of sexual reproduction becomes a consequence of the geographic mosaic of coevolution. Combining the Red Queen hypothesis with the arguments in previous chapters, the general hypothesis for sexual reproduction
driven by coevolution between parasites and hosts can be encapsulated as
follows:

> *The hypothesis of Red Queen mosaics.*
>
> *Local adaptation in parasite assemblages and selection favoring locally rare
> host genotypes together create continually shifting geographic selection mosaics
> and coevolutionary hotspots that are further modified through ongoing trait
> remixing among populations. Sexual individuals are more likely than asexual
> individuals to produce offspring with higher-than-average fitness amid such
> constantly changing coevolutionary selection across landscapes. Sexual repro-
> duction is therefore maintained by the geographic mosaic of coevolution.*

In effect, this is the current leading view of sexual reproduction as a result
of coevolutionary selection, but in this form the hypothesis emphasizes that
running with the Red Queen is not just a local game of chase. Local adaptation and local frequency dependence are only part of the raw material for the
overall coevolutionary process. Networks of parasites continue to evolve and
vary in their selection pressures on host populations. Gene flow from other

parasite and host populations fuels ongoing coevolution by introducing novel host genotypes that are resistant by virtue of their rarity, and it maintains a diverse array of fluctuating genetic polymorphisms in the host and parasite populations.

Sexual reproduction therefore should be especially favored in hosts that are sufficiently structured geographically to permit the evolution of locally adapted assemblages of parasites. At the extreme of geographic structuring, however, the role of parasites in the long-term maintenance of sexual reproduction becomes less clear. With extreme local adaptation and no gene flow in either the host or parasite populations, there are three possible evolutionary options. One is local extinction of both species, if locally adapted parasites with high levels of virulence create large oscillations in host abundance. Another possibility is persistence solely through local Red Queen coevolution. As for the general problem of the maintenance of local polymorphisms (chapter 9), such an outcome may be possible under some sets of local conditions, but it seems unlikely to explain the maintenance of sexual reproduction throughout much of the geographic range of most species over millions of years. Few species are composed of populations that remain locally stable and unconnected by gene flow to other populations over such long spans of time. The third option is the evolution of attenuated antagonism in local parasite populations, if lack of gene flow results in high within-species relatedness in parasites and very low local diversity of parasite species. Such conditions could decrease the strength of local selection for the maintenance of sexual reproduction.

With a few exceptions, current data and models do not provide guidance on the conditions under which coevolution is most likely to favor both the origin and the long-term maintenance of sexual reproduction along the continuum of geographic structures found in hosts and parasites. The importance of geographic structuring in shaping selection favoring sexual reproduction has only started to be explored, mostly through the demonstration of local adaptation in parasites and frequency-dependent selection in hosts. One clear general prediction, however, is that where asexual and sexual reproduction are both possible, asexual populations should be found only where coevolving parasites are rare or absent (Bell 1982; Lively and Jokela 2002). That general prediction has been supported by several studies of the geographic distribution of sexual and asexual populations, but recent analyses have emphasized the need for more research that pinpoints more precisely the geographic conditions under which sexual reproduction is favored (Lively and Jokela 2002).

The most convincing results on the geographic mosaic of Red Queen co-evolution are the long-term studies on *Potamopyrgus antipodarum* snails and *Microphallus* in New Zealand lakes (Lively 1987; Dybdahl and Lively 1996; Lively and Jokela 2002). The snails are found in all subalpine lakes in New Zealand, and the lakes differ in the frequency of diploid sexual snails and triploid parthenogenetic snails. Some lakes harbor only asexual clones, whereas others harbor a mixed population of sexual and asexual snails. The snails are attacked by at least fourteen trematode species, all of which typically sterilize their hosts (Jokela et al. 2003). The most common of these parasites is an undescribed species of *Microphallus,* and genotypes of the snails differ in their resistance to parasite species (Dybdahl and Lively 1995; Jokela et al. 1997). Research since the mid-1980s has concentrated on attack by this parasite.

Parasite infection and sexual reproduction covary among lakes in a manner consistent with a geographic mosaic of coevolutionary hotspots and coldspots. The lakes vary in the proportion of snails attacked by *Microphallus,* ranging from around 1 percent to over 20 percent. The proportion of males within the snail populations also varies among lakes, ranging from 1 percent to over 25 percent. (Some clonal populations produce up to 6 percent males.) Moreover, the frequency of infection and the frequency of males have remained fairly constant in these lakes for more than a decade (fig. 10.1). The proportion of males in a population is correlated both with the frequency of infection and with the proportion of sexual females (Lively and Jokela 2002; Jokela et al. 2003). The interactions between the snails and trematodes therefore suggest a complex mosaic of coevolutionary hotspots, warmspots, and coldspots across the subalpine of New Zealand.

Common snail genotypes differ among lakes, suggesting a possible geographic selection mosaic in which different genotypes are favored in different lakes or in different years among lakes. Moreover, studies of allozyme variation have shown that snail populations differ among lakes to a greater degree than trematode populations, indicating higher levels of gene flow in the parasites than in the snails (Dybdahl and Lively 1996). Because the trematodes alternate between the snails (their intermediate host) and waterfowl and wading birds (their final host), these parasites may be more readily transported among lakes than the snails. Gene flow among parasite populations may therefore continually add new genetic diversity into local parasite populations, fueling ongoing coevolution while preventing complete divergence of parasite populations into "host races" in different lakes (Dybdahl and Lively 1996).

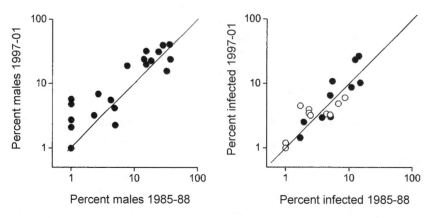

Fig. 10.1   Frequency of male *Potamopyrgus antipodarum* snails and percentage of snails infected with *Microphallus* sp. trematodes in New Zealand lakes in the 1980s and in the same lakes about a decade later. The solid line through the points indicates no change between the two sample dates. The points on the left-hand graph fall into two clusters: a lower cluster with few males within essentially clonal populations and an upper cluster with a higher percentage of males and a combination of sexual and asexual females. The graph on the right shows infection in asexual populations (open circles) and mixed populations (filled circles). After Lively and Jokela 2002.

Gene flow among lakes, however, is not so extensive as to prevent local adaptation in the parasites. When snails from different lakes are experimentally infected with trematodes from their normal lake, another lake, or a mixture of both populations, they show higher levels of successful infection on their sympatric hosts than on allopatric hosts (fig. 10.2). As expected, then, the mixed parasite populations show no differences in their ability to attack different snail populations.

Within each lake, the most common host clones suffer disproportionate attack, thereby favoring locally rare snail genotypes (Dybdahl and Lively 1998; Lively and Dybdahl 2000). Consequently, the frequency of snail genotypes changes over time, and the changes observed in one intensively studied single lake are consistent with time-lagged selection imposed by the trematodes (Dybdahl and Lively 1998). Infection frequency of snail genotypes increases with their relative commonness until they become disproportionately infected relative to other, less common clones within the population. As these clones then begin to decrease in frequency within the population, the parasites show a time-lagged effect by remaining more infective to recently common clones than to recently rare clones (Dybdahl and Lively 1998).

Hence, there is a continually changing geographic mosaic of Red Queen

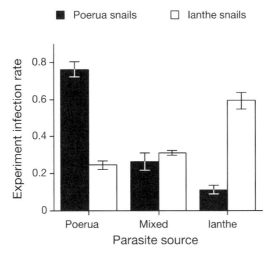

Fig. 10.2   Local adaptation in *Microphallus* sp. trematodes that infect *Potamopyrgus anti-podarum* snails in New Zealand. Snails from two populations, Lake Poerua and Lake Ianthe, were experimentally infected with trematodes from their local population, the other population, or a mixture of both populations. Local adaptation is indicated by significantly higher levels of successful infection (i.e., a higher proportion of snails attacked) on the sympatric combinations in comparison with nonsympatric combinations. The bars show the mean ± standard error of four replicates of seventy-five snails. After Lively and Dybdahl 2000.

coevolution between the snails and trematodes, driven by geographic selection mosaics, coevolutionary hotspots, and possibly gene flow. Exactly how ongoing gene flow contributes to the coevolutionary dynamics of this interaction is still unknown, but occasional infusion of new parasite genotypes into lakes is likely to fuel continual change in the local genotype frequencies of host populations. Genetically simple mathematical models of haploid sexual hosts and parasites suggest that different levels of gene flow, coupled with different levels of virulence, produce different patterns of oscillation in parasite and host fitness over time. For example, the dynamics of the genetic model shown in figure 10.3 suggest that highly virulent parasites can generate wide fluctuations in mean fitness of hosts and parasites, increasing the chance of local extinction of the host or the parasite. Moderate gene flow dampens those fluctuations slightly under high virulence and greatly under moderate virulence, which may potentially lower the chance of local extinction in either species. Adding additional host and parasite genotypes could further dampen the amplitude of oscillations.

It is currently unknown how the dozen or more other trematode species

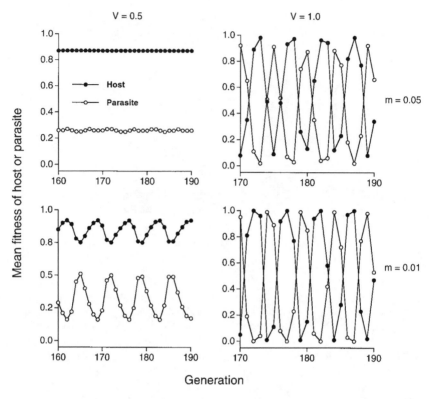

Fig. 10.3   Simulation of fluctuation in mean fitness of a host (filled circles) or parasite (open circles) population over time under two different levels of virulence (V) and gene flow (m). The model assumes two large haploid sexual populations each with one locus and two alleles. After Lively 1999.

contribute to the geographic mosaic of local adaptation and sexual reproduction in *Potamopyrgus* snails. The results for *Microphallus* sp., however, indicate either that this species is the most important selective force favoring sexual reproduction within snail populations, or that it is a useful index for the overall geographic pattern of selection imposed by all parasites. If most of the parasite species continually evolve toward local adaptation to the most common snail genotype, then that would place even more pressure for maintenance of sexual reproduction and the production of rare genotypes within each snail population in which parasite prevalence is high.

An important remaining question is why sexual reproduction does not become fixed within lakes with consistently high rates of parasite infection. Part of the answer may be that the lakes themselves are small-scale spatial mosaics. The snails are found at multiple depths within the lakes, and intensive sampling of one lake has shown that the percentage of infected snails decreases with water depth (Jokela and Lively 1995; Jokela et al. 2003). The proportion of sexual snails also decreases with water depth, ranging from greater than 70 percent in shallow-water habitats to less than 25 percent in habitats below four meters. Hence, selection may favor sexual snails in shallow-water habitats but asexual snails in deeper-water habitats. Nevertheless, the shallow-water habitats retain some asexual clones. Hence another part of the reason for lack of fixation of sexual reproduction may be that asexual clones are not at a selective disadvantage when rare.

Overall, the results for *Potamopyrgus* snails and their trematode parasites suggest a strong role for the geographic mosaic of coevolution in the distribution of sexual and asexual reproduction: sexual reproduction in hotspots and asexual reproduction in coldspots. Host populations differ at any moment in time in the alleles favored by frequency-dependent selection. Among rare genotypes, some are probably favored more than others simply because some alleles will confer inherently better defenses on hosts than other alleles. If so, this would further fuel the geographic mosaic, because the effectiveness of defense alleles can vary among environments. For instance, in the interaction between pea aphids, *Acrythosiphon pisum,* and the fungal pathogen *Erynia neoaphidis,* resistance and virulence among clones varies broadly with temperature, creating different resistance rankings in different environments (Blanford et al. 2003). Different allelic combinations of host defenses are therefore likely to oscillate in frequency over time in different populations of sexually reproducing species.

The only other major attempt so far to explore the coevolutionary consequences of geographic groups of sexual populations as compared with asexual populations are those on Australian wild flax and the rust *Melampsora lini.* These studies have concentrated on how the mode of reproduction affects the structure of genetic diversity of host resistance phenotypes and pathogen pathotypes within metapopulations. Populations of flax in the subalpine of southeastern Australia form asexual metapopulations, whereas populations in the interior plains to the west form sexual metapopulations (Burdon, Thrall, and Brown 1999). These two contrasting flax metapopulations are similar in their overall genetic diversity and resistance, but they differ considerably in how that diversity is distributed within and among local populations. Although the asexual mountain populations differ considerably from one another in their average resistance to a standard set of rust pathotypes, all these populations cluster together in their overall resistance structure. In contrast, the sexual populations on the plains differ considerably from one another in their overall resistance structure.

Less is known about the genetic structure of rust populations on these two contrasting flax metapopulations. A comparison of two mountain and two plains populations, however, has indicated that mountain populations of the rust may be fixed for more virulence/avirulence genes than plains populations (Burdon, Thrall, and Brown 1999). Future studies of the combined metapopulation structure of these two species will be invaluable in extending our understanding of how the geographic mosaic of coevolution builds hierarchically from local populations to metapopulations to broader regional and species-wide patterns.

Hierarchical Structure of Local Adaptation in Parasites

The relatively small size and fast generation times of many parasites adds another dimension to the problem of the evolution of sexual reproduction. In some cases, an entire subpopulation or deme of parasites may develop on a single host individual, and that deme may have to contend with an assemblage of coparasites and induced host defenses that change continually. If the parasite also has much shorter generation times than its host, then a parasite deme may evolve over multiple generations within or on that single host individual. Examples include interactions as different as the evolution of HIV within humans (Frost et al. 2001) and the evolution of local demes of phytophagous insects adapted to individual trees (Mopper 1996; Mopper and Strauss 1998; Van Zandt and Mopper 1998). In some cases, once a parasite population colonizes a host, its offspring have the potential to remain within or on that host for hundreds of generations. Each generation that a bacterial or viral population remains within a human body or an insect population remains on a single pine tree provides an opportunity for natural selection to fine-tune adaptation of that parasite lineage to that particular host individual.

Extreme local adaptation to host individuals creates two evolutionary problems. For the host, the problem is that there is now a parasite that would be very well adapted to attacking any asexual offspring the host produces. The evolutionary answer is sexual reproduction: produce offspring that are genetically different so that the parasite's fine-tuned adaptations are no longer so useful. By that argument, longer-lived host species should be under even stronger selection for sexual reproduction than shorter-lived species, all else being equal. The problem for the parasite deme is that, once the host sexually reproduces and dies, the process of adaptation to a new host must begin all over again.

Some parasites appear to solve the problem by reproducing through a combination of asexual reproduction while on a host and sexual reproduction when dispersing to new host individuals, but there is still much fertile ground for research on how parasites partition sexual and asexual reproduction to confront these problems. Just as with hosts, sexual reproduction provides a mechanism to produce the genetic diversity needed to stay ahead of host-induced responses and competition from coparasites. It is still unclear which conditions favor the evolution of sexual reproduction along the continuum of parasite life histories (Lythgoe 2000; Howard and Lively 2002). There are, in fact, multiple continua with extremes such as these: infections by single parasitic individuals as compared with large colonies of parasites; gene-for-gene interactions as compared with more complex genetic interactions involving antigens and antibodies; or similar host and parasite life spans as compared with highly disparate life spans.

Maintenance of genetic diversity for the parasite while on a host becomes even more pressing if the host is capable of producing antibodies following attack and if the parasite must contend with a continually changing assemblage of coparasites. The continual rebalancing of these combined selection pressures can be seen in the distribution of sexual reproduction within some phylogenetic lineages that include both parasites and free-living species. Parasitic nematode species are more commonly asexual than free-living species, but in some cases, parasitic nematodes are more likely to be sexual in hosts that are capable of strong immune response than in hosts incapable of such response (Maynard Smith and Szathmáry 1995; Gemmill, Viney, and Read 1997; Galvani, Coleman, and Ferguson 2003). Moreover, some parasitic nematodes are facultatively sexual, becoming sexual either in response to host immune defenses produced directly against that species or against co-occurring parasites (West et al. 2001).

Some mathematical models have suggested that sexual strains of relatively long-lived parasites are better able to prevent cross-infection by asexual strains (Galvani, Coleman, and Ferguson 2003). This advantage, however, may be reduced in parasites with very short generation times, because the high rate of appearance of novel mutations may maintain genetic diversity without sexual reproduction (Galvani, Coleman, and Ferguson 2003). In very short-lived parasites, some genes may be under selection to optimize mutation rates. One mathematical model on viral evolution has suggested that, in the absence of recombination, the optimal mutation rate for viruses responding to adaptive immune systems is one mutation within the time it takes for the immune system to adapt to a new viral form (Kamp et al. 2003). Although the model provides a way of thinking about optimization of mutation rates in parasites, it does not readily predict the order-of-magnitude differences found among different classes of virus, such as RNA viruses as compared with retroviruses or DNA viruses (Bonhoeffer and Sniegowski 2002). It does, however, emphasize the increasingly central importance of rapid evolution of parasites within hosts as a major component of coevolution between relatively long-lived hosts and short-lived parasites.

Parasitism, Parthenogenesis, and Reproductive Compatibility of Host Populations

If sexual reproduction in hosts is maintained largely by coevolution with parasites, it would seem at first pass that asexual reproduction should result from absence of coevolution. It is not turning out to be that simple. If a parasite can be transmitted directly from mother to offspring (i.e., vertical transmission, in the terminology of parasitology), natural selection can favor parasites that increase their own transmission by manipulating host reproduction either by suppressing host sexual reproduction completely or by killing males. Although it may involve a wide range of genetic processes and ecological outcomes, the general idea has been called "the Red Queen captured" (Clay and Kover 1996):

The hypothesis of the Red Queen captured.

Asexual reproduction in some species results from coevolution between parasites and host, such that selection on hosts favoring sexual reproduction is countered by selection on parasites favoring asexual reproduction in their hosts.

It has long been known that some parasites, such as trematodes, can prevent sexual reproduction in their hosts through mechanisms such as parasitic castration. During the 1970s and 1980s, however, microscopy and treatment of some invertebrate populations with antibiotics showed that some instances of parthenogenesis or sexual incompatibility among populations result from vertically transmitted endosymbionts. By the mid-1980s there were no more than a few dozen known examples from which to make inferences

(Thompson 1987a), but now there are many hundreds (Werren and Windsor 2000). The bacterium *Wolbachia* alone may occur within 20–75 percent of all insect species and is responsible for alteration of sexual reproduction and cytoplasmic incompatibility in many of the species in which it has been reported (Jeyaprakash and Hoy 2000; Werren and Windsor 2000; Koivisto and Braig 2003). This diverse genus of α-proteobacteria has developed symbioses with insects, isopods, mites, and nematodes (Werren and Windsor 2000).

Studies so far have concentrated on the effects of *Wolbachia* on hosts rather than on coevolutionary dynamics. The need for coevolutionary studies of these interactions is increasing, because, with the hundreds of new papers on *Wolbachia* now being published each year, the list of *Wolbachia*-harboring hosts continues to expand. *Wolbachia* may potentially affect sexual reproduction, sex ratios, and compatibility among populations in hundreds of thousands of species. It is therefore worthwhile to summarize here what is currently known about these interactions, because their importance to coevolutionary biology is bound to increase in coming years. Moreover, *Wolbachia* may be only one of multiple microbial lineages that shape reproduction in host species. A Cytophaga-like organism (CLO) that affects host reproduction was discovered in a mite species in 2001 and has already been reported from multiple species of true bugs (Hemiptera) and wasps (Weeks, Velten, and Stouthamer 2003).

Wolbachia occur as intracellular symbionts that are transmitted from host mothers to daughters through the egg cytoplasm. Consequently, symbionts within males are dead ends, and natural selection should favor *Wolbachia* genotypes that are able to manipulate hosts in ways that increase their chance of occurring in female host offspring. The known range of host manipulations is astonishing and includes feminization of genetic males, male killing, parthenogenesis, and cytoplasmic incompatibility (Stouthamer, Breeuwer, and Hurst 1999; Charlat, Hurst, and Merçot 2003; Poinsot, Charlat, and Merçot 2003). Sexual selection may possibly interact with coevolution to maintain a polymorphism in *Wolbachia* infection within some host populations. A few studies have shown that infected females have a lower probability of mating with males under some conditions (Jiggins, Hurst, and Majerus 2000; Moreau and Rigaud 2000). Studies of woodlice have shown that, when given a choice between the two female types, males prefer genetic females to males that have been feminized by *Wolbachia* (Moreau et al. 2001).

Wolbachia infection shows such a diversity of effects on its hosts that no simple statement is possible about coevolutionary trajectories in these inter-actions. There is some evidence that the phenotypic effects of *Wolbachia* infection can change within *Wolbachia* strains (Jiggins et al. 2002), and that diversity of effects is precisely what should be expected if coevolution between hosts and these symbionts depends upon the ecological and genetic structure of populations and the geographic mosaic of coevolution. Anything from antagonism to commensalism to mutualism should be possible in these interactions. In fact, *Wolbachia* and mitochondria are in the same α-proteobacteria lineage. Consequently, there is precedent within this lineage for establishment of not only parasitic relationships but also intimate mutualistic relationships. Nevertheless, most forms of manipulation of host reproduction induced by *Wolbachia* are detrimental to host fitness, especially under conditions where natural selection favors sexual reproduction in hosts.

Male killing, feminization of diploids, and parthenogenesis induction result in female-biased sex ratios within host populations (Charlat, Hurst, and Merçot 2003). In male killing, *Wolbachia* kill male offspring directly, and the interaction is likely to be highly antagonistic unless female offspring can benefit in some way from the killed males, such as cannibalism of brothers or decreased sibling competition (Charlat, Hurst, and Merçot 2003). Feminization in diploid hosts is achieved by turning males into females. In the pill woodlouse *Armadillidium vulgare,* it occurs in one of two ways: by preventing development of androgenic glands or by taking over sex determination completely (Bouchon, Rigaud, and Juchault 1998; Martin et al. 1999). Infected individuals develop into females, and uninfected individuals develop into males. Parthenogenesis is induced when *Wolbachia* infects haplodiploid hosts, in which females develop from diploid eggs and males develop from unfertilized haploid eggs. *Wolbachia* induce chromosome doubling in male eggs, thereby making them female. All these mechanisms of sex ratio distortion increase the spread of *Wolbachia* through host populations.

Cytoplasmic incompatibility induced by *Wolbachia* creates a yet more complex pattern of selection, leading to geographic mosaics that may shape coevolution. Mating between *Wolbachia*-infected females and infected or uninfected males produces offspring, but mating between infected males and uninfected females usually does not (fig. 10.4). This incompatibility is the most commonly reported effect of *Wolbachia* infection in insects. Unidirectional incompatibility can sweep through populations quickly, infecting one population after another, as happened in *Drosophila simulans* in California beginning in the 1980s (Hoffman, Turelli, and Simmons 1986; Turelli and Hoffmann 1991). Over longer periods of time *Wolbachia* populations have

Cytoplasmic Incompatibility

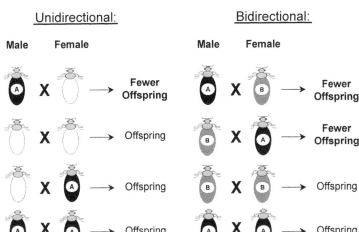

Fig. 10.4   Cytoplasmic incompatibility in invertebrate hosts induced by *Wolbachia*. In unidirectional incompatibility, uninfected females that mate with males infected with *Wolbachia* (A) produce fewer offspring than the other possible matings. In bidirectional incompatibility, crosses between host populations harboring different *Wolbachia* strains (A or B) result in fewer offspring than crosses between hosts harboring the same strain. After Telschow, Hammerstein, and Werren 2002.

diverged among regions, creating both bidirectional and unidirectional incompatibility between host populations harboring different symbiont strains. For example, populations of *Drosophila simulans* show a complex geographic pattern of reproductive compatibilities among populations worldwide. Different populations harbor different *Wolbachia* strains or combinations of strains, and a wide range of complete and partial incompatibilities occurs in crosses of fly populations harboring these different strains (fig. 10.5). The extent to which these kinds of mosaics of reproductive incompatibility are driven directly by geographic differences in coevolution between *Wolbachia* and their hosts is unknown, but it seems highly likely that coevolution is involved.

The pace of discovery of the phylogenetic distribution of *Wolbachia* among hosts, transmission mode, and effects on host reproduction is currently so fast that the only general conclusion possible at the moment about these organisms is that they are much more important to the evolutionary biology of invertebrates than anyone could have suspected a decade ago. Differences in the frequency of infection among populations suggest that these

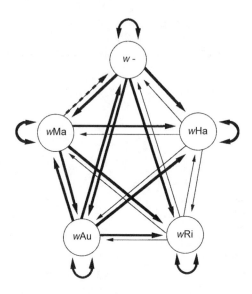

Fig. 10.5   Pattern of cytoplasmic incompatibility between *Drosophila simulans* lines that carry different *Wolbachia* strains or are uninfected (*w-*). Arrows are from males to females, with thicker arrows denoting greater compatibility. The broken line denotes variable expression of incompatibility. All infected lines used for this analysis harbored a single *Wolbachia* strain. *w*Au = Madagascar, *w*Ha = Hawaii and Tahiti, *w*Ma = Madagascar and Reunion, *w*Ri = Riverside, California, *w-* = California, Madagascar, and New Caledonia. After James and Ballard 2000.

and other interactions between vertically inherited symbionts and hosts are likely to develop complex geographic mosaics in which symbionts manipulate host reproduction and hosts resist those effects. That variation among populations and species may even involve large differences in the outcome of the interaction. There is now evidence that some *Wolbachia* that attack insects may not even be parasitic. In one set of experiments using antibiotic treatment, treated females of the parasitoid *Asobara tabida* from eleven tested populations were incapable of producing eggs (Dedeine et al. 2001). *Wolbachia* therefore appears to have become an obligate mutualistic symbiont in at least this one insect species.

Wolbachia co-occur with multiple genomes within their hosts, but the coevolutionary dynamics among all these genomes is unknown. Evidence of interactions among these genomes, however, is increasing. In the parasitoid wasp *Trichogramma kaykai,* for example, the biased sex ratio of 4–25 percent females within populations results at least partly from conflicting interactions between *Wolbachia* symbionts that favor asexual reproduction in their hosts and a parasitic B chromosome that causes females to produce only sons (Stouthamer et al. 2001). In addition, *Wolbachia* that infect insects are themselves infected with the bacteriophage WO (Masui et al. 2000, 2001). Such bacteriophages are known to affect virulence in symbiotic bacteria (Miao and Miller 1999). To add more genomic complexity, *Wolbachia* and similar symbionts that are for the most part vertically transmitted have the potential to evolve in concert with mitochondria because they are inherited together, leading to host populations that differ both in *Wolbachia* and mito-chondria haplotypes (Werren 1998). Consequently, the patterns of sexual reproduction, sex ratio, and parthenogenesis in some host populations may result from interactions among host nuclear genomes, mitochondrial ge-nomes, *Wolbachia,* various genetic elements, and bacteriophage WO.

Because *Wolbachia* and other bacteria within the Rickettsiaceae are so widespread and their effects so diverse, they may be a particularly important group for future work on the trajectories of coevolution. Even within *Wol-bachia* the relationship appears to differ fundamentally among host taxa. Al-though it has long been known that nematodes harbor bacterial symbionts, *Wolbachia* infection was confirmed only in 1995 through analysis of 16sRNA sequences (Sironi et al. 1995). The *Wolbachia* strains infecting nematodes belong to different clades (groups C and D) from those infecting arthropods and, unlike those that infect arthropods, these clades show strong phyloge-netic congruence with their host lineages (Bandi et al. 1998; Casiraghi et al. 2001). *Wolbachia* in nematodes therefore seem to be vertically transmitted much more consistently than *Wolbachia* in arthropods. These different *Wol-bachia* lineages are parts of ancient symbioses that have been coevolving and diversifying for many millions of years. The arthropod-infecting groups A and B are estimated to have diverged from the nematode-infecting groups C and D about 100 million years ago (Bandi et al. 1998). Moreover, ex-periments that eliminate *Wolbachia* by using antibiotics have shown that *Wolbachia* infection is essential to host reproduction in at least some nema-todes species (Stevens, Giordano, and Fialho 2001). The interaction between *Wolbachia* and nematodes may therefore often be mutualistic rather than antagonistic.

Antagonistic and mutualistic coevolution with *Wolbachia* and related taxa appear to favor different rates of evolutionary change on the outer-mem-brane proteins of these symbionts (Jiggins, Hurst, and Yang 2002). Outer-membrane proteins can act as antigens in interactions between hosts and parasites, favoring symbionts with novel membrane proteins. The *wsp* gene in *Wolbachia* and its homologue *map1* in the tick-borne parasite of ruminants *Cowdria ruminantium* code for these proteins. The rate of evolutionary change due to natural selection on the genes should be faster during antagonistic coevolution than during mutualistic coevolution. Studies of positive selection (ratio of nonsynonymous to synonymous nucleotide substitutions) indicate that *map1* in *Cowdria* and *wsp* in arthropod-infecting *Wolbachia* have, in fact, evolved faster through natural selection than *wsp* in nematode-infecting *Wolbachia* (Jiggins, Hurst, and Yang 2002). Moreover, selection has concentrated on the regions of the protein that may interact with the host. These studies are part of a new generation of coevolutionary approaches that will greatly improve our understanding of the molecular genetic structure of different forms of coevolution.

Conclusions

The evolution of sexual reproduction, parthenogenesis mediated by symbionts, and symbiont-mediated reproductive incompatibility among invertebrate populations are all mediated by the geographic mosaic of coevolution. Parasites become adapted to locally common host genotypes, creating geographic patterns in host response and reproductive compatibility among host populations. The hypotheses of Red Queen mosaics and Red Queen captured suggest ways in which coevolution creates these patterns. Variation among New Zealand snail populations in the percentage of sexual individuals and the complex patterns of unidirectional and bidirectional incompatibility among *Drosophila* populations infected with *Wolbachia* illustrate the ongoing geographic dynamics of Red Queen coevolution.

11 Antagonists III

Coevolutionary Alternation and Escalation

As the number and kinds of partners increase within antagonistic networks, so do the coevolutionary options. Oscillating polymorphisms, optimal allelic diversification, and sexual reproduction all appear to be viable ways in which hosts have coevolved with multiple parasites. These mechanisms, however, seem insufficient to explain the coevolved network structure of interactions between free-living predators and prey and between grazers and their victims. Within multispecific antagonistic networks, many predators and grazers preferentially choose among local victim species, and some parasites also have life-history stages in which individuals are capable of choosing among hosts (e.g., the free-living adult stages of some insects that feed parasitically as larvae on individual plants). Such preference hierarchies provide an additional mechanism by which some multispecific networks can coevolve in predictable ways. Coevolution continually changes the distribution of links among interacting species as preference hierarchies and relative levels of defense change over time and across space.

The hypotheses explored in this chapter are part of a continuing attempt to move coevolutionary biology beyond the old dichotomy between pairwise and diffuse coevolution toward a science of the structure and dynamics of reciprocal selection on interacting species. Toward that end, I consider two specific hypotheses on multispecific coevolution that may apply most directly to interactions in which predators, grazers, or parasites exhibit genetically based choice among available victim species: coevolutionary alternation, and coevolutionary alternation with escalation. These hypotheses differ in that one creates fluctuating defenses without escalation, whereas the other creates a coevolutionary arms race of escalating defenses and counterdefenses.

Background: Adaptation to Multiple Enemies

Each pairwise interaction is embedded in a community context that can potentially reshape local and regional coevolutionary dynamics, as is indicated by an increasing number of empirical studies (e.g., Mitchell-Olds and Bradley 1996; Benkman, Holimon, and Smith 2001; Thompson and Cunningham 2002; Zangerl and Berenbaum 2003) and mathematical models (e.g., Bronstein, Wilson, and Morris 2003; Gomulkiewicz, Nuismer, and Thompson 2003; Thomson 2003). These multispecific interactions may constrain the potential coevolutionary trajectories of pairwise interactions, but they do not necessarily prevent coevolution. We know, for example, that traits which evolved for one interaction can be co-opted for use in others (Armbruster 1997; Armbruster et al. 1997; Fellowes, Kraaijeveld, and Godfray 1999). Furthermore, natural selection can act in ways that optimize fitness amid trade-offs in interactions with two or more other species (Simms and Rausher 1993; Iwao and Rausher 1997; Juenger and Bergelson 1998). Co-opting of traits and balancing of selection pressures are simply part of the process of multispecific coevolution (Brody 1997; Cariveau et al. 2004).

Nevertheless, there is a lingering impression in some discussions of interspecific interactions that coevolution is greatly limited as a process, specifically because most species interact with multiple species during their lifetimes. The implication is that natural selection cannot shape genomes in ways that allow coevolution with multiple enemies and mutualists. That view, however, is equivalent to saying that natural selection can adapt populations to only one selective pressure at a time—say, temperature or salinity level or defense against a particular predator. Yet all populations are direct evidence to the contrary. No one would argue that natural selection is incapable of shaping the complexities of adapting a plant population to its herbivores, mycorrhizae, pollinators, and seed dispersal agents. Similarly, no one would argue that vertebrate species cannot display fine-tuned immune systems as defenses against parasites while maintaining morphological, chemical, and behavioral defenses against predators and competitors.

Yet the view persists that, even if populations can adapt simultaneously to, say, a mutualist and a particular enemy, it is somehow much harder to adapt simultaneously to multiple similar enemies. That view was part of a nonecological and nongeographic view of coevolution that hardened in the 1980s, restricting the word "coevolution" to interactions showing extreme reciprocal specialization across all populations (review in Thompson 1994). During those years, demonstration of networks of interaction among species sometimes became de facto evidence that a particular pairwise interaction was not coevolved. Interactions exhibited either pairwise coevolution or diffuse co-evolution. These phrases were initially developed to focus the attention of re-searchers on the need to demonstrate, rather than simply assume, reciprocal selection in interspecific interactions (Janzen 1980). Unfortunately, they quickly developed into caricatures of the coevolutionary process. The phrase "diffuse coevolution" became equivalent to saying that the structure and dy-namics of reciprocal selection could not be identified. By a jump in logic, this in itself became a way of arguing that coevolution was unimportant. Simi-larly, local populations within species that deviated from tight pairwise co-evolution became evidence that an interacting pair of species was not co-evolved. At the extreme, two species were either coevolved or they were not.

The artificial dichotomy between pairwise coevolution and diffuse coevo-lution created a typological view of the coevolutionary process that ignored the populational structure of species, the life histories of individuals, and the genetic structure of genomes. Viewing multispecific interactions a priori as hopelessly diffuse thwarts any scientific effort to evaluate how reciprocal se-lection can shape multispecific networks under different ecological condi-tions. By adopting a more biologically realistic view of the structure of spe-cies, life histories, and genomes, multispecific interactions simply become part of the raw material for the geographic mosaic of coevolution. As differ-ent combinations of species interact across landscapes, they collectively affect selection and fitness in different ways in different populations (Brody 1997; Benkman, Holimon, and Smith 2001; Thompson and Cunningham 2002; Zangerl and Berenbaum 2003).

The life histories of individuals provide a variety of mechanisms that al-low adaptation to multiple enemies. Perhaps the simplest is ontogenetic par-titioning of interactions. Most species pass through multiple stages during their lifetimes. The pelagic stages of marine invertebrates and the nymphal or larval stages of terrestrial arthropods are ecologically different organisms from their adult stages. The chewing mouthparts and feeding habits of cater-pillars contrast starkly with the sucking mouthparts of adult butterflies. Tens of thousands of parasite species have fine-tuned their interactions with other species to the point of having complex life histories in which they alternate among two or more hosts at different life-history stages. Ontogenetic shifts occur just as commonly with defenses against enemies, as natural selection favors hosts that deploy different defenses at different stages of ontogeny (Karban and Thaler 1999). By sequentially turning on different genes at dif-ferent life-history stages, species can confront the specific problems associated with enemies that are limited to each life-history stage. Partitioning interspecific interactions—and the expression of genes governing them—among life-history stages is clearly one of the fundamental ways in which species coevolve with multiple other species.

Even simple changes in body size during development create ontogenetic niches as growing individuals shift their use of habitats and interactions with other species (Werner and Gilliam 1984). For example, the generalist grasshopper *Schistocerca emarginata,* which is a common grazer in the grasslands of central North America, changes over its lifetime in the number of species on which it feeds. In Texas, early instar nymphs live parasitically on single host plants, and different populations use different plant species (Dopman, Sword, and Hillis 2002). One lineage feeds in early instars almost exclusively on *Rubus trivialis* (Rosaceae), and another feeds almost exclusively on *Ptelea trifoliate* (Rutaceae). Analysis of mitochondrial DNA has indicated that the populations that feed on these different hosts cluster as monophyletic groups. Adults are more generalized in their use of plant species, using plants from multiple families. Natural selection generally favors the use of multiple victims in species with a true grazing habit (Thompson 1982, 1994), but this example shows that, even within this form of interaction, ontogenetic partitioning allows significant specialization, thereby creating the potential for a species to coevolve with multiple other species.

Within a life-history stage, we also now know that adaptation to two or more other species can range from being genetically independent to being highly correlated (Simms and Rausher 1993; Hougen-Eitzman and Rausher 1994; Pilson 1996). Controlled studies of artificial selection imposed by two enemies have begun to sharpen our understanding of how local populations may respond to multispecific selection (Mitchell-Olds and Bradley 1996). In *E. coli,* for example, when a population is exposed to two bacteriophage species, it can sometimes evolve through epistasis among genes, conferring resistance to both bacteriophages and thereby reducing the cost of multiple resistance in some environments (Bohannan, Travisano, and Lenski 1999). In *Drosophila melanogaster,* artificial selection for defense against one parasitoid species can be either genetically independent of adaptation to another species or genetically correlated (Fellowes, Kraaijeveld, and Godfray 1999). Selection experiments such as these go beyond earlier correlational studies of multispecific defense and allow more powerful analyses of the genetic structure of defense and constraints on the evolution of defenses (Mitchell-Olds, Siemens, and Pedersen 1996).

Many organisms also have the ability to induce defenses against enemies when they are needed, rather than maintain a constitutive arsenal. Examples include the vertebrate immune system and the induced defense systems now known in plants and insects (Karban and Baldwin 1997). Some insects can even adjust their morphological defenses to the particular environments in which they grow. Caterpillars of the spring brood of emerald moths (*Nemoria arizonaria*) mimic oak catkins, whereas caterpillars of the summer brood mimic oak twigs (Greene 1996). Hence, the cost of defense against multiple enemies may not be as high as once thought.

These kinds of studies show how natural selection shapes genomes to confront, sometimes simultaneously and sometimes sequentially within lifetimes, selection pressures imposed by multiple enemies. Traits may fluctuate in frequency and become modified repeatedly over time within a lineage as they are reshaped by the combined interactions that shape selection on a species. Genes may be expressed continuously, induced only when an interaction occurs, or expressed only at particular life-history stages. Collectively, these studies suggest that coevolutionary selection can be highly focused in responses to the specific combination of enemies confronting a population.

Coevolutionary Alternation in Multispecific Networks

Most predators attack more than one prey species, but the total number of prey species used by any local predator population is still small. Consider the archetypal large predator—lions. Even lions eat only a small number of the animal species in their habitats. During the years of Schaller's (1972) classic study of Serengeti lions, the diets of those prides consisted of eighteen mammal species and four bird species. Most of the diet, however, was composed of only several mammal species. These observations imply that, even though lions have broad and flexible diets, they do not feed randomly among a large array of prey species. Moreover, lions in different parts of Africa concentrate on different prey species (Schaller 1972). Most of those differences undoubtedly reflect differences among regions in the relative abundance of prey species rather than genetic differences in the preference hierarchies of lion populations. These observations, however, highlight the fact that even some of the earth's largest predators rely upon a small number of prey species and differ geographically in their primary prey species.

Most predators are much smaller than large mammalian carnivores, and they are often divided into genetically differentiated populations. That creates the potential for multispecific coevolutionary networks between predators and prey to consist locally of a small number of predator species that differ in preference for a small number of prey species. The prey guild used by a particular set of predators may share a set of defense traits such as a thick calcareous shell or particular forms of running behavior to which the local predator has evolved effective counterdefenses. Multispecific networks of parasites and hosts are similar but not identical to networks or predators and prey. Most parasites are highly host specific, although some are capable of attacking a wide range of hosts. The general form of the network, however, can be similar to that found in predator-prey networks wherever individual parasites actively choose among multiple hosts. Examples include many phytophagous insects in which adults lay eggs on multiple plants or brood-parasitic birds that lay eggs in multiple nests. In these species one parasite interacts locally with several similar host species but differs geographically in its preference ranking among the hosts.

Scores of studies have now evaluated geographic differences in preference hierarchies for insect species that feed parasitically or as grazers on plants, and the range of taxa tested continues to expand (e.g., Fox 1993; Thompson 1993; Fox, Waddell, and Mousseau 1994; Lu and Logan 1994; Renwick and Chew 1994; Futuyma, Keese, and Funk 1995; Radtkey and Singer 1995; Nylin, Janz, and Wedell 1996; Scriber 1996; Singer and Thomas 1996; Craig, Horner, and Itami 1997; Wehling and Thompson 1997; Janz 1998; Thompson 1998b; Singer 2003). These studies are mostly based upon behavioral analyses of how females from different maternal families or populations rank plant species or develop on those species during controlled experiments. Almost all the studies show evidence of geographic differences in relative preference or performance among potential host species.

Moreover, breeding experiments on insects have suggested that only a few genes may sometimes be needed to alter preference hierarchies. Hybridization experiments on the *Papilio machaon* group swallowtail butterflies showed that major interspecific differences in preference ranking in this group are due to X-linked genes that may be modified by autosomal or Y-linked genes (Thompson 1988a). Subsequent experiments on other insects have shown sex linkage of oviposition preference or larval performance in some other species (Nielsen 1997b, 1999; Janz 2003). No one has estimated the minimum number of genes involved in creating geographic differences in preference hierarchies for prey or hosts among maternal families or populations, although some analyses are moving in that direction (Sezer and Butlin 1998; Via and Hawthorne 2002). The fact, however, that large differences in preference are sometimes localized onto a single chromosome suggests the possibility that single genes or gene complexes may sometimes control local and geographic differences in preference hierarchies in phytophagous insects.

Overall, antagonistic trophic interactions all have a structure that can favor multispecific coevolution through ongoing changes in preference hierarchies in predators or parasites and defenses within and among prey or host species. That structure is captured in the hypothesis of coevolutionary alternation:

The hypothesis of coevolutionary alternation.

Natural selection favors predators that preferentially attack prey populations with relatively low levels of defense, and it favors increased levels of defense in those prey populations. Meanwhile, selection favors the evolution of reduced defenses in unattacked prey populations, because defenses impose fitness costs in the absence of predation. As relative levels of defense change among prey species over time, selection favors those predators that preferentially attack the prey species that are currently least defended. The hypothesis is most likely to apply to predators and grazers and to those parasites that preferentially choose among multiple potential victims.

The hypothesis is stated for predators and prey to keep the terminology simple, but it applies equally to grazers and their victims and to parasites and hosts. Through coevolutionary alternation, local populations of predators continue to evolve over time in their relative preferences, and populations differ in how they rank potential prey species. At any moment within a local assemblage, one or two prey species will be the most preferred, although less-preferred species may continue to be attacked to some extent (fig. 11.1).

Strong selection imposed by a predator on the most-preferred species creates the conditions for coevolutionary alternation. Imagine a predator species capable of attacking at least several local prey species. All else being equal (e.g., similar levels of relative abundance among prey species), natural selection should favor those predator individuals that specialize on the least-defended prey species. Selection will then favor increased levels of defense in that prey species, making it less profitable with each passing generation than other local prey species. Eventually selection should favor mutant predator genotypes that specialize on other local prey species that are less defended. This, in turn, relaxes selection for defense on the original prey species. Over long periods of time, the local predator population will alternate among prey species as rela-

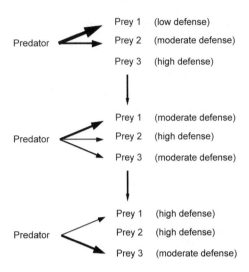

Fig. 11.1   Coevolutionary alternation between a predator and three local prey species. The thickness of the arrows represents the relative preference of predators for the prey species or the relative proportion of predator genotypes preferring each of the three prey species. The pattern of attack among prey species varies over time (top to bottom), and only one sequence of many possible patterns of shifts is illustrated. Coevolutionary alternation is most likely to apply to predators and grazers and to parasites that actively choose among multiple hosts. After Thompson 1994, based upon the hypothesis and results of Davies and Brooke 1989b.

tive levels of defense change among the prey species (fig. 11.2) and the preference ranking for prey species changes in the predator population.

Coevolutionary alternation should create dynamics similar in some respects to the dynamics of oscillating polymorphisms between parasites and hosts. Rather than a parasite population alternating among host genotypes, however, a predator population alternates among prey species. Neither class of coevolutionary dynamics, acting in the way presented here, creates an arms race of escalating levels of defense and counterdefense. Nevertheless, no formal mathematical models have yet explored the dynamics of the hypothesis of coevolutionary alternation. Any future analysis of the comparative dynamics of oscillating polymorphisms and coevolutionary alternative will be useful in helping us understand how these classes of coevolutionary dynamics differ from one another.

The possibility of coevolutionary alternation was first suggested as a specific explanation for interactions between cuckoos and their hosts in Britain (Davies and Brooke 1989a, 1989b) and later generalized as a hypothesis for time-lagged coevolutionary dynamics between predators, grazers, or par-

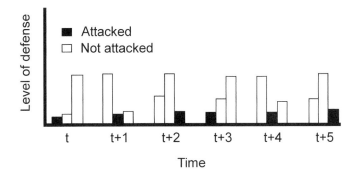

Fig. 11.2   Change over time in the relative levels of defense among three prey species, driven by the process of coevolutionary alternation with a predator. Time increments (e.g., *t* to *t*+1) are longer than a single generation. Note that levels of defense for each prey species fluctuate over time but overall levels of defense across the prey species do not increase over time.

asites and multiple victim species across geographic landscapes (Thompson 1994, 1999b). Davies and Brooke developed the hypothesis to explain the observed range of defenses found among host species and the patterns of egg mimicry found in the cuckoos. Although cuckoos lay their eggs in the nests of a wide range of birds in Britain and Europe, four species are currently their most common hosts. Several initially enigmatic aspects of this multispecific network are interpretable through the hypothesis of coevolutionary alternation. Some bird species in Britain are not currently used as hosts but exhibit high levels of defense against cuckoos. In addition, three of the commonly used hosts actively eject cuckoo eggs from their nest, but the fourth, the dunnock, does not. By the hypothesis of coevolutionary alternation, the birds that are active ejectors but nonhosts may be former hosts that, because they have evolved high levels of defense, have been temporarily abandoned by cuckoos. The dunnock is a new host that has not yet evolved high levels of defense.

Recent molecular work on the genetic structure of cuckoo populations has provided some additional insights into the structure of these and other interactions between brood parasites and hosts (Sorenson and Payne 2002). Some of the results are consistent with the hypothesis of coevolutionary alternation, but much work remains, because the genetic basis for host-race formation in brood parasites is unknown. Cuckoo chicks sampled from the nests of different host species in Britain and Japan differ in mitochondrial haplotypes, suggesting that some female lineages remain faithful to particular hosts long enough to produce host races (gentes, in the terminology of

brood-parasite biology) with novel haplotypes (Gibbs et al. 2000). The mitochondrial haplotypes within each host race, however, are not monophyletic,
potentially implying that switches onto these hosts and subsequent evolution
of egg mimicry may have occurred multiple times. That would be consistent
with the hypothesis of coevolutionary alternation.

In contrast, host races do not differ in their nuclear DNA haplotypes.
Male cuckoos appear to mate with females without regard to host race (Marchetti, Nakamura, and Gibbs 1998). Consequently, a current working hypothesis on the genetics of these interactions is that host race formation is
governed by the W chromosome in females, while random mating among
host races by the males holds the species together. Females are the heteroga-
metic sex in birds, and inheritance through the W chromosome would allow
these matrilineal lineages. The combination of female host races governed by
the W chromosome and random mating of males relative to female host races
would therefore be conducive to coevolutionary alternation, allowing maintenance of multiple host races within a single cuckoo species over time.

This kind of multispecific network should exhibit a strong geographic
mosaic, because most species involved in multispecific coevolution will differ
from one another in their geographic ranges. Moreover, the rate at which coevolutionary alternation occurs is likely to differ geographically even among
assemblages of the same species, because the intensities of selection and the
reproductive rates of species will differ geographically. The resulting geographic mosaic is evident in traits that differ among populations. In the case
of cuckoos and hosts, Davies and Brooke (1989a, 1989b) found that white
wagtails ejected eggs in Britain, where cuckoos are present, but not in Iceland, where cuckoos are absent. Other studies have found that magpie populations differ across Europe in their probability of ejecting cuckoo eggs
(Soler et al. 1999; Soler and Soler 2000).

The geographic mosaic found in these multispecific interactions provides
an indirect way of testing the hypothesis of coevolutionary alternation, because it allows assessment of multiple snapshots of the process across broad
geographic regions. The main predictions of the hypothesis are:

> 1. Local populations of predators (or grazers or parasites) have ge-
> netic preferences for the least defended of their potential local prey
> species.
> 2. Predator preference hierarchies vary geographically, providing in-
> direct evidence of alternation.
> 3. Some currently unattacked prey populations show high levels of

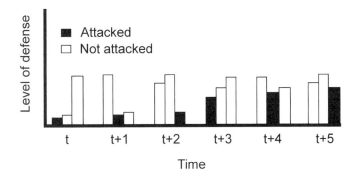

Fig. 11.3   Coevolutionary alternation in an interaction between a predator and three prey species, accompanied by escalation in the overall levels of prey defense over time. Unlike in fig. 11.2, the predator population evolves so quickly in preference for the least-defended prey species that defense levels in unattacked prey species never relax completely. Consequently, overall levels of defense in all victim species ratchet up over time. Time increments are longer than a single generation.

defense, providing correlative evidence of past defense that has not yet been completely lost.

4. Some currently attacked prey populations show very low levels of defense. These low levels of defense may indicate that the species is a completely new prey, or it may indicate that the species is a previously used prey that lost its defenses during the period of time that the predator focused its attack on other species.

Coevolutionary Alternation with Escalation

If genetically based switching among prey or hosts occurs with only short time lags, then an alternative form of coevolutionary dynamics is possible (fig. 11.3). Stated as a hypothesis, the process links coevolutionary alternation with coevolutionary escalation.

The hypothesis of coevolutionary alternation with escalation.

Coevolutionary escalation results from coevolutionary alternation when predator, grazer, or parasite populations evolve their relative preferences among victim species faster than previously preferred, but now unattacked, victim populations completely lose their defenses. Under these conditions, the overall levels of defense and counterdefense ratchet upward over time, but the geographic mosaic of coevolution prevents relentless escalation.

Coevolutionary alternation with escalation involves true reciprocal selection and change favoring higher levels of defense and counterdefense. It produces coevolutionary arms races in the sense that each species must devote more resources to the interaction to be successful at defense or counterdefense. Use of the word "escalation" here differs from the broader use of the word in interactions envisioned by Vermeij (1987, 1994). By that hypothesis, overall selection for defense has increased, diversified, and become so pervasive over eons and so diffusely distributed across species that escalating defenses have become an inherent part of the background of evolutionary change. In contrast, coevolutionary alternation with escalation hypothesizes a specific coevolutionary process by which a multispecific network coevolves. Although described here as one predator coevolving with multiple prey, it could just as readily apply to networks in which multiple similar predators coevolve with multiple prey.

Coevolutionary alternation with escalation may explain the exaggerated traits of snails and crabs in Lake Tanganyika. The shells of the snails are unusually thick, and the chelicerae of the crabs that feed on them are unusually large and forceful (West, Cohen, and Baron 1991; West and Cohen 1996). The snail shells more closely resemble those of marine snails than those of other typical lake-dwelling snails from other regions. Moreover, the thickened shells appear to have arisen multiple times and have microscopic changes in the number of cross-lamellar layers, suggesting overall escalation among snail species in their resistance to shell-crushing predators (West and Cohen 1996). Through the process of coevolutionary alternation with escalation, one can imagine how levels of defense and counterdefense could have ratcheted up over time, as the crabs concentrated at any point in time on the least-defended snail species. The same process could potentially explain the more complex geographic patterns of shell thickness and design in marine invertebrates worldwide.

The main predictions for coevolutionary alternation with escalation are the same as those for coevolutionary alternation without escalation, with two differences.

1. Prey assemblages differ geographically in their overall levels of defense, indicating different rates of defense ratcheting among different assemblages.
2. Only prey populations truly new to a local assemblage have very low levels of defense. That is, once established within a local assemblage, a prey population does not cycle between high and low levels of defense.

Coevolutionary alternation with escalation is unlikely to lead to relentless escalation, because it will be tempered by the geographic mosaic structure of antagonistic trophic networks. As the geographic ranges of interacting species continue to change, simple escalation of current defenses and counterdefenses will not always be the option favored by natural selection. Ongoing mutation and gene flow from other populations will continue to provide host or prey populations with various options, including escalating the expression of one defense mechanism, substituting a qualitatively different form of defense, or some intermediate combination of the two. The selective advantage of developing qualitatively different and cheaper defenses should increase as current defenses escalate within local populations. Over evolutionary time, populations are likely to accumulate a variety of defenses—and counterdefenses—even against single enemy species. The level of investment in any one defense is therefore likely to wax and wane over time within and among populations as natural selection favors different combinations of defenses in prey species.

Coevolution within antagonistic trophic networks will therefore often be driven by a constantly changing selective tension between escalation of current traits and evolution of novel traits. In addition, gene flow among coevolutionary hotspots and coldspots across landscapes and adaptive compromises imposed on populations within multispecific networks will themselves slow the rate of relentless escalation. Part of the future challenge is to understand how coevolution proceeds amid that tension within and among populations having different network and geographic structures and different phylogenetic constraints. Studies in functional morphology and evolutionary developmental biology have made it clear that evolution along some morphological axes is more likely than evolution along other axes. Analyses of structures as different as the evolution of butterfly eyespots (Brakefield 2001; Brakefield, French, and Zwaan 2003) and sunfish feeding mechanisms (Wainwright 1996) show that some trajectories of evolutionary change are more likely than others. These evolutionary biases in trajectories could potentially bias the pattern of evolution of network structure.

Coevolution in Broader Predator-Prey Networks: The Structure and Dynamics of Food Webs

As the number of species in local antagonistic networks expands and contracts over time, coevolutionary selection will continually reshape the number of links among species. Most food webs depicted in ecological publications are representations of the structure of local specialization in predators and grazers based upon known food items in diets. These complex networks usually show lines connecting each predator to two or more prey species, implying that extreme specialization to a single prey species is rare. The same applies to grazers that feed on multiple victims without killing them outright, and the subsequent discussion uses predators as a shorthand for predators and grazers. Within these webs, predators do not attack prey at random. They often specialize on groups or guilds of prey that share similar traits, including defensive traits. The result can be strong compartmentalization within food webs (Krause et al. 2003).

These compartmentalized food webs are rarely static and can range from relatively simple trophic cascades to complex multispecific networks (Polis and Strong 1996; Warren 1996; McPeek 1998; Jeffries 1999; Strong 1999). Within each network there is always the potential for moderate to strong, but continually varying, indirect effects of species on one another (May 1973; Wootton 1994; Miller and Travis 1996; Schoener and Spiller 1999; Holt et al. 2003; Trussell, Ewanchuk, and Bertness 2003). These indirect effects may continually alter the structure of natural selection within food-web compartments (Miller and Travis 1996). Local patterns of specialization on other species are therefore unlikely to remain static over time as prey defenses and predator preferences continue to evolve. The dynamic ecological structure of food webs may reflect a dynamic coevolutionary structure.

The structure of food webs therefore has the potential to be shaped by coevolutionary alternation, with or without escalation, whenever three conditions are met. Predator populations must exhibit genetic variation in preference hierarchies for prey, so that natural selection could vary over time in the predator genotypes that are favored. Prey populations must exhibit genetic variation for defenses against particular predators and perhaps suites of predators. Moreover, prey species must continue to differ from one another in their levels of defense as perceived by the predators. Once established, food webs would vary in compartmental structure over time as preference hierarchies continually evolve in the predators and relative levels of defense continually evolve in the prey species.

Although there are some studies on genetic variation for prey preference or foraging in predators and victim preference in grazers (e.g., Arnold 1981; Hedrick and Riechert 1989; Sanford et al. 2003; Sotka, Wares, and Hay 2003), the numbers are still few relative to the number of studies that have been conducted on parasitic species such as insects that complete development on a single host plant (Janz 2003; Singer 2003). Conclusions on preference hierarchies from studies of parasites cannot simply be applied to predators and grazers, because hierarchies in these species may be more complicated than parasite hierarchies. Natural selection on predators and grazers can favor a greater diversity of learned behavior than probably occurs even in parasite species that show active host choice. Moreover, selection on predators and grazers can favor individuals that attack different combinations of species at different ages and sizes. Genetic studies of preference hierarchies within and among predator populations are needed to clarify the geographic mosaic of coevolution in predator-prey and grazer-victim networks.

Such studies would also go a long way toward resolving differing perceptions about natural selection on predators. There is a long-standing perception among some evolutionary biologists that there is an inherent asymmetry in predator-prey interactions, with prey evolving much faster than predators (see discussion in Abrams 2000a). That view, however, does not take into account the overall network structure of coevolving predator-prey interactions. Most coevolutionary responses of predators probably involve evolutionary shifts in preference hierarchies rather than major changes in morphology or complete shifts to different prey. These subtle but crucial shifts in preference are easily missed in any study of coevolution that relies upon analysis of morphology or the number of arrows connecting species within food webs. Predators that miss fewer dinners will be favored by natural selection over those that miss more dinners, and the major mechanism favoring fewer missed dinners is probably evolutionary adjustment of preference hierarchies. The hypothesis of coevolutionary alternation with or without escalation captures this view of how coevolution between predators and prey is likely to proceed within complex food webs.

Nevertheless, we currently lack a mathematical theory that would predict how the structure of food webs would appear if driven by the process of coevolutionary alternation, with or without escalation, as compared to some other coevolutionary process of network formation. More than twenty years ago, Pimm (1982) noted that we lack any real understanding of how evolution molds the overall shape of food webs, and that is still true today. Current food-web theory remains mostly ecological rather than evolutionary in approach and does not invoke particular hypotheses on the form of multispecific coevolution. There are a small number of coevolutionary food-web models (e.g., Matsuda and Namba 1991; Freeland and Boulton 1992; Caldarelli, Higgs, and McKane 1998; Drossel, Higgs, and McKane 2001). Some evolutionary network models have begun to incorporate phylogenetic constraints in ways that create guilds of interacting species (Cattin et al. 2004), and network models based upon work on artificial intelligence are also getting closer in design to multispecific coevolutionary models (Potter and De Jong 2000). None of the current models, however, has a structure that could be used directly to evaluate the effects of coevolutionary alternation and alternative hypotheses on the structure of multispecific coevolution in predator-prey networks. Multiple competing models should become available in the next few years as network algorithms continue to develop, inspired collectively by the need to answer questions about the structure of the Internet, artificial intelligence, neural networks, genome structure, hierarchical population structure, and biodiversity structure.

Current coevolutionary models of pairwise predator-prey interactions cannot provide direct insight into how processes such as coevolutionary alternation with escalation could shape the multispecific structure and dynamics of predator-prey interactions. As Abrams (2000a, 2001) has emphasized in reviews of predator-prey models, the range of coevolutionary models is still very narrow, and almost all models have been developed in the absence of their food-web context. Most models have focused either on the local dynamics and stability of pairwise interactions or on the evolution of general properties such as prey capture rate or escape ability in pairwise interactions (e.g., Dieckmann and Law 1996; Marrow, Dieckmann, and Law 1996; Doebeli 1997; Gavrilets 1997; review in Abrams 2000a). Nevertheless, these studies of pairwise interaction have been very useful in providing hints as to what we might expect in multispecific predator-prey networks. Through these models, we now know that coevolution between local pairs of predator and prey can generate local population cycles, if there is gene matching in the interacting species. Under these conditions, prey escape if they have a trait that deviates in either direction from the current mean. If prey escape only through escalation of traits, however, then stable local cycling is unlikely (Abrams 2000a, 2001). Hence, local pairwise coevolution between predators and prey can create a wide range of potential outcomes. If these pairs of populations are connected to others across a geographic gradient of prey productivity (i.e., prey birth rates), then coevolutionary hotspots and coldspots develop, which in turn produce geographic differences in the levels of defense and counterdefense (Hochberg and van Baalen 1998).

Collectively, results from pairwise models beg the question of what would happen if predators and prey coevolved not as pairs of species with matched or escalating traits but rather as multispecific groups in which local prey species increase and decrease their defenses over time in response to varying local levels of predator attack and constantly changing network structure. Models of adaptive switching among prey by a predator population have already shown that addition of predator switching can have major direct and indirect effects on the dynamics of predator-prey interactions (Abrams 1999). There is, then, wide scope for the next generation of predator-prey models that incorporate multispecific network structure, coevolutionary alternation, and geographic structure.

Whether motivated by evaluation of coevolutionary alternation or by other coevolutionary mechanisms, further research on predator-prey coevolution can proceed best if developed in the context of antagonistic networks that vary geographically as species shift in their distributions across landscapes. The hypothesis of coevolutionary alternation provides a specific hypothesis on the coevolutionary process in multispecific interactions between predators and prey. Even if this particular hypothesis turns out to be incorrect as a description of how coevolution shapes multispecific predator-prey interactions, it shows that there are ways of moving beyond general descriptions of "diffuse" coevolution that provide no specific hypotheses on how the actual multispecific coevolutionary process works. The way forward is to understand how different forms of interaction—such as those between predator and prey, parasite and host, or free-living mutualists—develop different network structures through a predictable coevolutionary process.

Unresolved Problems in Multispecific Coevolution with Parasites, Grazers, and Predators

Among the unresolved problems on coevolution within antagonistic trophic networks, two stand out. We need more explicit hypotheses on how indirect effects of interspecific interactions shape coevolutionary dynamics within and among populations. Studies of multispecific networks are beginning to make inroads into the problem by replacing the assumption of diffuse selection with careful analyses of the actual structure of selection. We now know that some indirect effects may even link antagonistic and mutualistic networks. Herbivory on leaves or flowers, for example, can influence the pattern of visitation by pollinators and selection on floral phenology (Strauss, Conner, and Rush 1996; Strauss and Armbruster 1997; Juenger and Bergelson 1998). Progress in our understanding of the role of indirect effects in coevolving antagonistic networks will demand detailed study of a few model interactions in which the structure of multispecific selection can be evaluated within and among populations.

We also need to understand better how long-lived taxa lacking immune systems confront fast-evolving parasites. The evolution of sexual reproduction, discussed in the previous chapter, is part of the solution. Alone, however, it is unlikely to explain how redwoods living for a thousand years remain remarkably healthy when a pathogen population could take up residence and evolve over thousands of generations on a single large redwood tree. One possibility for such plants and also for clonal animals is genetic mosaicism (Whitham 1981; Gill 1986; Gill et al. 1995). By this hypothesis, species with modular body plans may coevolve with multiple fast-evolving parasites through somatic mutations. As host individuals grow and produce new modules—tree limbs, coral branches—any somatic mutation at the base of the branch will affect all subsequently formed tissue on that branch. A long-lived modular organism can therefore become over time a genetically diverse individual as it continues to produce new modules. Even without genetic mosaicism, modular organisms could possibly eliminate small modules (e.g., root tips) that are attacked by parasites (Hoeksema and Kummel 2003).

Alternatively, or additionally, some species with long-lived individuals or colonies may cope with fast-evolving parasites by enlisting assemblages of mutualists. Some long-lived social insect colonies use specialist mutualistic microorganisms in their battles against pathogenic microorganisms (Currie, Bot, and Boomsma 2003), sometimes requiring the colonies to tend the mutualists so that these microbes do not themselves become antagonists (Matsuura, Tanaka, and Nishida 2000; Matsuura 2003). Long-lived trees may take this process a step further and enlist mutualistic assemblages. The tropical tree species cacao (*Theobroma cacao*) harbors a diverse assemblage of horizontally transmitted fungal endophytes that reduces leaf necrosis and mortality in leaves challenged with a major fungal pathogen (*Phytophora* sp.). Endophyte species diversity changes throughout the lifetimes of cacao individuals, and overall endophyte composition varies geographically across Panama (Arnold et al. 2003). Fungal endophytes are common in many tree species (Arnold et al. 2000). They therefore deserve careful study as a mechanism by which long-lived tree species may cope with a broad array of rapidly evolving parasites. If trees exhibit an ever-changing genetic mosaic of fungal endophytes as they grow and produce new modules, then pathogens would be less able to adapt fully to individual trees over hundreds of years.

Conclusions

Coevolution through coevolutionary alternation and coevolutionary alternation with escalation are testable hypotheses of multispecific coevolution. These hypotheses suggest how predators, grazers, or parasites may coevolve with multiple victims and how those victim species may collectively escalate defensive traits. The hypotheses suggest that identifiable processes of reciprocal selection, rather than diffuse coevolution, may shape multispecific antagonistic interactions. Moreover, these hypotheses suggest how coevolution may contribute to compartmentalization within food webs.

Coevolutionary alternation with escalation creates coevolutionary arms races in defense and counterdefense. Although the term "arms race" is sometimes used loosely for any form of coevolution, the analogy only makes sense when the reference is to true escalation in levels of investment in defense or counterdefense. Coevolutionary alternation with escalation provides a mechanism by which multispecific coevolution may shape coevolutionary arms races. In contrast, coevolutionary alternation without escalation creates varying levels of defense amid changing patterns of preference. Together, these hypotheses highlight that escalating arms races are not an inevitable outcome of antagonistic coevolution.

Any process that involves evolutionary switching among hosts or prey is almost bound to fall apart occasionally through stochastic events at the local level over long periods of time. Similarly, any defense system that involves the evolutionary gamble of having the right mix of alleles to confront all local forms of parasites will occasionally fail at the local level. Consequently, it seems likely that long-term persistence of antagonistic trophic networks must often depend to some extent upon the constantly shifting geographic mosaic of defense and counterdefense in these interacting networks.

The current working hypotheses on coevolution in antagonistic trophic networks are best viewed as testable starting points. The long-term goal of work on multispecific coevolution is to understand how the network structure of interspecific interaction shapes reciprocal evolutionary change—and how reciprocal evolutionary change, in turn, shapes network structure within local assemblages and across broader geographic landscapes. Achieving that goal will require novel approaches to understanding how reciprocal selection is weighted among species and how selection changes over time within these networks.

12 Mutualists I
Attenuated Antagonism and Mutualistic Complementarity

Mutualisms are inherently selfish interspecific interactions that increase the fitness of both species. Mutualistic interactions often derive from antagonistic interactions, and mutualisms may in turn produce descendant lineages that are commensalistic or antagonistic (Thompson 1994; Pellmyr and Leebens-Mack 2000; Dale et al. 2002). Mutualism and antagonism are therefore ends of a continuum. Interacting species can differ among populations in their placement along that continuum as investment in the interaction and the ecological outcomes vary across landscapes. That variation allows the local origin of mutualism (Doebeli and Knowlton 1998), and it may fuel the persistence of mutualisms over time as mutualistic networks accumulate species.

This chapter discusses how coevolution may shape the transition from antagonism to mutualism in symbiotic interactions, and the following chapter explores geographic mosaics in symbiotic mutualisms. Symbiotic mutualisms are those that involve sustained, intimate interactions between individuals of two species. Examples include the hundreds of thousands of interactions between endosymbiotic bacteria and fungi harbored by plants and animals as well as more complicated forms of intimate interaction, such as those between leaf-cutter ants and the fungi they grow and harvest within their nests. Chapter 13 then evaluates coevolving mutualisms among free-living species, such as plants and their pollinators. The differences in the structure of selection in symbiotic and nonsymbiotic mutualism parallel the differences considered earlier in the structure of selection on parasitic interactions and those involving grazing and predation. This chapter focuses on three hypotheses: coevolution with symbionts toward attenuated antagonism, coevolved monocultures and complementary symbionts, and symbiont partitioning.

To begin though, it is worthwhile reviewing how far we have come in recent years in our appreciation of the importance of mutualistic symbioses in the organization of the earth's biodiversity.

Coevolved Symbiotic Mutualisms and the Organization of Life

We now understand that mutualisms and the geographic structure they produce are probably an inevitable consequence of the genetic organization of biodiversity. If an individual can use not only its own genes but also those of other species to increase its chance of survival and reproduction relative to others in a population, it may be favored by natural selection. Most such interactions will be antagonistic, producing parasites, grazers, or predators. But a subset of the interactions among genetically variable species is bound to be mutualistic, and those mutualistic effects are just as inevitably bound to vary among environments. As research has expanded over the past decade both on microbial symbionts and the structure of interactions between free-living species, we have learned that mutualisms are not only common but also much more fundamental to the organization of life than we had previously imagined.

Mutualistic interactions are so pervasive and central to the organization of communities they are easy to ignore. In fact, one of the oddest aspects of the history of the sciences of ecology and evolutionary biology was the lack of attention to mutualistic interactions during much of the twentieth century in comparison to the large body of work on competition and predation (Bronstein 1994b; Douglas 1999). Yet even a cursory scan of the pervasiveness of mutualisms shows them to be central to the organization of the earth's biodiversity. Eukaryotic species depend upon genes from ancient mutualistic symbioses (Dyall, Brown, and Johnson 2004). In the absence of the ancient interactions that have become mitochondria and chloroplasts, the fundamental structure of the earth's biodiversity would be entirely different. Without the mutualism between corals and dinoflagellates, coral reefs, with their biological complexity, would not have appeared. At greater ocean depths, deep-sea vent animals have developed obligate endosymbiotic associations with sulfur-oxidizing bacteria to obtain nutrition (Peek et al. 1998). In terrestrial environments, plant communities and the process of succession are built upon the underlying lichen, mycorrhizal, and rhizobial associations, and colonization of land by plants may even have resulted from the development of mycorrhizal associations (Blackwell 2000; Redecker, Kodner, and Graham 2000). In turn, the animals that feed upon these plants are able to do so through the microbial assemblages they harbor in their digestive systems. The two dominant animal groups in many tropical environments, ants and termites, include ecologically pervasive species that rely obligately upon agricultural symbioses with fungi within their nests (Mueller et al. 2001; Aanen et al. 2002). The expanding list of examples of symbiotic mutualisms central to community organization spans all kingdoms and phyla and permeates every biological community on earth (Margulis and Fester 1991; Douglas and Raven 2003).

At its core, then, the organization of the earth's biodiversity is built upon coevolved mutualistic symbioses that are both ancient and pervasive. Take away these mutualisms and most species would simply disappear. In fact, multiple mutualistic symbioses are a fundamental part of the life histories of almost all multicellular organisms. An individual plant is usually a combination of four or more genomes—its own and those of its mitochondria, chloroplasts, and the multiple fungal species with which it has formed mycorrhizal associations. Many animals harbor at least three obligate genomes— nuclear, mitochondrial, and one or more symbiont species that aid in nutrition. Studies of an increasingly wide range of species have indicated that multigenomic associations are the norm rather than the exception. Eukaryotes are multigenomic, and symbioses are therefore not simply add-ons to the diversification of life. They are the very basis for much of the innovation and diversification that has occurred over the past several billion years (Margulis and Fester 1991; Maynard Smith and Szathmáry 1995; Margulis and Sagan 2002).

Our growing appreciation of the pervasiveness of mutualistic endosymbiotic relationships is a result of the great proliferation of studies of many forms of symbiosis since the early 1990s. These studies continue to accelerate (Margulis 2000; Moran and Baumann 2000; Margulis and Sagan 2002; Moran 2002; Saffo 2002), and much has happened very recently, especially in our understanding of some of the earth's most important symbioses. For example, the nod factors in rhizobia that begin the complicated series of events leading to nitrogen-fixing symbioses with legumes were discovered only in 1990, and the corresponding plant receptors were discovered only in 2003 (Limpens et al. 2003; Madsen et al. 2003; Radutoiu et al. 2003). Studies of local adaptation in mycorrhizae, molecular analyses of the diversity of zooanthellae in corals, calculations of the rates of evolution and lateral gene transfer in endosymbionts, and theoretical and experimental examination of the coevolutionary consequences of vertical and horizontal transmission of symbionts are all research programs and approaches that began to proliferate in the 1990s.

Attenuated Antagonism in Symbioses

Increased research on symbiotic interactions has helped us begin to understand why some symbiotic interactions are parasitic, whereas others are commensalistic or mutualistic. That research has changed major assumptions about the evolution of attenuated antagonism and mutualism. In fact, the sharp focus on sexual reproduction as a coevolved response to the relentless and rapid evolution of parasites and appreciation of the complicated roles of *Wolbachia* in host reproduction and coevolution are both part of that paradigm shift. Until development of the field of evolutionary ecology in the 1960s and 1970s, many evolutionary biologists and parasitologists would have argued that highly antagonistic dynamics between parasites and hosts were an indication of the evolutionary youth of an interaction. The common expectation was that interactions between parasites and hosts eventually coevolve toward reduced antagonism. Parasites that kill their host population would themselves become extinct, and natural selection would therefore favor less aggressive parasites. Over evolutionary time, parasites on a host species would become less virulent, and antagonism would become attenuated. In a sense, the lions would lie down with the lambs, although the argument applied only to parasites and not to predators.

This view was so prevalent that, when Roy Anderson and Robert May first started publishing their landmark papers on the mathematical theory of evolutionary epidemiology (Anderson and May 1979; May and Anderson 1983, 1990; Anderson and May 1991), in one of their papers they placed their major conclusions in italics: the coevolutionary pathways between parasites and hosts allow for many trajectories, and there is nothing inevitable about evolution toward reduced antagonism or commensalism (Anderson and May 1982). Their models and later models by others showed that views that assumed reduced virulence were based upon multiple assumptions about the relationship between host density, rates of parasite transmission, and mortality patterns in infected as compared with uninfected individuals. (Note that "virulence" here is used in the sense of any reduction in host fitness caused by parasites and pathogens, not in the special sense used in the earlier discussion of gene-for-gene coevolution in chapter 9). Changing those assumptions changes the evolutionary outcomes (Lenski and May 1994; Day and Burns 2003; Ganusov and Antia 2003). Evolution toward a high level of virulence results under some ecological conditions, whereas evolution toward attenuated antagonism results under other conditions.

Experimental and phylogenetic studies have now shown quite convincingly that interactions between parasites and hosts have no inherent directionality relative to the age of the interaction itself. Increasingly, empirical studies have made use of the geographic structure of interacting species to show the lack of inherent directionality. For example, the four species that make up the *Drosophila testacea* species complex vary in geographic distribution and in their probability of being attacked by nematodes in the *Howardula aoronymphium* species complex. Cross-infections of the nematodes to allopatric and sympatric populations and species of *Drosophila* show no overall pattern of virulence related to sympatry or phylogenetic position within a clade. Levels of virulence are at least as variable in normal interactions between parasite and host populations as in novel interactions (Perlman and Jaenike 2003b). The overall pattern of variation in virulence in these species complexes appears to be due to a combination of population differences in parasite and host traits (Perlman and Jaenike 2003a).

Identifying the conditions that favor escalating virulence or attenuated antagonism has become a central focus of research in evolutionary epidemiology. Some models show that spatial structure can favor reduced levels of virulence in parasites, but different levels of spatial structure and the potential for acquired immunity can create alternative stable states of levels of virulence (Boots, Hudson, and Sasaki 2004). Current models differ in how they incorporate genetic architecture, demographic structure, natural selection, mutation, acquired immunity, geographic selection mosaics, and coevolutionary hotspots. A tentative general working hypothesis of the conditions favoring attenuated antagonism, based on current results, can be stated as follows:

The hypothesis of coevolution with symbionts toward attenuated antagonism.

Coevolution between parasites and hosts leads to attenuated antagonism when local symbiont genetic diversity is low, the rate of infectious spread of symbionts low, and the geographic structure of species allows high rates of local adaptation in hosts and parasites.

This is stated only as a general hypothesis, because the interrelationships among the factors contributing to the evolution of attenuated antagonism are only partially understood. The Anderson-May models emphasize the effects of host density and parasite transmission rates on the dynamics of parasite-host interactions, and these two components of the models have become predictive tools in studies of natural selection on parasites. In two influential books, Ewald (1994, 2000) extended these arguments by suggesting that the mode of parasite transmission and host mobility are the two aspects of life histories most important in determining how host density and parasite transmission rates shape the evolution of virulence. If parasites can be transmitted infectiously among hosts (that is, through horizontal transmission, in the terminology of parasitology) even when the host is immobilized by the disease, then natural selection will favor highly virulent parasites. Ewald has argued that human diseases such as malaria, dengue fever, and yellow fever, which are transmitted by mosquitoes, fall into this category, because the mosquitoes can transmit these diseases in many human cultures even after a person becomes ill and immobilized. Water, too, can act as a vector of human diseases, including cholera, typhoid, and dysentery. In cultures with poor water sanitation, these virulent diseases can thrive. If, instead, parasites rely upon direct contact between hosts, then selection will favor genotypes of the parasite that do not result in acute disease that would immobilize their host.

By this view, vector-transmitted diseases should generally be more lethal or incapacitating than diseases transmitted by direct contact between hosts. Host population density, however, also has an effect. High host population densities will favor high levels of virulence in parasites transmitted by direct contact, provided that the number of contacts between hosts is high. In contrast, low population densities will often favor low levels of acute disease in parasites transmitted by direct contact.

The range of possible evolutionary outcomes is even greater if symbionts are transmitted directly from mother to offspring. Obligate vertical transmission of a symbiont is possible only if the maternal host stays alive long enough to reproduce. Hence, for a vertically transmitted symbiont to spread, it must either maintain the ability to spread through horizontal transmission as well, manipulate host reproduction in ways that increase its spread through a population, or confer some direct benefit on host fitness. Consequently, vertical transmission is probably the main way by which attenuated antagonism and commensalism and mutualism may evolve. The diversity of ecological outcomes between *Wolbachia* and hosts (see chapter 10), however, indicates that there is nothing inevitable about particular coevolutionary trajectories in vertically transmitted symbionts.

As the theory of the evolution of virulence has progressed, it has moved increasingly toward a coevolutionary approach by evaluating how parasite adaptation and host resistance, both individually and together, produce outcomes ranging from increased virulence to attenuated antagonism. Current mathematical models and some empirical studies of parasites and hosts suggest that transmission mode, transmission rate, the geographic mosaic, and symbiont diversity within local host populations are the most important foci of coevolutionary selection affecting virulence (Anderson and May 1991; Lockhart and Thrall 1996; Gandon 2002a, 2002b; Blanford et al. 2003; Day and Burns 2003). In general, current theory suggests that natural selection on hosts should favor suppression of multiple symbiont genotypes, which may be easier for vertically transmitted symbionts than for horizontally transmitted symbionts. Intrahost competition among symbiont genotypes can favor more virulent genotypes (Bull 1994; Frank 1995; van Baalen and Sabelis 1995; Frank 1996a, 1996b; Ebert 1998), although exactly how competition shapes the evolution of virulence in parasites can depend upon the form of competition (May and Nowak 1994, 1995). Exploitation competition, interference, and apparent competition have potentially different effects on selection for virulence in the absence of a coevolutionary response by the host (Read and Taylor 2001). From the host side of the interaction, then, the potential for attenuated antagonism may depend upon the form of parasite transmission, its ability to control parasite diversity, and the form of competition among parasite genotypes and species.

In turn, natural selection favors parasite genotypes that are best able to spread across host populations. Such spreading is mediated by the structure of competition with other parasites, the potential for local adaptation within hosts and within host populations, and the genetic structure and coevolutionary response of hosts. Current theory suggests that, all else being equal, selection on parasites should favor some degree of infectious spread, because dispersal reduces competition with similar genotypes (Hamilton and May 1977). Hence, infectious genotypes may sometimes spread faster through a population than vertically transmitted parasites. But all else is never equal, and the broad range of transmission modes, from purely infectious to purely vertical and everything in between, creates opportunities for natural selection to favor highly attenuated antagonism under some ecological conditions. Some parasites even maintain phenotypic flexibility in their capacity for vertical or horizontal transmission. The bacterium *Holospora undulata* is vertically transmitted when growth rates are high in its protozoan host *Paramecium caudatum*. At low growth rates, the bacteria switch to infectious spread (Kaltz and Koella 2003).

Variation across landscapes in selection on parasites and hosts must inevitably produce a geographic mosaic of coevolutionary response, tipping the balance toward different forms and levels of antagonism, or even commensalism or mutualism, in different populations. Where symbiont genotypic diversity and infectious spread are high, selection can favor high levels of virulence in parasites and prevent hosts from favoring single strains of mutualistic, vertically transmitted symbionts. Evidence for infection by multiple genotypes is increasing for some diseases as molecular probes make it easier to identify separate symbiont genotypes, although most current data are for human diseases (Read and Taylor 2001). For other taxa, the available data range from experimental inoculations of local populations to a few broad geographic surveys of many populations (e.g., Kondo et al. 2002; Hood 2003). For example, surveys of *Wolbachia* infection have shown infection of hosts by two or more *Wolbachia* genotypes or strains (Rousset, Braid, and O'Neill 1999; Werren and Windsor 2000). In Japanese populations of azuki bean beetles, *Callosobruchus chinensis,* more than 90 percent of beetles harbor three *Wolbachia* genotypes, with multiple infection levels almost identical in females and males (fig. 12.1). Within a broad geographic sample of 410 individuals from nine populations, all but one individual harbored two or more *Wolbachia* strains, based upon analysis of the *wsp* gene (Kondo et al. 2002). In New Caledonia, 85 percent of *Drosophila simulans* individuals from one population were infected with two *Wolbachia* strains (James et al. 2002), and in Thailand one tephritid species is known to harbor five *Wolbachia* strains (Jamnongluk et al. 2002).

Multiple *Wolbachia* infections within hosts are sometimes from different *Wolbachia* lineages, suggesting horizontal transmission. One of the three *Wolbachia* strains in Japanese bean beetles falls into *Wolbachia* group A, and two are in group B (Kondo et al. 2002). Surveys of other insect groups have also found instances of *Wolbachia* groups A and B in the same host (Werren and Windsor 2000). These results, together with others showing lack of strong phylogenetic congruence between arthropod hosts and their *Wolbachia* symbionts, suggest that horizontal transmission of these symbionts occurs at least occasionally, with transfer through parasitoids as one possible mechanism (Heath et al. 1999; Vavre et al. 1999).

Considerable uncertainty remains over how parasite transmission mode, transmission rate, parasite competition, host density, and spatial structure interact to shape the continuum between aggressive virulence and attenuated antagonism. Most of the uncertainty comes from the fact that coevolutionary models of the evolution of virulence and attenuated antagonism are still

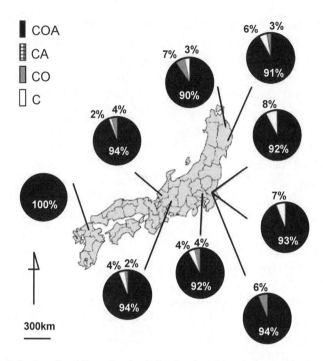

Fig. 12.1   Infection of azuki bean beetles *Callosobruchus chinensis* in Japan by three geno-types (wBruCon = C, wBruOri = O, and wBruAus = A) of *Wolbachia*, identified using highly specific primers for their *wsp* genes. Values are the percentage of individuals harboring the C genotype alone or a combination of the C genotype and other genotypes. The CA combination of genotypes is rare and not visible in the pie charts. In all, 410 beetles were sampled from the nine beetle populations (36–50 beetles per population). After Kondo et al. 2002.

in the early stages of development (e.g., Koella, Agnew, and Michalakis 1998; Sasaki and Godfray 1998; Hochberg et al. 2000; Roy and Kirchner 2000; Koella and Restif 2001; Restif, Hochberg, and Koella 2001; Restif and Koella 2003). Many mathematical models on the evolution of virulence have evaluated only the evolution of virulence in the parasite, while assuming that the host population does not evolve. Other models have evaluated only the effects of parasite infection on host evolution.

Coevolution and geographic structure, however, create novel routes within these models to changes in the levels of virulence. If parasites reduce host survival, then natural selection could favor the evolution of increased reproductive effort in hosts at the expense of future survival. If host populations are allowed to evolve life histories in response to parasite evolution in a model with an epidemiological structure, then the life histories of the two species coevolve across environmental gradients (Gandon 2002b). The perceived level of virulence becomes an outcome of the coevolutionary process, as it shapes allocation of resources to survival and reproduction in both species across different environments. Other general models of the geographic mosaic of coevolution have shown that gene flow between coevolutionary hotspots and coldspots or between environments where interactions differ in outcome can tip the balance among potential trajectories and equilibrium states of local interactions between parasites and hosts (Gomulkiewicz et al. 2000; Nuismer, Thompson, and Gomulkiewicz 2000, 2003).

Looked at another way, the current state of virulence theory is one version of our changing views of the coevolutionary process. The old view of interactions between parasites and hosts concentrated on adaptation in one of the interacting species within one population, or it treated the overall interaction as a large panmictic unit. The current view is becoming more geographic in perspective. It suggests that parasite virulence and host defenses will evolve in different ways among populations under different ecological conditions. Endemic and epidemic cycles of disease create constantly changing geographic structures in parasites and hosts. No epidemic affects all populations equally. Hence, levels of parasite virulence and host defense may fluctuate within and among populations over time, creating geographic mosaics in polymorphisms that develop in tandem with the overall frequency-dependent selection that drives much of coevolution between parasites and hosts. General patterns in the evolution of virulence and attenuated antagonism will therefore ultimately be interpretable only when we understand the overall geographic selection mosaic on virulence and host defense.

Coevolutionary Selection on Symbiotic Mutualisms

At the evolutionary moment when a reciprocal fitness benefit occurs in a local interaction between a symbiont and host, the potential coevolutionary dynamics change fundamentally from those found in parasitic interactions. Rather than fluctuating polymorphisms, coevolutionary alternation, or escalation of defenses, natural selection favors complementarity of traits. Such complementarity does not in any way imply a reduction of the inherent Darwinian selfishness of these interactions.

A hypothetical sequence of events could progress in the following way, by piecing together theoretical and empirical results from the past decade. Some individuals in a host population happen to acquire a symbiont species that makes, say, an amino acid in short supply in the host's diet, and some genotypes of that symbiont species survive and reproduce better in those hosts than in alternative hosts or as free-living individuals. The initial fitness benefits vary greatly among partners, creating a distribution of ecological outcomes ranging from antagonism to mutualism among various combinations of host and symbiont genotypes (see fig. 3.3). Coevolution could shift the interaction back toward increased antagonism, if the fitness costs of the interaction were too great for most combinations of host and symbiont genotypes. In this environment, however, the fitness gains for individuals of both species outweigh the costs, allowing establishment of the mutualism and driving further coevolution by favoring host and symbiont genotypes best able to extract the greatest fitness benefits from the interaction at the lowest costs.

During this process, natural selection favors hosts that preferentially interact with the most beneficial genotypes of the symbiont. In this hypothetical example, it occurs through selection favoring those hosts that pass on the beneficial symbionts directly to their offspring through vertical transmission. Alternatively, the partners could acquire complex genetic signaling mechanisms that allow both partners to differentiate among genotypes of the other species. As the mutualism continues to coevolve, selection favors individuals lacking the ability to accomplish certain functions themselves, because keeping those functions imposes a fitness cost. The result is fixation of a core set of complementary traits in the interacting populations. Ongoing selection, however, continues to favor individuals with new mutations that produce even greater fitness gains at lower costs.

Alternative scenarios are possible. This hypothetical example simply illustrates that the evolution of mutualistic symbioses can result from selection acting on the distribution of ecological outcomes by favoring highly mutualistic combinations of host and symbiont genotypes. This scenario also illustrates how mutualistic coevolution develops from the ongoing biased fixation of complementary traits that confer the greatest fitness gain for the lowest cost in each species. Throughout this process, asymmetries in fitness benefits may add further selective complexity to mutualisms (Bever 1999), and the combination of traits conferring the greatest mutualistic benefits may vary among habitats.

In some cases, selection for mutualistic complementarity may eventually develop into a complementary network of symbionts within hosts. Hosts may acquire multiple symbionts that accomplish different functions without competing with one another. The process is analogous to acquiring new beneficial genes at different loci. At the same time, selection on symbionts favors elimination of competing symbionts, thereby reinforcing selection for complementarity. Overall, then, complementarity in symbiotic mutualisms has two components: between hosts and symbionts, and among symbionts.

More precisely, the working hypothesis on coevolutionary selection in symbiotic mutualisms can be summarized in the following way:

The hypothesis of coevolved monocultures and complementary symbionts.

Symbiotic mutualisms have evolved from the subset of symbiotic interactions in which the interacting species have complementary traits that confer reciprocal fitness benefits and in which hosts are able to restrict the interaction locally to single symbiont genotypes or complementary sets of noncompeting symbionts.

The hypothesis follows from the arguments earlier in this chapter, which suggest that infection of hosts by competing symbionts may favor the evolution of aggressive parasitic genotypes that have detrimental effects on host fitness. Competing symbiont genotypes or species create a coevolutionary environment that may make it impossible for natural selection to favor directional selection and fixation of mutualistic traits in the host and the most beneficial symbiont (Frank 1996a).

Some studies have shown highly restricted genotypic diversity of symbionts within individual hosts or host colonies of some species (Herre et al. 1999; Bot, Rehner, and Boomsma 2001; Tsuchida et al. 2002) and highly stable and conserved genotypes in some obligate symbionts (Tamas et al. 2002). If restriction of symbiont genetic diversity is a central focus of selection on symbiotic mutualisms, then these mutualisms should exhibit a common set of attributes:

- single genotypes of a symbiont species within host individuals (i.e., genotypic monocultures of symbiont species)
- often vertical rather than horizontal transmission of symbionts
- complex reciprocal signaling during formation of symbioses between hosts and horizontally transmitted symbionts
- local coevolution toward fixation of coevolving traits

These attributes should, in turn, favor an overall predictable structure to coevolved symbiotic mutualisms:

- inherently complementary, rather than competing, networks of symbiont species within host individuals (i.e., complementary cultures)

- within each host population, a small number of vertically transmitted symbiont species that compete minimally with each other and complement each other, and an even smaller number of horizontally transmitted mutualistic symbionts
- strong phylogenetic conservatism in the hosts used by vertically transmitted symbiont lineages
- a geographic mosaic of mutualistic symbionts harbored by each host species
- a geographic mosaic of coevolving host and symbiont traits

The genetic diversity of mutualistic symbionts within individual hosts has been analyzed for few species, although that is changing quickly with the increased availability of molecular markers. Current results suggest that host species with small body size and short life histories commonly harbor single genotypes of mutualistic symbiont species. Moreover, these hosts often harbor only one symbiont species that functions in a particular way. That is, individual hosts appear rarely to harbor multiple, similar mutualistic symbionts that could compete with one another, although some major exceptions to this observation in large organisms will be considered later in this chapter.

The best-studied example of what appears to be strong selection for restricted genetic diversity in symbionts is that of coevolution of attine ants and their fungal gardens. Selection on this interaction appears to be focused on maintaining single-genotype monocultures of a fungal lineage that has been associated with these ants for more than fifty million years (Mueller et al. 2001). The attines are a monophyletic group of more than two hundred species that obligately cultivate fungi in their nests as an agricultural symbiosis. The ants grow the fungus for food and in return provide the fungi with a mostly competition-free and parasitism-free environment for growth (Mueller, Rehner, and Schultz 1998). Different attines cultivate different fungi, but there is broad phylogenetic congruence in the diversification of the two lineages (Chapela et al. 1994; Currie et al. 2003). Basal attines cultivate a group of closely related fungi in the family Lepiotaceae, although some of these attines have domesticated some other fungi within another family. The more derived attines, which include the well-known leaf-cutter ants in the genera *Acromyrmex* and *Atta*, cultivate fungi within the Lepiotaceae that are derived from the fungi used by more basal attines. The fungi cultivated by derived attines have unique nutrient-rich hyphal swellings that are easily harvested by the ants (Mueller et al. 2001).

Reciprocal transplant experiments have shown that *Acromyrmex* ants can differentiate between fungi grown in their own nest and those grown in other nests, even though the fungi are very closely related, as analysis of molecular markers indicates (Bot, Rehner, and Boomsma 2001). Ants kill genetically different novel fungal cultivars if the fungi are transplanted into a nest. They therefore actively maintain a monoculture of their own fungal cultivar. These monocultures are maintained within ant lineages, because each attine ant queen carries a fungal fragment in a small cavity beneath the esophagus (infrabuccal pocket) when she leaves the colony in which she developed and starts a new one (Hölldobler and Wilson 1990; Wirth et al. 2003).

Cultivation of a single fungal cultivar creates problems similar to those caused by cultivation of single cultivars in human agriculture. It makes the nest vulnerable to invasion by competitors and catastrophic loss caused by highly adapted parasites, and the ants must therefore spend considerable time tending these gardens (Currie 2001a). Among the serious potential problems is attack by the microfungus *Escovopsis,* which has evolved as a specialized parasite of leaf-cutter fungus gardens (Currie, Mueller, and Malloch 1999; Currie, Bot, and Boomsma 2003). *Acromyrmex octospinosus* ants combat this parasite by enlisting yet another symbiont, a filamentous bacterium in the Actinomycetes (Currie et al. 1999; Currie, Bot, and Boomsma 2003). Reproductive females carry this bacterium within them when they start a new colony (Poulsen et al. 2003). Hence, both the mutualistic fungus and the mutualistic bacterium are vertically transferred as a complementary unit. Major ant workers acquire the bacteria on their cuticle a few days after eclosion, and the bacteria have strong antibiotic effects on *Escovopsis* (Currie, Bot, and Boomsma 2003). Infection rates and virulence of *Escovopsis* are known to vary among attine species (Currie 2001b), and Poulsen et al. (2003) have predicted that future work will show strong geographic mosaics in coevolution across populations of the three mutualists (ants, the mutualistic fungi, and Actinomycetes bacteria) and the parasitic *Escovopsis.*

Occasional horizontal transmission may occur among ant colonies, especially among lower attines but also in more derived taxa. Two closely related species of lower attines in the genus *Cyphomyrmex* are known to exchange cultivars (Green, Mueller, and Adams 2002), and the broader phylogenetic history of attines and their fungal cultivars shows evidence of past horizontal transmission (Mueller, Rehner, and Schultz 1998; Bot, Rehner, and Boomsma 2001). In addition, lower attines appear to have repeatedly domesticated novel cultivars. Although the higher attines (*Acromyrmex* and *Atta*) show strong fidelity to the asexual lineages with which they have coevolved, horizontal transmission may sometimes occur. A nest can sometimes lose its fungal cultivar through attack by *Escovopsis,* and the ants under those conditions may usurp a neighboring fungal colony (Adams et al. 2001). Horizontal transmission could also occur if a colony is founded by multiple queens (Bekkevold, Frydenberg, and Boomsma 1999). These occasional horizontal transfers may be possible because the ability of the ants to distinguish their local fungus cultivar from alien fungi may be due to chemical recognition of their cultivar rather than to genetic differences among ant nests in preference for particular cultivars. After ants are force-fed a novel fungus cultivar for about ten days, they no longer kill these fungi (Bot, Rehner, and Boomsma 2001). The overall structure of these mutualisms, however, suggests that attine ant colonies are structured to maintain monocultures of any fungus they cultivate.

Controlling Symbionts during Vertical and Horizontal Transmission

If symbiotic mutualism generally imposes intense selection on host populations to minimize the genetic diversity of a symbiont species and maximize the complementarity of co-occurring symbiont species, then the most efficient selective mechanism is to favor hosts that transmit the favored symbiont genotype or complementary genotypes directly to their offspring. Not only can vertical transmission minimize symbiont genetic diversity, it can also keep intact particular combinations of symbiont and maternal host genotypes. Because the benefits and costs of an interaction are likely to differ among host and symbiont genotypes, vertical transmission provides a powerful evolutionary mechanism for maintenance of mutualistic interactions.

Many species have evolved elaborate ways of transmitting mutualistic symbionts directly to offspring, although the genotypic diversity of those symbionts has been analyzed for few species. Some true bugs (Heteroptera), for instance, coat their eggshells with a bacteria-rich excretion that is eaten by their offspring after they hatch, whereas others deposit symbiont capsules that are eaten by hatchlings (fig. 12.2). Nymphs of the Japanese plataspid stinkbug that are prevented from ingesting symbiont capsules experience retarded growth and development and abnormal coloration (Fukatsu and Hosokawa 2002). The symbiont capsules contain mostly a single γ-proteobacteria species, which is harbored in the adult insects in a specialized area of the midgut. The active behaviors of some hosts in vertical transmission of

Fig. 12.2   Symbiont capsules (dark clumps) deposited together with eggs by an ovipositing female of the Japanese plataspid stinkbug, *Megacopta punctatissima*. Scale bar: 0.5 mm. From Fukatsu and Hosokawa 2002.

symbionts therefore provide a mechanism of direct control of hosts over their symbionts.

In contrast to vertical transmission, the spread of symbionts by infection makes it more difficult for hosts to control symbiont genotypes. Some well-studied interactions suggest that hosts can solve this coevolutionary problem, but it requires complex signal exchange between potential partners during formation of the interaction. Perhaps the best-studied example is the mutualism between legumes and rhizobial bacteria, which involves a multistep process of genetic handshaking. Rhizobia are a group of genera of soil bacteria, including *Azorhizobium, Bradyrhizobium, Mesorhizobium, Rhizobium,* and *Sinorhizobium,* among others. These bacteria are capable of infecting the roots of legumes and causing formation of characteristic root nodules. In fact, most of the earth's natural and agricultural terrestrial communities rely upon the nitrogen fixation produced by rhizobial interactions (Sessitsch et al. 2002). The bacteria multiply within the nodules and differentiate into a bacteroid form that is capable of fixing atmospheric nitrogen, thereby making it available to the plant host. The bacteria, in turn, use the products of photosynthesis provided by the plant to obtain the energy they need for biosynthetic processes. Molecular phylogenetic studies suggest that rhizobial lineages have not simply coevolved with different legume lineages. Lateral gene transfer among rhizobial taxa appears to have occurred repeatedly, thereby creating novel symbioses as these interactions have diversified over hundreds of millions of years (Turner and Young 2000; Sessitsch et al. 2002; Qian, Kwon, and Parker 2003).

Formation of nitrogen-fixing root nodules involves a series of reciprocal signals between the plant and the rhizobia (Miklashevichs et al. 2001), which have sometimes been compared to a series of locks and keys (Broughton, Jabbouri, and Perret 2000). Each signal elicits characteristic developmental changes in both the plant and the rhizobia. Plant roots secrete specific flavonoids that are recognized by the rhizobial nod protein, leading to activation of rhizobial *nod* genes and synthesis of nod factors. When these nod factors are recognized by two receptor genes on a root hair (Radutoiu et al. 2003), the tip of the root hair curls around the bacteria. Mutation in either of these receptor genes can prevent infection by rhizobia. The cradled rhizobia invade the plant through an infection thread constructed by the plant, possibly through interaction with rhizobial genes (Broughton, Jabbouri, and Perret 2000). Simultaneously, the plant begins to develop a nodule primordium in the root cortex below the base of the root hair. The infection thread grows toward the primordium, and the bacteria are released into the cytoplasm of the host cells (fig. 12.3).

Nodulation in the plant involves multiple nodulin genes that are expressed at different times during development of the nodule. Failure in reciprocal signaling at any of these stages can prevent successful nodule formation. The complexity of the process, coupled with lateral gene transfer of plasmids among rhizobial taxa, creates the opportunity for a range of specialization in hosts and rhizobia. Some combinations of keys may open multiple locks.

The lock-and-key metaphor can be useful in envisioning how formation of mutualisms could evolve between hosts and horizontally transmitted symbionts, but it can also be misleading, because it can conjure up an artificially restrictive mental image of the genetics of this coevolutionary process. Genetic locks and keys function in different ways depending upon whether the interaction is determined by gene-for-gene relationships, matching genes, multiple additive genes, multiple genes exhibiting a range of dominance and epistatic interactions, or lateral transmission of plasmids. The interactions between plants and rhizobia involve a wide range of genetic architectures,

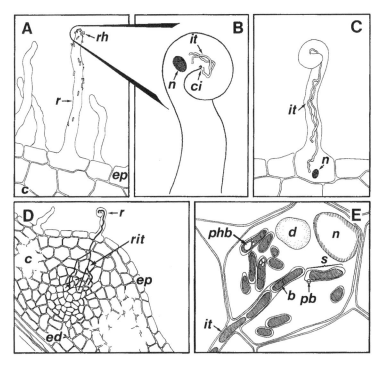

Fig. 12.3   The multistep process of establishment of an interaction between *Rhizobia* and a legume root. (A) *Rhizobia* (*rh*) attach to a root hair (*r*) growing from the epidermis (*ep*) above the root cortex (*c*). (B) Nod factors induce root hair curling, allowing the bacteria to penetrate the root hair at a center of infection (*ci*) and travel along its length through growing infection threads (*it*) preceded by the plant nucleus (*n*). (C) At least one infection thread carrying the bacteria eventually reaches the base of the cell. (D) The infection thread ramifies (*rit*) near the developing nodule primordium, which is composed of cortical cells located between the epidermis and the endodermis (*ed*). (E) Bacteroids (*b*) are released from the infection thread and form symbiosomes (*s*) within the nodule cells. In addition to harboring a nucleus and digestive vacuole (*d*), these symbiosomes accumulate granules of poly-β-hydroxybutarate (*phb*) within the peribacteroid membrane (*pb*). From Perret, Staehelin, and Broughton 2000.

with different genes and gene combinations interacting in different ways during nodule formation. This complex structure appears to have resulted from further modification of more ancestral symbioses between plants and mycorrhizal fungi (Provorov, Borisov, and Tikhonovich 2002; Radutoiu et al. 2003). The probability of nodule formation between particular host and rhizobial genotypes can also differ among environments. The combination of genetic architectures, lateral transfer, and genotype-by-environment interactions makes varying degrees of reciprocal specificity almost inevitable in interactions between plants and rhizobia as they diversify across landscapes.

Molecular and ecological studies among *Bradyrhizobium* isolates in four widely separated geographic regions have highlighted the complex network structure of specificity that can result in these mutualisms. In California, a third of a sample of legumes with multiple nodules on their roots harbored more than one molecularly identifiable *Bradyrhizobium* strain (Simms and Taylor 2002). In eastern North America, the legumes *Apios americana, Desmodium glutinosum,* and *Amphicarpaea bracteata* share a set of identical or nearly identical *Bradyrhizobium* multilocus genotypes, but among the three plant species these shared genotypes vary in their ability to function as effective symbionts (Parker 1999b). The plant species also harbor some genotypes that are not shared among the species. A similar mix of many shared and few unique *Bradyrhizobium* genotypes occurs among legume species on Barro Colorado Island in Panama (Parker 2003). In Australia, *Bradyrhizobium* strains are able to form successful symbioses with multiple *Acacia* species, but there are no simple geographic or phylogenetic patterns. Some, but not all, *Bradyrhizobium* genotypes (that is, strains or isolates) produce higher plant growth on their own host plants than on other *Acacia* species that have been tested (Murray, Thrall, and Woods 2001), and some, but not all, *Acacia* species will produce nodules with a wide range of *Bradyrhizobium* genotypes (Thrall, Burdon, and Woods 2000). A test of sixty-seven populations spanning twenty-two *Acacia* species found that the average local acacia-rhizobia combination was about 70 percent as effective as the best local combination in stimulating plant growth as measured by dry matter production. Many combinations were much poorer, and the worst combinations produced plants after three to four months that were less than one-tenth the size of the best combinations (Burdon et al. 1999).

Mutualism between rhizobia and plants appears to be maintained by amino acid complementarity within the biosynthetic pathway created by the interaction (Lodwig et al. 2003). Each species controls a vital amino acid needed for the overall process and therefore has some control over the other species. This complementarity may prevent either the host or the rhizobia from exploiting the interaction at the expense of other species. For example, if air is replaced with a nitrogen-free atmosphere, thereby preventing rhizobia from fixing nitrogen (i.e., the equivalent of cheating on the mutualism), rhizobial reproductive success is itself reduced by about half (Kiers et al. 2003). Such ability to control exploitation by the other species through mutual dependence has become a part of most mathematical and empirical formulations of how symbiotic mutualisms are maintained over the long term (Pellmyr and Huth 1994; Herre et al. 1999; Denison 2000; West et al. 2002; Sprent 2003).

The genetically complex interactions between legumes and rhizobia suggest that mutualisms with horizontally transmitted microbial symbionts often require two layers of checks and balances on the participants. One is multistage reciprocal signaling during formation of the interaction, permitting invasion only by potentially mutualistic symbionts. The other is integration of biosynthetic pathways that prevents either species from gaining the upper hand in the interaction.

Some symbiotic mutualisms with animals, however, may instead use true lock-and-key mechanisms that rely upon the active behavior of animals. The studies of *Leonardoxa* ants and their host plants provide the clearest potential example of how a symbiotic mutualism between plants and animals coevolves in this way as a geographic mosaic (McKey 1984; Chenuil and McKey 1996; Brouat et al. 2001). Three of the four subspecies of the African rainforest tree *Leonardoxa africana* harbor ants in structures called domatia. One of these subspecies is associated with a specialist ant in the genus *Aphomomyrmex,* another with a specialist ant in the genus *Petalomyrmex,* and the third with a diverse group of ant species, with one ant species monopolizing each tree. Specificity for ants within *L. africana* subspecies is maintained by a morphological plant structure, the prostoma, that constrains the shape of the opening into the domatium. The mutualistic ants chew through the prostoma while remaining within its morphological boundaries. These host-specific ants have a morphological shape that matches the shape of the prostoma, creating a lock-and-key mechanism that maintains reciprocal specificity (fig. 12.4).

Overall, most mutualisms with horizontally transmitted symbionts probably originate through multistep genetic handshaking between a host population and one or a small number of mutualistic symbiont genotypes. As populations diversify across landscapes, the locks and keys become modified, often creating geographic mosaics. Those mosaics, however, may sometimes retain a high degree of phylogenetic conservatism. The evolution of mutualism between termites in the Macrotermitidae and the fungi they garden provide some of the best, although still preliminary, data. Molecular phylo-

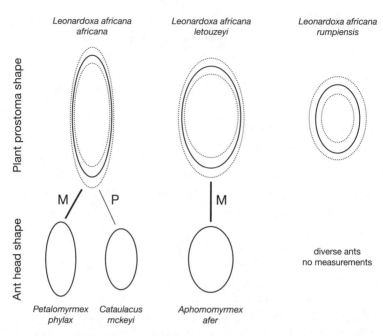

Fig. 12.4   Means (solid ellipses) and extremes (dotted ellipses) of the shapes of prostomata of three subspecies of *Leonardoxa africana* inhabited by ants. Also shown are the mean cross-sectional head shapes of the associated ants for the two *Leonardoxa* subspecies specialized to one ant species and the head shape of the ant (*Cataulacus mckeyi*) that parasitizes the mutualism with *L. africana africana*. M = mutualistic and P = parasitic relationship between that particular combination of plant and ant populations. After Brouat et al. 2001.

genetic analyses suggest that this mutualism arose from a single African origin (Aanen et al. 2002). Although host switching has occurred repeatedly, the descendant termite and fungal taxa show a high level of phylogenetic congruence. As the interaction has further diversified, vertical transmission of the fungi has evolved twice in two highly derived lineages. The extent to which termites are capable of directly controlling fungal genotypes is unknown, but the overall phylogenetic congruence of these termites and their fungi suggest that either direct or indirect control occurs. Aanen et al. (2002) have hypothesized that the evolution of vertical transmission in termites has been constrained by the fact that males and females together found colonies, and both sexes may carry fungi with them. In contrast, vertical transmission of a single fungal genotype is easier in ants, because males do not survive beyond mating and only females found colonies.

Limits to the Hypothesis of Coevolved Monocultures and Complementary Symbionts: Large and Modular Hosts

The hypothesis that mutualism favors the maintenance of single genotypes of symbiont species and complementary symbiont species through vertical transmission or complex genetic handshaking will undoubtedly undergo much revision as molecular analyses provide data on more populations and species. Important qualifications are already evident. Large or modular organisms are able to maintain complex microbial communities in ways that are still poorly understood. These complex assemblages go beyond anything that can be explained by simple complementarity among a small number of symbiont species within host individuals. Some trees harbor a diverse array of endophytic fungi (Arnold et al. 2003), and molecular analyses of vertebrate gut symbionts have suggested that these microbial assemblages are even more genetically diverse than previously thought (Simpson et al. 2002). Colonies of termites, which function collectively as large superorganisms, harbor complex assemblages of gut symbionts. Analysis of 16s RNA sequences have suggested that individual termite species harbor multiple *Treponema* spirochetes as part of their gut microbial assemblage and that, together, these microbial assemblages may contribute to a combination of the termites' carbon, nitrogen, and energy requirements (Breznak 2002).

A reasonable hypothesis, then, is that large or modular organisms may be able to control their symbiont assemblages by partitioning the symbionts into subassemblages.

The hypothesis of symbiont partitioning.

Large and modular organisms maintain complex assemblages of symbiotic mutualists by partitioning them among body parts, maintaining "controller" symbionts, selectively eliminating modules, or altering assemblage structure at different ages and in different environments.

Arguing simply that large organisms or colonies create multiple potential niches for symbionts solves only part of the problem. The real challenge is to understand how these hosts are able to maintain assemblages of mutualistic symbionts within individuals, within populations, and among populations over thousands of generations. One possibility is that natural selection on hosts favors maintenance of a "controller" symbiont genotype that is capable of preventing spread of nonmutualistic symbionts in other species. Another possibility is that coevolution toward a network structure among symbionts may sometimes minimize the potential for the evolution of cheating. Yet another possibility for modular organisms, suggested by Hoeksema and Kummel (2003) for mycorrhizal associations, is that host individuals could eliminate modules colonized by nonmutualistic genotypes. No current mathematical or empirical studies deal directly with these coevolutionary possibilities.

Some hosts may change symbiont genotypes at different developmental stages, allowing coevolution with multiple similar symbionts. Many organisms interact with different species at different ages, sizes, or life-history stages (Werner and Gilliam 1984), and this ontogeny of interactions is an important way by which organisms can coevolve with multiple species (Thompson 1994). Hosts may also change symbionts under different environmental conditions. Reef-building corals harbor endosymbiotic dinoflagellates in a single genus, *Symbiodinium*, but, in some coral species, individuals living under different light conditions within a local population differ in the *Symbiodinium* genotype they harbor (Rowan and Knowlton 1995; van Oppen et al. 2001). Some corals also harbor multiple *Symbiodinium* genotypes, with each genotype restricted to a particular photic position on the coral (Rowan et al. 1997).

The full extent of genotypic diversity in *Symbiodinium* and the ability of corals to choose repeatedly among these genotypes under different ecological conditions are poorly known, but molecular analyses indicate that this symbiotic genus harbors more genetic diversity than is indicated by the current morphologically based descriptions of species. There are between four and ten recognized *Symbiodinium* species, but molecular analyses have now identified over one hundred molecular types, distributed among approximately four clades (Baker 2003). It is not yet known how much of this molecular variation reflects corresponding functional and ecological variation, but there is clearly more potential for local and geographic differentiation in these interactions than previously suspected.

The regional genetic structure of the interaction between the Caribbean gorgonian coral *Pseudopterogorgia elisabethae* and its *Symbiodinium* sp. clade B symbionts suggests the kind of symbiont assemblage structure that may be found in other interactions between corals and *Symbiodinium* (Santos et al. 2003). Microsatellite analysis of twelve coral populations along an approximately 450-kilometer transect identified twenty-three unique genotypes among 575 individuals. Most colonies harbored only a single *Symbiodinium* sp. clade B genotype, but about 4 percent of coral colonies harbored two genotypes. Nine of the twelve populations harbored a genotype that was either unique or uncommon in the other populations, suggesting a strong geographic structure in this interaction.

The unfolding complexity of genetic structure in the interactions between corals and *Symbiodinium* highlights the potential for ongoing coevolutionary dynamics in symbiotic mutualisms. Such studies are surely the tip of the iceberg in this almost completely unexplored area of research. Many marine taxa other than corals harbor symbionts. Anemones, jellyfish, and giant clams all include species with symbiotic algae, and juveniles of many marine taxa rapidly acquire symbionts early in ontogeny (Baker 2003). These taxa, and similar symbioses in terrestrial environments, offer tremendous opportunity for future studies on the ecological and coevolutionary dynamics of mutualistic symbioses.

Conclusions

Our understanding of coevolving symbiotic mutualisms has increased enormously in recent years. The models and the molecular and ecological data have suggested several hypotheses on the structure of coevolutionary selection and the patterns that result. Selection appears to favor mechanisms in hosts that restrict symbiont genotypic diversity to a single monoculture or a small network of complementary symbiont species. At the same time, selection on vertically inherited symbionts should also act to minimize the presence of competing symbionts, while retaining the potential for horizontal transmission. As host and symbiont genomes become increasingly complementary and integrated, horizontal transmission may eventually become impossible in some mutualisms, making selection for parasitism in symbionts less likely. Among symbionts that retain horizontal transmission, maintenance of mutualism appears to require complex genetic handshaking between the participants.

These arguments, however, seem insufficient to explain the complex assemblages of symbionts found in some large or modular organisms. Hosts in these interactions may possess additional mechanisms for partitioning of symbionts among body parts and at different ages and sizes. As techniques in molecular ecology and microbiology continue to improve, and as these studies become part of the mainstream of coevolutionary research, it should become possible to develop more specific hypotheses on the structure of coevolutionary selection in these interactions.

13 Mutualists II

The Geographic Mosaic of Mutualistic Symbioses

Coevolutionary complementarity and symbiont partitioning should favor the development of geographic mosaics in interactions between hosts and mutualistic symbionts. Until recently, it has been difficult to analyze these mosaics, but new molecular tools and an increasing number of experiments designed to test for local adaptation are helping to uncover previously unsuspected geographic complexity in these interactions. Beginning with three predictions that follow from coevolutionary complementarity and symbiont partitioning, this chapter discusses our current understanding of these coevolving mosaics and the various approaches that are developing to assess these mosaics.

Geographic Mosaics with Horizontally Transmitted Mutualistic Symbionts

Prediction: Geographic mosaics in coevolved traits and outcomes should occur in both ancient and more recently derived interactions with horizontally transmitted symbionts, because natural selection can act in each generation on newly formed genotypic combinations of hosts and symbionts. The same should apply to interactions with symbionts capable of both vertical and horizontal transmission, because these interactions retain the potential for natural selection to test new genotypic combinations of hosts and symbionts.

Some of the best evidence for geographic mosaics in interactions with horizontally transmitted symbionts comes from studies of legumes and rhizobia. For most Australian acacias that have been tested, the mean effectiveness of any isolate chosen at random is 15–20 percent lower than that of the most effective isolate (Burdon et al. 1999). This means that there is a strong potential for natural selection to favor particular combinations of host and

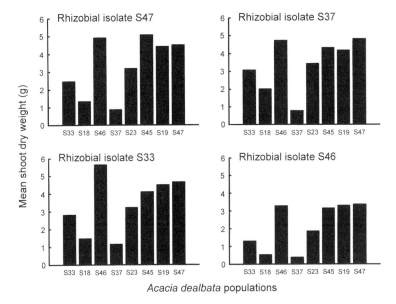

Fig. 13.1 Differences among eight populations in mean early growth of *Acacia dealbata* from southeastern and southern Australia when inoculated with different *Bradyrhizobium* isolates. The four isolates came from acacia populations with the corresponding number (e.g., acacia population S33 and isolate population S33). After Burdon et al. 1999.

symbiont genotypes. More direct evidence of a selection mosaic comes from experiments in which seedlings from different populations of Australian acacias have been inoculated with the same rhizobial isolate (Burdon et al. 1999). Populations of *Acacia dealbata* differ significantly in early growth when inoculated with the same isolate (fig. 13.1). These differences are likely to have a genetic basis, because half-sib genetic families of *A. dealbata* differ in growth when inoculated with the same isolate (Burdon et al. 1999). Each acacia species must face continually shifting geographic differences in selection, resulting from its own genetic structure, the local mix of rhizobia genotypes, and the mix of sympatric legumes that contribute to the local relative abundance of rhizobial genotypes.

Further evidence of a geographic mosaic in interactions between legumes and rhizobia comes from studies of the North America annual legume *Amphicarpaea bracteata*, which differs geographically in the degree to which individuals are adapted to particular genotypes of *Bradyrhizobium* sp. (Parker 1996; Parker and Wilkinson 1997; Parker 1999a). Plant genotypes differ in their breadth of compatibility with two known *Bradyrhizobium* genotypes.

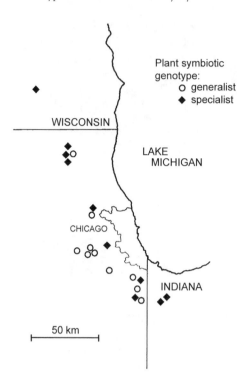

Fig. 13.2   Geographic mosaic of symbiotic genotypes of the legume *Amphicarpaea bracteata* in the midwestern United States. The specialist genotype is compatible with one *Bradyrhizobium* genotype, whereas the generalist genotype is compatible with two *Bradyrhizobium* genotypes. After Parker 1999a.

Plants tested in the midwestern United States are compatible with one bacterial genotype (specialists), but some populations are compatible with two bacterial genotypes (generalists). Compatibility with the second bacterial genotype is determined by alleles at a single locus (Parker and Wilkinson 1997). *Amphicarpaea* populations form a geographic mosaic in which local populations are fixed for either specialists or generalists (fig. 13.2). In some cases, the boundaries between populations fixed for specialist genotypes are within a kilometer of populations fixed for the generalist genotype. A similar geographic mosaic of these same genotypes occurs about a thousand kilometers away in New York. Cross-inoculation experiments across this thousand-kilometer region indicates a yet broader geographic mosaic in which combinations of plants and bacterial genotypes native to the same habitat exhibit growth rates 9–48 percent higher than nonnative combinations (Wilkinson, Spoerke, and Parker 1996).

Simple genetic models of symbiotic mutualism can produce sharp geo-

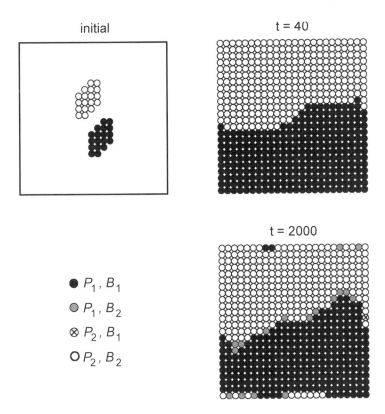

Fig. 13.3   Development of sharp geographic boundaries between genotypic combinations of two mutualists. In this simulation P_1 and P_2 represent the densities of host genotypes capable of forming associations with one symbiont genotype P_1 or P_2, and B_1 and B_2 represent the effect on plant fitness caused by interaction with bacterial mutualists. The plant species compete in a manner described by traditional Lotka-Volterra competition models. In this particular example, different regions were initially colonized solely by one or the other combination of plant and bacterial genotypes. The stable geographic mosaic breaks down in this model under higher levels of gene flow and large overall fitness differences favoring particular genotypic combinations across all geographic regions. After Parker 1999a.

graphic boundaries similar to that found between *Amphicarpaea* and *Bradyrhizobium* populations (fig. 13.3). If selection favors the numerically or proportionally dominant local genotype, and if local regions are initially colonized by only one host genotype, then selection can favor local fixation through positive frequency and density dependence (Parker 1999a). Other models that exhibit relatively strong coevolutionary selection and low levels of gene flow can produce similarly sharp boundaries among genotypes (Nuismer, Thompson, and Gomulkiewicz 1999; Gomulkiewicz et al. 2000; Nuismer, Thompson, and Gomulkiewicz 2000).

Moving up to a worldwide scale, the interactions between *Bradyrhizobium* and legumes show even more complex mosaics. Phylogeographic analysis of *Bradyrhizobium* from Asia, Australia, Central America, and North America shows significant incongruence between the 16S rRNA gene and the *nifD* gene, suggesting lateral transfer of genes during evolution and geographic diversification of this genus (Parker et al. 2002). Analysis of the 16S rRNA gene suggests that two lineages spread across these continents, with both lineages present in almost all geographic areas. In contrast, the *nifD* gene shows clear differentiation among continents. The overall phylogenetic analysis suggests initial worldwide colonization by multiple 16s rRNA lineages followed by lateral gene transfer within regions, leading to greater regional homogeneity for that gene. The *nifD* gene resides close to the *nod* gene cluster within *Bradyrhizobia* and may therefore potentially be transmitted as part of that cluster through lateral gene transfer (Parker et al. 2002).

Evidence for geographic differences in selection is accumulating from studies of other horizontally transmitted mutualistic symbioses, although much of it is still indirect. For example, the highly diverse interactions between *Crematogaster* ants and the *Macaranga* plants they inhabit show complex geographic patterns of morphological and genetic diversification and association (Feldhaar et al. 2003). The interactions between *Acacia, Cecropia,* and *Cordia* plants and their ant inhabitants all show strong potential for geographic differences in species composition, as individual plant species draw throughout their geographic ranges on subsets of competing ant species specific to particular habitats or environmental conditions (Yu and Davidson 1997; Yu, Wilson, and Pierce 2001; Stanton, Palmer, and Young 2002). Discussion sections of published papers on these complex relationships have increasingly emphasized the need for a better understanding of the geographic structure of these relationships (e.g., Brouat et al. 2001; Davies et al. 2001).

Reduced Potential for Geographic Mosaics in Ancient Vertically Transmitted Symbionts

Prediction: Ancient interactions with a vertically transmitted symbiont species should show less of a geographic mosaic in coevolved traits and outcomes than more recent symbioses.

This prediction follows from results suggesting small effective population sizes, reduced genome sizes, and ongoing random genetic drift in ancient vertically transmitted symbionts (Funk, Wernegreen, and Moran 2001; Abbott and Moran 2002; Wernegreen 2002), leading potentially to reduced genetic variation on which selection can act. There is a potential caveat to this prediction, which only future research can clarify. If a vertically transmitted symbiosis is established from a single infection, as molecular analyses have suggested for several interactions (Chen, Li, and Aksoy 1999), then interacting species may initially show little geographic variation in traits and outcomes, increase in geographic variation as selection and drift act on the symbiont genome at intermediate time scales, and eventually become more geographically homogeneous as the symbiont genome is reduced to the minimum number of genes that form the basis for the symbiosis.

The interactions between aphids and their *Buchnera* symbionts show a much clearer pattern of codivergence than those between legumes and rhizobia, resulting from obligate vertical transmission and tight genomic integration over millions of years. These well-studied interactions provide insight into the rare set of conditions under which coevolution results in strict parallel geographic and phylogenetic divergence in hosts and symbionts. *Buchnera* are γ-proteobacteria that reside within special organs in an aphid's body cavity and synthesize amino acids that are limiting in the plant tissues fed upon by their aphid hosts. The relationship is reciprocally obligate, and the endosymbionts are maternally inherited through infection of aphid eggs or embryos (Baumann et al. 1995; Douglas 1998).

Buchnera is one of a group of obligate nutritional endosymbioses found in a diverse array of insects, including some cockroaches, bedbugs, lice, tsetse flies, and some beetles and ants (Douglas 1998; Wernegreen 2002). These insects tend to feed on nutritionally unbalanced foods, such as wood, vertebrate blood, or phloem sap. In many of the symbioses, the symbionts are proteobacteria and are related to other bacteria that cause pathogenesis in other organisms (fig. 13.4). Collectively, these symbionts, both mutualistic and parasitic, belong to a set of organisms whose ancestors appear to have had a set of genetic mechanisms permitting symbiotic survival and replication (Goebel and Gross 2001). The obligate mutualistic symbionts of these insects and other invertebrates generally show evidence of strong parallel speciation with their hosts resulting from a single ancient infection (Bandi et al. 1995; Bandi et al. 1998; Peek et al. 1998; Chen, Li, and Aksoy 1999; Clark et al. 2001).

The interactions between *Buchnera* and aphids are ancient and appear to

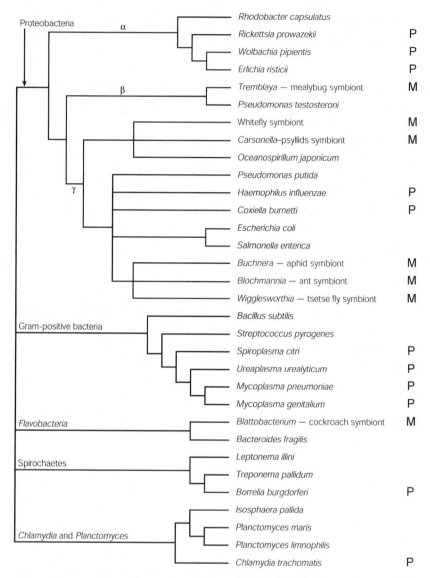

Fig. 13.4   Phylogenetic relationships among some known symbiotic Eubacteria based upon 16S rRNA sequences. P indicates pathogenic species, whereas M indicates mutualistic symbionts of insects. α, β, and γ are the commonly used designations for the three major subdivisions within the Proteobacteria. After Wernegreen 2002.

have originated from a single infection event 84–164 million years ago (Moran et al. 1993). Since then, *Buchnera* has undergone a major decrease in genome size, and estimates of the numbers of coding genes are among the lowest reported for bacteria (Gil et al. 2002). The retained genes include those involved in amino acid synthesis and form the basis for the mutualism (Tamas et al. 2002). Sequence divergence and genome reduction continues among some *Buchnera* genomes (Gil et al. 2002), but these genomes are now so restricted in size that analysis of two completely sequenced *Buchnera* genomes that separated 50 million years ago shows no subsequent chromosome rearrangements or gene acquisitions (Tamas et al. 2002). *Buchnera* are now, in effect, obligately mutualistic organelles with phylogenetic and geographic patterns of divergence that completely match those of their host species.

The highly reduced genomic structure in *Buchnera* and its vertical transmission among hosts has created genomes that show evidence of low effective population sizes, accelerated sequence evolution through random genetic drift, and very low levels of genetic polymorphism. The intensively studied interaction between the aphid *Uroleucon ambrosiae* and *Buchnera aphidicola* shows little geographic variation across the United States at three *Buchnera* loci. This low genetic diversity, together with genetic bottlenecks in their aphid hosts, appears to be the cause of geographic homogenization in the genetic structure of this interaction (Funk, Wernegreen, and Moran 2001). Analysis of *Buchnera* polymorphism on a distantly related aphid with a very different life history has shown similar patterns of low polymorphism (Abbott and Moran 2002).

Hence, the genomic structure of the interaction between aphids and *Buchnera* seems to have reached a point at which the major evolutionary processes are further reduction in genome size and sequence divergence through random genetic drift on most genes, rather than significant ongoing adaptive change across landscapes (Abbott and Moran 2002). Selection is focused on amplification of symbiont genes that aid in host nutrition or other host functions (Moran and Baumann 2000). These appear to be common features of ancient and obligate vertically transmitted mutualistic symbioses as genomic integration between hosts and symbionts proceeds (Gil et al. 2003; Martin 2003). In fact, Moran and Wernegreen (2000) have argued that irreversible loss of genes and restricted functional capabilities in obligate mutualistic symbioses may preempt any shift back to parasitism, thereby guaranteeing the long-term stability of mutualism.

Geographic Mosaics in Multispecific Mutualistic Symbioses

Prediction: The overall geographic mosaic of coevolution with *assemblages* of vertically transmitted symbionts should increase over time as ancient symbioses with individual species degrade in mutualistic effects through random genetic drift and hosts acquire additional symbionts to compensate for this loss of symbiont function. Douglas and Raven (2003) have argued that the transfer of mutualistic function to secondary symbionts may be a common mechanism for the maintenance of mutualism in multicellular hosts. In general, mutualistic interactions involving complementary symbiont species should show more pronounced geographic mosaics in traits and local adaptation than interactions with single symbionts, because multispecific selection allows more coevolutionary options.

If hosts coevolve with complementary mutualistic symbiont species, the potential for geographic mosaics increases further. As genotypes of hosts and symbionts vary in expression across environments, each pairwise interaction between a host and symbiont may range from mutualism to commensalism or antagonism. In addition, interactions among symbiont species may either increase the mutualistic effects of symbiosis due to indirect effects or decrease those effects due to competition among symbionts and subsequent selection for increased virulence. Indirect effects and competition among symbionts are also likely to vary across environments, adding to the geographic selection mosaic. If this argument holds, then most host populations should commonly harbor a small to moderate number of mutualistic symbiont species that compete minimally with one another. The number and combination of symbionts should vary geographically, as different genotypes, environments, and life histories favor different symbiont combinations.

There are still few data on the geographic structure of multispecific mutualistic symbioses. Current data consist mostly of a few intensive studies on local and geographic variation in the number of phylogenetically or functionally similar symbiont species harbored by hosts. For example, pea aphids (*Acyrthosiphon pisum*) can harbor up to five vertically transmitted facultative symbionts, including a *Rickettsia*, a *Spiroplasma*, and three proteobacteria (Tsuchida et al. 2002), in addition to their obligate *Buchnera* symbionts. Populations of pea aphids throughout Japan differ geographically in their combination of four of the facultative symbionts (fig 13.5). The fifth facultative symbiont has not been found in Japan. Analysis of hundreds of pea aphids throughout Japan has shown that individual aphids rarely harbor more than one of the facultative symbionts, even though between two and four sym-

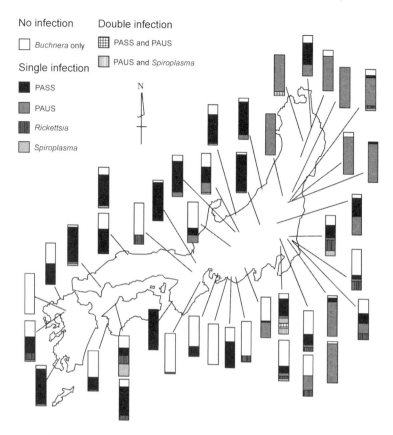

Fig. 13.5   The proportion of pea aphids in Japan harboring *Buchnera* and any of four known facultative symbionts. Populations range from harboring only *Buchnera* to harboring *Buchnera* and various combinations of facultative symbionts. Each bar shows the proportion of the individuals in that population harboring *Buchnera* only or *Buchnera* and one of the facultative symbionts. Only a few individuals harbor *Buchnera* and more than one facultative symbiont. After Tsuchida et al. 2002.

biont species are commonly present within local pea aphid populations (Tsuchida et al. 2002). *Buchnera* occurs in all individuals and populations, but populations differ greatly in the proportion of individuals harboring one or more additional symbionts. The secondary symbionts are usually maternally transmitted, but phylogenetic analyses of DNA sequences indicate that horizontal transmission has contributed to their distribution among aphids (Russell et al. 2003).

The causes of these differences among host individuals and populations in the symbiont combinations they harbor are unknown. Studies on other

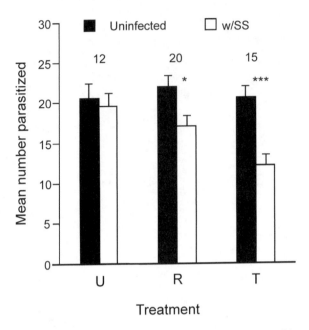

Fig. 13.6   Effect of facultative symbionts (sometimes called secondary symbionts) on parasitism of pea aphids by the wasp parasitoid *Aphidius ervi*. Bars show the mean ($\pm$ standard error) number of aphids successfully parasitized out of a possible thirty in each trial. The same aphid clone was used in all trials to evaluate parasitism on aphids lacking facultative symbionts (= uninfected) and those harboring symbionts (= w/SS). The trials evaluated three symbiont species, designated U, R, and T. The numbers above the bars are the number of replicates. The asterisk (*) indicates significant differences between neighboring bars, and the number of asterisks indicates the level of statistical significance. After Oliver et al. 2003.

pea aphid populations, however, have suggested that geographic differences in host plants used by the aphids and attack by parasitoids may produce geographically variable selection. When a pea aphid is infected with one of its *Rickettsia* symbionts, the effect of the interaction varies with the host plant on which the aphid feeds. *Rickettsia* infection has no influence on fecundity of aphid clones grown on alfalfa, but it reduces that of clones grown on sweet pea (*Lathyrus odoratus*) and that of clones grown on bur clover (*Medicago hispida*) even more (Chen, Montllor, and Purcell 2000). These antagonistic effects, however, are only part of the selection mosaic. The symbionts can also affect interactions between pea aphids and the wasp parasitoid *Aphidius ervi* (fig. 13.6), and the symbionts differ in their effects. In one set of laboratory trials that used the same aphid clone for all treatments, aphids that harbored the *Rickettsia* (= R) symbiont or the T-type symbiont (sometimes called PABS) were more successful at defeating parasitoid attack than symbionts lacking these symbionts. Aphids harboring the U-type (sometimes called PAUS) symbiont, however, were not more successful than control aphids at preventing successful parasitism. Because pea aphids (*Acyrthosiphon pisum*) also differ genetically in their ability to resist attack by the wasp parasitoid *Aphidius ervi* (see chapter 3), there is a strong potential for the development of a geographic mosaic of coevolution among pea aphids, parasitoids, and aphid symbionts, driven in part by host plant differences.

Constraining Divergence in Geographic Mosaics

Taken to its extreme, the geographic mosaic of coevolution between hosts and mutualistic symbionts could lead to locally fixed coevolved traits across all populations—that is, extreme geographic divergence caused by local coevolutionary convergence. Such extreme divergence among all populations, however, is likely often counterbalanced by the continually shifting geographic mosaic of coevolution. Synthesizing the above arguments, those counterbalances include several ecological aspects of mutualism. First, selection on mode of transmission of symbionts can vary among populations, which can create a geographically dynamic structure to mutualisms. The tension between selection on hosts to favor vertical transmission and selection on symbionts to retain the capacity for infectious spread may maintain some degree of gene flow among symbiont populations. Even a small amount of gene flow among symbiont populations maintained by horizontal transmission could prevent local fixation of mutualistic genotypes, as shown in geographically structured models of mutualism (Nuismer, Thompson, and Gomulkiewicz 1999; Gomulkiewicz et al. 2000; Nuismer, Thompson, and Gomulkiewicz 2000). Moreover, lateral gene transfer among symbiont species may create a continually changing geographic structure in the traits of symbiont species.

Second, the demographic dynamics of populations and interactions may maintain a shifting geographic mosaic in many mutualisms. Stochastic loss of symbionts by local host populations may contribute to a constantly changing geographic mosaic of symbiont and host genotypes, again provided that symbionts can at least sometimes be acquired through horizontal transmission among hosts. Different symbiont genotypes or species also may be favored at different host ages or life-history stages. Because the demographic structure of host populations is likely to vary over time, selection may act in ways that maintain some symbiont genetic diversity within and among host populations. Temporally variable selection on local populations could also act to prevent or slow fixation of locally selected mutualistic traits. General mathematical models that allow selection on matched alleles to vary locally from mutualism to antagonism can maintain polymorphisms in locally coevolving traits over long periods of time (Nuismer, Gomulkiewicz, and Morgan 2003). As in models of single-species evolution, the greater the variance in selection, the greater the likelihood that selection will maintain local polymorphism.

Finally, varying productivity across landscapes is likely to contribute to the shifting mosaic of coevolving traits and outcomes. One set of mathematical models has suggested that environments that are demographic sinks for host populations may, in the absence of symbioses, favor the formation of mutualistic symbioses (Hochberg et al. 2000). If host populations are distributed across an environment with two competing symbionts, one virulent and one relatively avirulent, the virulent form in these models is favored only in environments in which the host population is highly productive (i.e., a demographic source), whereas the avirulent form is favored in environments where the host teeters on the edge of population viability. Under such conditions, any mutant symbiont that contributes to host survival and reproduction would increase simultaneously its own fitness and that of the host genotype in which it occurs. That combination of symbiont and host genotypes would spread at the expense of others within these populations. If the results of these models are broadly true, then demographic sinks in host populations may sometimes become coevolutionary hotspots for the evolution of mutualistic symbioses. These results reinforce the conclusions of a related set of mathematic models, which show that productivity gradients (i.e., variation among environments in the intrinsic rate of natural increase) can produce selection mosaics and coevolutionary hotspots and coldspots (Hochberg and van Baalen 1998).

No empirical studies have yet evaluated the distribution of mutualistic symbioses and mutualistic selection across gradients of host productivity. Any future studies, however, will be valuable to our understanding of how demography, metapopulation structure, geographic ranges, and coevolutionary hotspots interact to shape the coevolutionary structure and dynamics of mutualistic symbioses. Moreover, such studies will be important for work on the conservation of coevolutionary processes. Much effort on conservation is directed toward conserving the most viable populations of species, but the geographic edges of species' ranges may themselves be important as cradles for the evolution of novel outcomes in interspecific interactions.

Lineage Codiversification, Secondary Contact, and the Geographic Mosaic in Mutualistic Symbioses

As coevolving lineages diverge and interactions diversify, mutualistic symbioses should exhibit patterns of codiversification that depend upon the geographic structure of interacting populations, transmission mode and frequency of symbionts, the potential for lateral gene transfer among symbionts, and complementarity among symbiont species. At the simplest level, strict vertical transmission should result in parallel cladogenesis, and horizontal transmission should show more complex patterns that reflect the ongoing possibility of host switching in symbionts and symbiont choice in hosts. For the most part, the phylogenetic structure of mutualistic symbioses follows these expectations (Chen, Li, and Aksoy 1999; Funk et al. 2000).

Nevertheless, some intriguing exceptions suggest there is still much to understand. Sepiolid squid species harbor luminescent bacteria, which they acquire from the environment when newly hatched. Despite horizontal transmission, the squid and bacterial species show strong evidence of long-term parallel cladogenesis. Nuclear and mitochondrial phylogenies of the squid mostly match the molecular phylogeny obtained for the bacteria (Nishiguchi, Ruby, and McFall-Ngai 1998). When newly hatched *Euprymna scolopes* are presented individually with any of the seven strains used in the phylogenetic analysis, all strains are capable of colonizing this one squid species. When the squid are presented with their normal strain and a novel strain, both strains colonize the host in approximately equal numbers, but after forty-eight hours the native strain shows a twenty-fold advantage over any of the other strains. Hence, this interaction appears to have maintained parallel cladogenesis across multiple oceans despite horizontal transmission.

More typically, a mix of vertical and horizontal transmission creates phylogenetic patterns that become more reticulate as lineages diversify and come repeatedly into secondary contact as geographic ranges expand and contract over eons. The evolution of lichens illustrates the complex patterns that are probably common in many mutualistic symbioses that have coevolved and diversified over millions of years in this way. Some lichens propagate and disperse as a single fungus-symbiont unit (sometimes called mycobionts and photobionts, respectively), either through vegetative fragmentation or through specialized dispersal organs (Romeike et al. 2002). In other cases, the fungi and symbionts disperse separately, forming new associations after dispersal. Through both mechanisms, lichens have become a spectacularly successful lifestyle. Almost one-fifth of known fungal species form obligate symbioses with cyanobacteria, green algae, or both (Lutzoni, Pagel, and Reeb 2001), and the resulting estimated fourteen thousand lichens are major components of terrestrial environments worldwide (Brodo, Sharnoff, and Sharnoff 2001). More than 98 percent of the lichenized fungi are within the phylum Ascomycota, within which they compose about 42 percent of known species. These lichenized species are distributed, depending upon the classification system, among fifteen to eighteen orders, and between eight and eleven of those orders include both lichenized and nonlichenized species. On the surface, it would appear that the evolution of lichens was phylogenetically conserved at higher taxonomic levels (i.e., restricted mostly to ascomycetes) but highly dynamic within that phylum.

Large-scale phylogenetic analyses, however, have suggested a simpler picture of the phylogenetic dynamics. Lichens appear to have evolved early within the Ascomycota, and the lichen-forming habit may have arisen only three times at most, with one or two being most likely (Lutzoni, Pagel, and Reeb 2001). In contrast, multiple lineages appear to have lost the lichen-forming habit over time. Bayesian statistical analysis of 19,900 possible phylogenetic trees suggests that the rate of loss of lichenization has greatly exceeded that of gain (fig. 13.7). Hence, one of the most important forms of symbiotic mutualism on earth may have resulted from one or two initial colonizations of fungi many eons ago. The current phylogenetic patterns reflect subsequent diversification in lichen-forming species and loss of the habit in other species as the mutualism spread over time across landscapes worldwide.

The partners in lichens show similar phylogenetic conservatism. About 1500 lichen species contain cyanobacteria, and most of these cyanobacteria are in the genus *Nostoc,* which form a monophyletic clade (Rikkinen, Oksanen, and Lohtander 2002). The lichens that contain eukaryotic green algae rather than cyanobacteria usually contain algae of a single genus, *Trebouxia* (Romeike et al. 2002). Current data suggest that lichen-forming fungi usually associate with the same algae or cyanobacteria throughout their geographic range, but some algal symbiont species associate with many different lichen fungi (Brodo, Sharnoff, and Sharnoff 2001). Less than 5 percent of algal symbionts, however, have been formally identified to species, and molecular

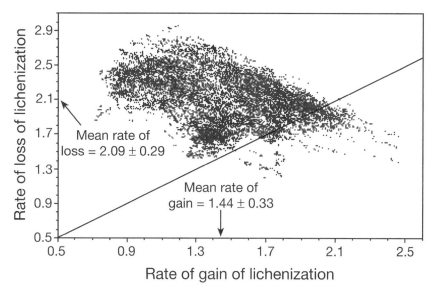

Fig. 13.7   Rates of gain and loss of lichenization during the diversification of lichen-forming fungal lineages. The line indicates equal rates of gain and loss. The observed pattern suggests that the rate of loss of lichenization exceeds the rate of gain. The analysis is based upon a continuous-time Markov model of trait evolution. After Lutzoni, Pagel, and Reeb 2001.

analyses are beginning to resolve patterns of geographic differentiation and diversification in these taxa. Hence, although lichens fit the view that hosts should be more restrictive than symbionts in mutualistic symbioses, much additional work is still needed.

Both the fungi and the cyanobacteria are selective in their choice of partners. As cyanolichens have diversified, however, different combinations of species have come into secondary contact in different parts of their geographic ranges. The result is a geographic mix of partnerships that has developed through a combination of phylogenetic conservatism and secondary switching, but how that complex geographic structure develops is only now starting to become clear. Molecular studies have indicated that cyanolichen symbioses are more species specific and geographically stable than previously suspected (Paulsrud, Rikkinen, and Linblad 2001). Molecularly similar or identical cyanobacteria have been found in the analysis of some lichens from Sweden and the northwestern United States (Paulsrud, Rikkinen, and Linblad 2000). Hence, there is strong phylogenetic conservatism in these interactions. One monophyletic group within *Nostoc* is associated with cyanolichens found only within old-growth forests. These epiphytic lichens are a diverse group of species distributed in similar habitats worldwide. In contrast, many terrestrial cyanolichens share a different monophyletic group of *Nostoc* (Rikkinen, Oksanen, and Lohtander 2002). There is, however, no simple one-to-one relationship of specificity between fungi and cyanobacteria. Some closely related sympatric cyanolichens harbor different *Nostoc* strains, but several local fungal species may share cyanobacterial strains (Paulsrud, Rikkinen, and Linblad 2000). The community of interacting fungi and cyanobacteria is therefore structured both phylogenetically and ecologically, forming local networks of mutualistic interaction. In addition, this sharing of symbionts may itself have produced novel lifestyles that depend upon the ability of identical cyanobacteria to use different fungal hosts. Cyanolichen guilds may rely upon core species that produce large numbers of symbiotic diaspores (Rikkinen, Oksanen, and Lohtander 2002). Some species produce only fungal spores and may depend upon the core species for dispersal of the appropriate cyanobionts.

These studies on the phylogeny and geography of lichen association highlight the inevitability of geographic mosaics in mutualistic symbioses as they diversify in species and overlap in various combinations of species over time as climates and geographic ranges change. So long as some horizontal transmission is possible, the geographic mosaic of coevolution is likely to produce geographically variable degrees of reciprocal specificity in coevolving mutualistic symbioses.

Conclusions

Molecular and ecological studies over the past decade have suggested that mutualistic symbioses show much more geographic structure than previously suspected. Too few mutualistic symbioses have been studied in multiple populations to make robust assertions about the geographic mosaic of coevolution in these interactions. The current predictions and results, however, suggest useful approaches for future analyses. The outcomes of interactions between particular host and symbiont genotypes differ among environments. Vertically transmitted symbionts may eventually lose alleles through random genetic drift, possibly creating differences among populations in their mutualistic effects on hosts and favoring hosts that accumulate complementary symbionts. In mutualisms with horizontally transmitted symbionts, the complex genetic handshaking between the host and symbiont creates ongoing opportunities to test new partners as mutations occur in the hand-shaking genes. Meanwhile, lateral gene transfer among horizontally trans-mitted symbionts continues to fuel the geographic mosaic of symbiont traits and acceptable symbiont partners for hosts.

Over time, some local populations with vertically transmitted symbionts may become extinct, if random genetic drift in the symbionts is not coun-tered by other selective processes. Some parasitic symbionts may crack the handshaking code and decimate a local host population or even drive it to ex-tinction. Other populations, however, will thrive, and the overall geographic mosaic of coevolution is likely to stabilize mutualisms over the long term. The sheer ubiquity of mutualistic symbioses suggests that, although they may sometimes collapse locally, as can occur in all interactions, they are no more evolutionarily fragile than other interspecific interactions, and their long-term stability appears to be just as likely as that of other forms of interaction. If anything, the diversity of life appears to have been made possible because mutualistic symbioses have been a major part of the evolutionary process.

14 Mutualists III

Convergence within Mutualistic Networks of Free-Living Species

Even though symbiotic mutualisms are pervasive, our most common images of mutualism are of free-living species because these are among the most easily observed interspecific interactions. In fact, pollination by animals and Müllerian mimicry rings were the first interactions to be studied from a coevolutionary perspective following publication of *The Origin of Species*, long before we had the molecular tools necessary to understand symbiotic mutualisms (Thompson 1994). The structure of coevolution at this end of the continuum differs from symbiotic mutualisms in two important ways.

The most immediately evident difference is that free-living mutualists interact during their lifetime with multiple individuals of the other species. As in symbiotic mutualisms, one species may serve as a host—for example, fruiting trees, fish tended by cleaner fish—but each interaction with a mutualistic visitor is short term rather than sustained. Mutualisms between free-living species are therefore often more similar to antagonistic interactions between grazers and their victims than to those between parasites and hosts. Like antagonistic interactions with grazers, these mutualisms rarely favor high levels of extreme reciprocal specialization between pairs of species (Thompson 1994).

The other difference is that free-living mutualisms inherently form interspecific networks. Mutualisms proliferate through the diversification of species harboring the mutualistic traits and through convergence of traits in unrelated species. That proliferation, in turn, favors the evolution of lifestyles that rely directly upon a local diversity of mutualists (Thompson 1982, 1994). The lifestyles of most highly frugivorous birds became possible only after multiple plant taxa had diverged within lineages to produce many fleshy-fruited species while converging among lineages in similar fruit traits. The resulting diversity of fruits created the opportunity for natural selection to favor avian species that feed on fruits year round or throughout much of the year. Similarly, dependence on nectar by hummingbirds and eusocial bees that maintain large colonies (e.g., honeybees) became possible only after multiple plant taxa offered similar floral resources that attracted these taxa.

These two common aspects of mutualism between free-living species favor specialization in the form of interaction (e.g., exploitation of nectar from tubular flowers), but these mutualisms rarely favor extreme specialization to one species and even more rarely favor extreme reciprocal specialization. Instead, coevolution of free-living mutualists favors networks that expand to encompass an ever-widening range of specialization in the interacting species. As a mutualism draws in more species, asymmetries in specialization increase, because the range of evolutionary options increases.

This chapter uses these two components of mutualistic selection to explore the geographic mosaic of coevolution among free-living mutualistic species. It builds upon the observation that networks of mutualistic and commensalistic interactions are widespread among taxa (Hacker and Bertness 1996; Waser et al. 1996; Hacker and Gaines 1997; Palmer, Stanton, and Young 2003; Stanton 2003). The major argument of the chapter is that, even though species composition within these mutualistic networks changes geographically, coevolutionary selection appears to shape these networks in predictable ways. The final part of the chapter explores interactions that straddle the continuum between symbiotic and free-living mutualisms. The chapter focuses on the hypothesis of convergence within mutualistic networks.

The Coevolving Structure of Mutualistic Networks

By evaluating mutualistic networks as coevolved structures rather than as diffuse multispecific interactions, we are now on track to understand better how these networks develop and undergo continual rearrangement across landscapes. The mathematical models and empirical results together suggest the following general hypothesis and ecological outcomes.

The hypothesis of convergence within mutualistic networks.

Reciprocal selection on mutualisms between free-living species favors genetically variable, multispecific networks in which species converge and specialize on a core set of mutualistic traits rather than directly on other species.

This selective regime should result in a structure to coevolved mutualisms among free-living species with the following attributes:

- a wide range of asymmetries in specialization to other species within the mutualistic network
- an overall high level of local genetic diversity and local species diversity within the network
- high geographic interchangeability in mutualistic participants and low phylogenetic constraints on the species composition of the mutualistic network
- variable outcomes among habitats in pairwise interactions among species, including cheating on the network by some species in some populations

Mutualisms appear generally to coevolve through manipulation of a core set of traits in other species. Plants offer floral rewards and fleshy fruits to animals that, in turn, evolve to exploit those rewards. Cleaner fish offer parasite-cleaning services that are exploited by larger fish. As coevolution proceeds, selection acts on each species to maximize the fitness gain and minimize the fitness loss during these interactions. As the network grows, the core set of traits becomes, in effect, an evolutionarily stable strategy.

The evolution of "pollination syndromes" in plants is the classic example of such convergence (Faegri and van der Pijl 1979), which becomes explicable when placed in the context of coevolutionary selection within mutualistic networks. Hawkmoth-pollinated plants, for example, occur in multiple plant families, and their flowers take many shapes. Most species, however, share various combinations of traits that increase the likelihood of visitation and pollination by hawkmoths, including white flowers, tubular flowers, sucrose-based nectar, and the opening of flowers or pollen dehiscence at dusk. Individual plant species mix and match these core traits in different ways and consequently attract different combinations of hawkmoth species (Haber and Frankie 1989). Networks of interacting species with similar traits also develop during diversification of genera. *Penstamon* species form guilds of species with traits that attract mostly bumblebees, and other guilds with traits that attract mostly hummingbirds (Thomson et al. 2000; Castellanos, Wilson, and Thomson 2003; Wilson et al. 2004).

Within these networks, mutualistic interactions between free-living species can result in extreme specialization to one other species, but it should be the exception (reviewed in Thompson 1994). Such extreme specialization in interactions among free-living species requires life histories and ecological conditions that allow a species to depend solely upon one other species throughout the year or throughout the part of the year in which these repeated interactions take place. Once a multispecific mutualistic network begins to develop, however, alternatives to extreme specialization to a single species can often provide equal or higher fitness.

Because the ecological basis of some mutualistic interactions and the traits under convergent selection are similar worldwide—such as pollination of flowers or dispersal of fruits—the participants often become geographically interchangeable. North American hummingbirds readily feed in North American gardens on imported plants from Australia that are adapted to honeyeaters or on South African plants adapted to sunbirds. In the same gardens, European honeybees efficiently collect nectar and pollen from native North American plants.

This flexibility, however, does not imply that coevolved mutualistic networks lack structure. Analyses of some plant-pollinator networks have shown more extreme specialists and generalists than those predicted by null models (Vásquez and Aizen 2003), and a small but growing number of studies of mutualistic networks suggest that these networks may evolve toward a predictable structure (Olesen and Jordano 2002; Jordano, Bascompte, and Olesen 2003). Interaction networks have the potential to coevolve toward any of three general structures: nested, random, or compartmentalized. A nested network is one in which the core set of generalist species interact with one another and the most specialized species interact only with the most generalist species (fig. 14.1A). The network therefore forms proper nested subsets of interacting species (Atmar and Patterson 1993). By contrast, a random network is one in which each species interacts with a random subset of the total pool of species (fig. 14.1B). A compartmentalized network is one that is built on tight reciprocal specialization in the interacting species (fig. 14.1C). Compartmentalized networks therefore correspond at the extreme to universal pairwise interactions or, more generally, to small clusters of species that interact only with each other within a community, as often occurs within symbiotic mutualisms. Real networks, of course, are bound to form more complex patterns (fig. 14.1D), but comparing multiple networks across broad geographic areas can allow an assessment of whether mutualistic networks tend to coevolve toward any of these three possible structures.

An analysis of twenty-seven plant-frugivore networks and twenty-five plant-pollinator networks by Bascompte et al. (2003) has suggested that local mutualistic networks are generally nested rather than random or compartmentalized (fig. 14.2A). Moreover, these mutualisms have levels of nested-

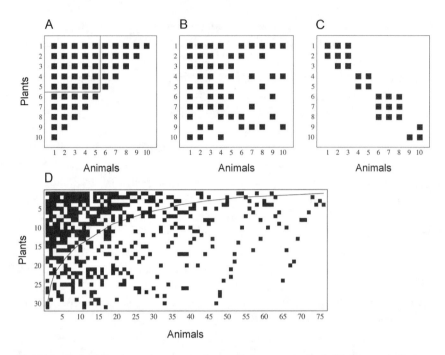

Fig. 14.1   Potential and actual structure of mutualistic networks. Each filled square indicates an interaction between a particular animal species and a corresponding plant species. (A) A perfectly nested mutualistic network between ten animal species and ten plant species. The internal box drawn around the plant and animal species 1–5 highlights a core group of species that all interact with each other. (B) A hypothetical example of a random network of interacting species. (C) A hypothetical example of a compartmentalized network. (D) An example of an actual local plant-pollinator network. In a perfectly nested interaction, all the observed points would occur to the left of the curve. After Bascompte et al. 2003.

ness higher than that found in more general food webs (fig. 14.2B), suggesting that mutualisms between free-living species coevolve toward a particular form of network structure. Very high values of nestedness are especially common in mutualistic networks that involve large number of species.

Trophic mutualistic networks may be more likely to be nested than compartmentalized, because mutualistic networks are not subject to coevolutionary alternation (see chapter 11). The hypothesis of coevolutionary alternation suggests that parasites and predators continue to evolve in their relative preference for host and prey species as local host and prey population undergo continual evolution in their levels of defense. Coevolutionary alternation therefore provides one mechanism for coevolution of compartmen-

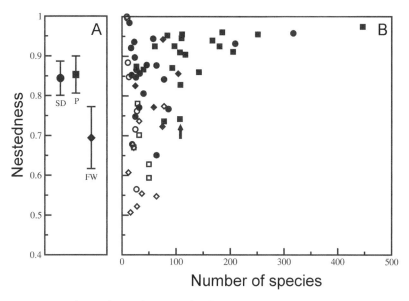

Fig. 14.2   Nestedness values within networks of interacting species. (A) Mean nestedness value ± standard error for seed dispersal (SD) mutualisms between plants and frugivores, pollination (P) mutualisms, and, by comparison, more general resource-consumer food webs (FW). (B) Nestedness values for all data in (A). The arrow points to the data set shown in fig. 12.1C. Filled symbols in (8) are significantly nested. After Bascompte et al. 2003.

talized structure in antagonistic networks, with nothing comparable occurring in mutualistic networks.

Instead, local mutualistic networks may often be based upon a cohesive core of generalist species that interact with one another. The core of generalist species may grow over time, as it provides a stable resource that other species can exploit. In plant-frugivore and plant-pollinator interactions, the number of links among species increases slightly faster than the number of species, and the network becomes more structured as this complexity increases (Bascompte et al. 2003). This stable core of species may, in turn, allow the evolution of a subset of specialists that exploit single hosts or visitors. Specialists become attached to the core, but, as suggested by the analyses of Bascompte et al. (2003), the pattern of specialization is often asymmetric. Specialists interact with generalists rather than with one another. Through differential sorting of species among networks and coevolutionary selection within local networks, individual species will vary among communities in how they fit within mutualistic networks.

Geographic differences in the ways in which species fit within networks are becoming increasingly apparent as more mutualistic interactions are studied across multiple communities. Even the most highly specialized mutualistic associations on islands can coevolve as a small network of closely related species that vary in patterns of reciprocal specialization across the geographic range of the interaction. Among the most visually dramatic are the geographic differences in the morphology of purple-throated carib hummingbirds (*Eulampis jugularis*), which are the sole pollinators of two geographically variable species of *Heliconia* on Caribbean islands (Temeles and Kress 2003). The interaction varies among islands and also among populations within the islands. The birds feed on multiple plant species, but their principal food plants are these species of *Heliconia*.

On St. Lucia, *H. caribaea* has red-bracted flowers that match the straight bills of the relatively large males, and *H. bihai* has green-bracted flowers that match the long, curved bills of the females (fig. 14.3). At sites on St. Lucia where *H. caribaea* is rare or absent, a red-and-green-bracted form of *H. bihai* with shorter, straighter flowers and overall intermediate morphology has evolved. At these sites males tend to visit the red-and-green-bracted morph of *H. bihai*.

On Dominica, the two *Heliconia* species are mostly allopatric, but they come into contact in some parts of the island. *Heliconia bihai* has only one morph, with bracts that are red with a yellow stripe and flowers that are long and curved. *Heliconia caribaea* has two morphs, one red and one yellow. As on St. Lucia, these two plant species are the principal food sources for *E. jugularis*, and the males and females specialize on different *Heliconia* morphs. Throughout the island, females are the sole pollinator of *H. bihai*. The interaction with *H. caribaea*, however, is more geographically variable. In non-contact zones between the two species, male and female hummingbirds do not distinguish between the colors of *H. caribaea*. In contact zones, the flowers of the red morph are longer and more curved and differentially attract females. Males in these contact zones are restricted mostly to the yellow morph.

Additional geographic variation is built into the interaction on Dominica. Because males do not visit *H. bihai*, they feed from plants other than *Heliconia* in places where *H. bihai* is the more common *Heliconia* species. Consequently, even in this tight interaction between hummingbirds and *Heliconia* the mutualistic network coevolves as a geographic mosaic, with patterns of specialization and local adaptation varying across landscapes.

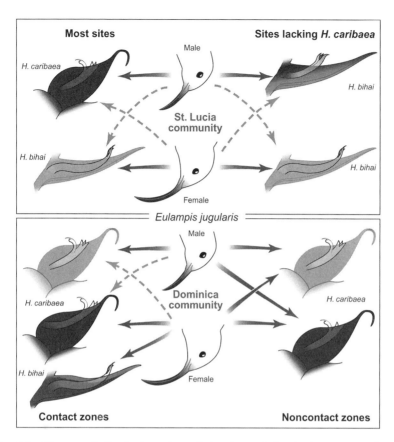

Fig. 14.3   Divergence in floral color and morphology in two *Heliconia* plant species within and between the Caribbean islands of St. Lucia and Dominica relative to use by males and females of the purple-throated carib hummingbird (*Eulampis jugularis*). Solid arrows point to the plant forms most commonly used by the local male and female hummingbirds; broken arrows point to forms used only occasionally or rarely. The more darkly shaded flowers correspond to the red morph, and the more lightly shaded flowers correspond to either the green morph (St. Lucia) or the yellow morph (Dominica). The two-toned bract of *H. bihai* at sites lacking *H. caribaea* on St. Lucia is red and green. From Altshuler and Clark 2003, based on results of Temeles and Kress 2003.

Within larger mutualistic networks, the shifting geographic ranges of species will create a geographic mosaic of local matches and mismatches of coevolved traits. Studies of geographic differences in the network structure of interactions between the plant *Helleborus foetidus* and the ants that disperse its elaiosome-bearing seeds in Spain illustrate the first step in studies of matching and mismatching within networks (Garrido et al. 2002). Seed traits in this species vary geographically, as does the distribution of body sizes in local ant assemblages. Variation in seed traits, however, matches the distribution of body sizes in local ant assemblies in only some localities. Ants of different body sizes differ in the seeds in they can handle. Therefore, geographic variation in the pattern of seed sizes and ant sizes can become a focus of studies on how selection may shape network structure in these interactions. Because this plant species is only one component of the overall mutualistic network, it is not possible in this particular study to evaluate the overall causes of the local mismatches. Nevertheless, these results are a step toward understanding the overall geographic network structure of interactions between ants and elaiosome-bearing seeds.

Future analyses of local mutualistic networks will require several kinds of study. Additional studies on other forms of mutualism must test the generality of the observation that mutualistic networks of free-living species coevolve toward a nested rather than a compartmentalized structure. These studies also need to include an evaluation of how facultative and obligate generalists within mutualistic networks contribute to the structure of coevolutionary selection within networks. In addition, we need a better understanding of the network conditions that allow the evolution of extreme specialization to one other species within a network. Finally, we need studies that evaluate closely how the structure of reciprocal benefits shapes patterns of coevolutionary convergence. Asymmetries in fitness benefits among genotypes may add to the complexity of selection within mutualisms, thereby contributing to the maintenance of genetic diversity within these interactions (Bever 1999). Together, these analyses will help us understand how coevolution stabilizes mutualistic networks over time even as it continually creates local mismatches in coevolved traits. Formal comparison of network structures among regions that differ in degree of trait matching will help us pinpoint the components of network structure that are under the strongest coevolutionary selection.

The Geographic Mosaic of Müllerian Mimicry Networks

Networks of species that share warning colors as signals of distastefulness to potential predators are common in terrestrial and marine environments. These networks differ in coevolutionary structure from trophic mutualisms among free-living species, because coevolution is indirect among the species sharing the warning signals. Evolution of shared warning colors within these networks requires that predators avoid prey based upon the prey's degree of resemblance to the locally common warning signals. In Müllerian networks, selection mediated by predators leads to coevolutionary convergence of distasteful prey species on a common set of warning signals. Suggested examples include *Heliconius* butterflies in Central and South America (Mallet and Gilbert 1995; Gilbert 2003) and mimicry rings of wasps and unpalatable wasplike tiger moths (Simmons and Weller 2002). In Batesian mimicry networks, palatable species evolve to mimic unpalatable species. Examples include syrphid flies (i.e., bee-flies or hover flies) that evolve to mimic unpalatable wasps, or palatable butterflies that evolve to mimic unpalatable butterfly species (Scriber, Hagen, and Lederhouse 1996).

Müllerian and Batesian mimics should differ in their coevolutionary dynamics. Coevolution among Müllerian mimics should be driven by coevolutionary convergence, whereas Batesian mimicry should produce a coevolutionary chase. In Batesian mimicry selection should favor individuals in the palatable species that resemble the unpalatable model, but selection should simultaneously favor individuals of the distasteful species with warning signals that are effective against predators but allow the predators to distinguish between the distasteful models and the palatable mimics.

Between these two coevolutionary extremes are a variety of evolutionary possibilities. An abundant distasteful species may act as a model for a rare species that evolves to mimic it. Under these conditions, true coevolution between the abundant model and rare mimic may not occur, if the presence of the rare species has no effect on selection on the abundant species (Gilbert 1983). In other interactions, one species may be more distasteful than another. If predators can distinguish among these species, then these interactions could evolve in a manner resembling Batesian mimicry (Speed and Turner 1999; Speed et al. 2001). Species could also differ geographically in their relative unpalatability to predators (Ritland 1995), and the relative abundances and relative palatabilities of the prey species could vary over hundreds or thousands of years, shifting the roles of model and mimic across space and time within these networks.

Distinguishing among the various possibilities has proven to be difficult. Although mimicry has a long tradition of study in evolutionary biology, few studies have approached an unequivocal demonstration of Müllerian mimicry (Mallet 1999; Kapan 2001). With that caution in mind, the following discussion assumes that the classic cases of shared warning colors among distasteful species are sometimes driven by coevolutionary convergence. Considering these interactions in this way provides a chance to explore how convergence, rather than trait complementarity and convergence together, may shape the geographic mosaic of coevolution.

The best-studied Müllerian mimicry rings are those among *Heliconius* butterflies in Central and South America, which develop clear geographic mosaics in the signaling patterns on which species converge (Gilbert 2003). Genes that control the wing patterns in these species show steeper clines across hybrid zones than do other genes, providing strong evidence for geographic differences in selection for local convergence of traits in these butterfly species (Mallet 1986; Mallet et al. 1990; Jiggins et al. 1997). Multiple mimicry rings may coexist within a local area, and the same species often differ geographically in the color patterns on which they converge (Turner and Mallet 1996). For example, nine *Heliconius* species coexist within Corcovado Park in Costa Rica, and those species form three distinct mimicry rings (Gilbert 2003). Those networks include some widespread species such as *H. melpomene* and *H. erato*, both of which exhibit visually dramatic geographic differences in wing color and pattern. Breeding and hybridization experiments have shown that, although many genes are involved in these wing color patterns, single allelic changes can have large effects on these patterns (Sheppard et al. 1985; Nijhout 1991; Gilbert 2003).

Moreover, introgression of genes from neighboring geographic areas can create novel phenotypes. Gilbert (2003) has argued that such introgression may be the primary source of new local mimicry rings, because, unlike mutation, it is renewable and creates repeated opportunities for these "macromutants" to become established. More commonly, tension zones produced by introgression between color morphs develop and shift across landscapes over time. In Panama, *H. erato* and *H. melpomene* form parapatrically distributed color morphs that are separated by hybrid zones. In one region, populations have a strong yellow hindwing bar, whereas populations in a neighboring region do not. The yellow bar is determined by a single Mendelian allele, with heterozygotes showing either a lightened or a broken bar (Mallet 1986). The hybrid zone is phenotypically evident in *H. erato* as a cline in the frequency of individuals showing yellow bars that are intermediate in size and shape (fig. 14.4).

In 1982, the cline in Panama was about eighty kilometers wide (Mallet 1986). During the seventeen years between 1982 and 1999, the center of the cline shifted westward approximately forty-seven kilometers (fig. 14.5). The

Fig. 14.4 Dorsal wing color pattern in the butterfly *Heliconius erato* in Panama. The wing background is black, the gray band on the forewing is red in actual butterflies, and the white band on the hindwing is yellow. The wings on the far left and right are normal morphs from two different regions, and the other two wings are intermediate forms within the hybrid zones. From Blum 2002.

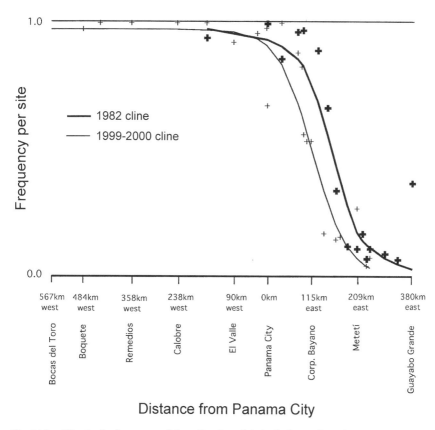

Fig. 14.5 Cline in the frequency of the yellow bar allele in the butterfly *Heliconius erato* in Panama. The cline has moved between the 1982 and 1999–2000 samples of these populations. After Blum 2002.

shift is most likely driven by changing spatial patterns of strong selection against individuals with intermediate colors within this *Heliconius* hybrid zone. Hybrids between the two regions do not exhibit sterility, inherently lower viability, or hybrid breakdown. Hence, these potential alternative interpretations of selection against hybrids within the tension zone have been ruled out (Blum 2002).

Population-genetic models of Müllerian mimicry suggest that shifting geographic mosaics of mimicry rings can develop as clines intersect among species and experience geographic differences in the direction of positive frequency-dependent selection (Sasaki, Kawaguchi, and Yoshimori 2002). Geographic differences in selection on a warning pattern on one species can create a cline that moves across a spatially heterogeneous landscape. Convergence of another species on that pattern can then create combined clines in the warning pattern for the two species. The combined clines spread more slowly than for one species alone and may reach a boundary in regions of low population density. Geographic tension zones between mimicry rings may therefore tend to develop in regions of low population density or low dispersal. Depending upon the strength and geographic structure of the combined positive frequency-dependent selection on these species, the result can be either geographic mosaics with no coexisting Müllerian mimicry rings or multiple coexisting mimicry rings as one species but not the other crosses the tension zones.

Mutualistic Networks along the Continuum from Symbiotic to Nonsymbiotic Mutualisms

Some interactions are a mix of symbiotic and nonsymbiotic relationships, creating a coevolutionary tension between the selective forces driven by symbiotic mutualism and those driven by mutualism between free-living species. Examples include mycorrhizal networks, pollinating floral parasites, and ant-plant and ant-butterfly interactions, which mix components of the continuum in different ways. The geographic mosaic of network structure has not yet been studied in most of these interactions. Future studies of these interactions, however, should help refine our understanding of the conditions favoring nestedness within mutualistic networks. Toward that end, the following sections synthesize what is currently known about the structure of specialization in these interactions.

MYCORRHIZAL NETWORKS

The structure of the interactions between plants and mycorrhizal fungi may differ from other known forms of interaction, because a single genotypically complex mycorrhizal fungus individual may be able to attach simultaneously to multiple host individuals and change host combinations over time. Like the words "lichen" and "rhizobia," the word "mycorrhiza" itself is a testament to the commonness of these interactions, because these are three of a very small number of words that have been coined to describe particular interspecific interactions between taxa. Mycorrhiza refers neither to the plant nor the fungus, but rather to the interaction itself. Most plants routinely form mycorrhizae on their roots (Smith and Read 1997; van der Heijden and Sanders 2002), and even plants within the most isolated archipelago on earth, the Hawaiian Islands, depend highly on mycorrhizae. Over 90 percent of endemic Hawaiian plant species form arbuscular mycorrhizae (Koske, Gemma, and Flynn 1992). Mycorrhizae are known from fossils that are 400 million years old, and these relationships may have made possible the colonization of land by plants (Remy et al. 1995).

Some natural history background can make more evident the uniqueness of mycorrhizae and their importance as tools for understanding mutualistic coevolution along the continuum from symbiosis to interactions among free-living species. More than 60 percent of the earth's vascular plants form arbuscular mycorrhizae (Smith and Read 1997). Arbuscular mycorrhizal (AM) fungi are obligate symbionts of plants that germinate in the soil from presumed asexual spores and form hyphal connections with plant roots soon after spore germination. No free-living AM fungi are known (Sanders 2002). Once in contact with a root, hyphae penetrate the plant epidermis, grow toward the root cortex, and then invade cortical cells, forming with the cell's membrane a dual membrane called the arbuscule. Therefore, as with rhizobia, formation of the symbiosis requires a complex series of steps. The arbuscular membrane forms the interface for transport of nutrients between the two species, providing the fungus with fixed carbon and the plant with higher phosphorus intake, greater access to water and organic nitrogen, or protection from root pathogens (Miyasaka, Habte, and Matsuyama 1993; Newsham, Fitter, and Watkinson 1995; Hodge, Campbell, and Fitter 2001; van der Heijden and Sanders 2002).

The soil of many landscapes is laced with a loose mycorrhizal mat, sometimes called the "wood-wide web," as fungal hyphae weave through one another. In some cases, these networks connect the roots of unrelated plants, creating, either directly or indirectly, local changes in plant nutrient dynamics (Zabinski, Quinn, and Callaway 2002). This network of connections with multiple plant individuals blurs the line between symbiotic and nonsymbiotic interactions. A single mycorrhizal net has the potential to make simultaneous connections with multiple hosts, and a single plant genotype has the potential to form multiple mycorrhizal connections on different parts of its root system (Allen et al. 2003). But that is only part of the genomic richness of these interactions. A single AM fungal spore can contain variable ribosomal DNA sequences, and the growing hyphae are multigenomic (Sanders et al. 1995; Lanfranco, Delpero, and Bonfante 1999; Sanders 2002). Some evidence suggests that this genetic diversity results at least partly from the stable maintenance of multiple genomes contained within separate nuclei in these presumed asexual individuals (Kuhn, Hijri, and Sanders 2001), but more recent studies have indicated that highly divergent rDNA variants may be maintained within each nucleus, indicating a relaxation of concerted evolution of rDNA in these species (Pawlowska and Taylor 2004). Adding to the complexity, hyphae can anastomose and exchange protoplasm (Giovannetti, Azzolini, and Citernesi 1999; Giovannetti et al. 2001).

The combination of genomic complexity and ability to form simultaneous connections with multiple plant species is the perfect recipe for the development of mutualistic networks favoring a wide range of specificity in both the plants and the fungi. Nevertheless, despite the pervasiveness of arbuscular mycorrhizal associations in terrestrial communities, detailed studies of the local, geographic, and phylogenetic structure of these and other mycorrhizal interactions are only now underway (Taylor 2000; Streitwolf-Engel et al. 2001). Fewer than two hundred morphospecies of AM fungi have been described, based primarily on spore morphology, and the links between these morphospecies and the ecology and geography of mycorrhizal specificity are mostly unknown (Sanders 2002). The complex multigenomic structure of mycorrhizae suggests that simple species designations will become more, rather than less, difficult as molecular analyses proceed.

Community-wide characterizations of network structure are underway, but most of these analyses have used morphological criteria and grouped some fungi into species complexes. Most use spore counts that may correlate only roughly with local patterns of specialization. One analysis of a temperate grassland with about forty plant species found thirty-seven arbuscular mycorrhizal (AM) fungi (Bever et al. 2001), and a study of the fourteen most common plant species at La Selva, Costa Rica, identified thirteen AM fungi associated with these species (Lovelock, Andersen, and Morton 2003). Most collected spores at La Selva were from various morphospecies of the fungal genus *Acaulospora*. The mycorrhizal assemblage varied among tree species and habitats, but most of the variation was in frequency of fungal species rather than in clear restriction of fungi to particular plant species. The overall mycorrhizal network at La Selva is certainly much richer, since there are 1864 vascular plant species recorded from that site. This initial subsample, however, suggests that the mycorrhizal network will show a complex interaction structure in this lowland tropical forest. A similar analysis of another lowland tropical forest, Barro Colorado Island in Panama, used restriction fragment length polymorphisms and sequencing and found variation in the mycorrhizal assemblage across plant species, plant ages, and environments (Husband et al. 2002). Yet other studies show that different combinations of plant and AM fungi species result in different effects on plant and fungal growth (van der Heijden et al. 1998; Eom, Hartnett, and Wilson 2000), and experimental evidence suggests local adaptation in at least some of these interactions (Klironomos 2003). Together these studies suggest that AM mycorrhizal networks are much more geographically and temporally dynamic than previously suspected.

Complex species networks also develop with ectomycorrhizae, which are the other major form of mycorrhizal interaction. These phylogenetically diverse fungi commonly form mushroom fruiting bodies and are usually associated with large woody plant species, especially in temperate zones (Hibbett et al. 1997; Bruns, Bidartondo, and Taylor 2002). Like AM fungi, ectomycorrhizal (EM) fungi form intimate and obligatory associations with plant roots. These associations have arisen multiple times, mostly within the Ascomycota and Basidiomycota, and molecular phylogenetic evidence suggests that some lineages have produced species that have returned to the free-living state (Hibbett, Gilbert, and Donoghue 2000). Hence, these appear to be highly dynamic associations between plants and fungi.

Like AM fungi, EM fungi form networks that connect multiple plant individuals. Seedling plants become inoculated when they come into contact with hyphae in the soil Through these instances of horizontal transfer, a single tree or patch in a monoculture forest can include connections with a dozen or more EM fungal species (Bruns 1995), and a tree species may form connections with a wide range of phylogenetically distant EM fungi throughout its geographic range (Molina, Massicotte, and Trappe 1992).

Specificity to particular plant species varies among EM fungi, as does specificity of plant species to particular fungi. Among the specialist EM fungi, for example, several suilloid fungal genera form a monophyletic lineage restricted almost exclusively to plants in the Pinaceae, and most are restricted to single genera or species groups within that plant family (Bruns et al. 1998). In contrast, molecular analyses of mycorrhizal fungi in a region of mixed Douglas fir and pine have shown that other EM fungi associate with both these conifers (Horton and Bruns 1998). Patterns of variation in specificity are still being sorted out (Hoeksema 1999; Bruns, Bidartondo, and Taylor 2002), but local plant and EM fungal diversity is positively correlated in some regions (Kernaghan et al. 2003). Moreover, mycorrhizal assemblage structure changes during plant succession. The complex network structure in these interactions has yet to be fully analyzed at the community level for nestedness or compartmentalization.

The extent to which individual populations of plants and EM mycorrhizae are locally adapted and form geographic mosaics in their interactions is mostly unknown. In fact, the only known examples are for some orchid species and some parasitic plant species in the Monotropoideae that are known to be specialists on particular fungal groups. These are the only species for which the question has been assessed directly. The first demonstration of a geographic mosaic in mycorrhizal interactions involved two nonphotosynthetic orchids in the genus *Corallorhiza* and the fungi in the family Russulaceae on which they specialize (Taylor and Bruns 1999; Taylor, Bruns, and Hodges 2004). These two orchid species are patchily distributed across western North America. They differ in the EM fungi with which they form mycorrhizae even when they grow intermixed at the same site, suggesting a genetic basis to these differences in mycorrhizal association. Both species also show a strong geographic mosaic in their mycorrhizal associations, using different combinations of Russulaceae in different regions. Some of these geographic differences may directly reflect differences in geographic ranges between the plants and their fungal associates. Similar geographic differences in mycorrhizal associations are also known for nonphotosynthetic orchids in the *Hexalectris spicata* complex (Taylor et al. 2003).

Parasitic plants in the Monotropoideae show similar evidence of phylogenetic and geographic specialization in mycorrhizal associations. The family occurs in forests throughout the Northern Hemisphere, and molecular phylogenetic analyses indicate that each lineage within the family is specialized either to one EM fungal species or to a group of species within a single genus (Bidartondo and Bruns 2001, 2002). Among the plant taxa that have been studied, no fungal lineage is shared by any sympatric plant lineages, even though those sympatric combinations include sister taxa. The *Monotropa hypopithys* group, which is distributed across the Northern Hemisphere as differentiated populations or sibling species, shows parallel geographic divergence with fungi in the genus *Tricholoma*.

The results for *Corallohiza* orchids and plants in the Monotropoideae are probably not representative of mycorrhizal interactions, because they are for nontypical plant species. These plants are nonphotosynthetic exploiters of mycorrhizal networks. They parasitize other plants through their mycorrhizal connections and rely exclusively upon these mycorrhizal associations for fixed carbon. This exploitative lifestyle may require a greater degree of specificity than occurs in mutualistic mycorrhizal associations. Still, one study of photosynthetic tropical orchids has shown evidence of differences in mycorrhizal specificity among sister genera (Otero, Ackerman, and Bayman 2002). Further understanding of mycorrhizal networks will require a combination of molecular analyses and ecological experiments on local adaptation.

These studies will also require a better understanding of how mycorrhizae interact indirectly with herbivores, pathogens, and pollinators, and how mycorrhizal interactions differ among environments. Some of the conditions necessary for a geographic mosaic of coevolution in mycorrhizal networks have already been indicated from two kinds of analysis. Molecular and ecological studies show differences in metal tolerance in the EM fungus *Suillus luteus* on a polluted and nearby nonpolluted site (Colpaert et al. 2000), reinforcing the results from earlier ecological studies of AM and EM fungi (Taylor 2000). Although heavy metals impose perhaps unusually strong selection on populations, these results show the potential for mycorrhizal differentiation across landscapes. The other indirect evidence for a geographic selection mosaic in mycorrhizal networks comes from studies on the indirect effects of interactions between mycorrhizae and insect herbivores on pines discussed in chapter 3.

POLLINATING FLORAL PARASITES

Mutualisms between plants and pollinating floral parasites are the classic textbook cases of coevolution. In these interactions, insects lay eggs in the flowers of the same plants they pollinate. The mutualism depends upon free-living pollinators visiting multiple plants to pick up and deposit pollen, but it comes at a cost. The larvae feed on the flowers that the adult pollinated. Developing seeds eaten by the larvae are part of the cost of the mutualism.

Hence, part of the interaction is between free-living species, and part of it is symbiotic. The interaction is usually mutualistic only when it is simultaneously antagonistic, because pollination generally occurs only during oviposition into a flower.

Most pollinating floral parasites that have been studied in detail are highly host specific. Many of the textbook examples of obligate coevolved mutualisms are of these interactions—especially figs and fig wasps, and yuccas and yucca moths. New examples continue to be discovered, including globeflowers and globeflower flies (Pellmyr 1992), senita cactus and senita moths (Fleming and Holland 1998; Holland and Fleming 2002), and *Glochidion* plants and their *Epicephala* moths (Kato, Takimura, and Kawakita 2003). Phylogenetic analyses of the origins of some of these mutualisms, however, have suggested that the extreme specialization in these interactions is driven by the antagonistic part of the interaction rather than by the mutualism. In the few cases that have been studied in detail, the close relatives of these insects are highly host specific parasites of plants. The mutualism evolves as a serendipitous consequence of floral feeding and oviposition into the floral corolla by the ovipositing insects.

The most complete phylogenetic analyses are for moths in the family Prodoxidae, which includes a wide range of species that feed parasitically on different plant tissues. Only a few phylogenetically derived species in the family pollinate the hosts on which they live as larvae. These include one lineage within the genera *Parategeticula* and *Tegeticula* that pollinates yuccas and two lineages within the genus *Greya* that pollinate plants in the Saxifragaceae. Host specialization within the Prodoxidae is similar between the purely parasitic species and the pollinating floral parasites. In fact, all but a few prodoxids are highly host-specific (Davis, Pellmyr, and Thompson 1992; Pellmyr et al. 1996; Pellmyr, Leebens-Mack, and Thompson 1998), and even the species that are recorded from multiple hosts show geographic variation in the plants they use as hosts (e.g., Thompson 1997). Hence, the high degree of host specialization and the geographic variation in host use found in the mutualistic prodoxids cannot be attributed directly to mutualistic selection.

Earlier views that these interactions evolve toward obligate reciprocal extreme specificity across all populations are changing quickly. Now that more is becoming known about the local and geographic network structure of these interactions, it seems clear that these interactions sometimes form networks of closely related species. Of the eleven yucca moth species within the *T. yuccasella* group, seven of the pollinator species have only one known host, but one has two hosts, two have three hosts, and one has six hosts (Pellmyr 2003). Some globeflowers (*Trollius* spp.) are pollinated by *Chiastochaeta* flies, but globeflower populations may harbor multiple *Chiastochaeta* species that both pollinate and oviposit into the flowers (Pellmyr 1992). And the most commonly cited example of obligate reciprocal specificity—figs and fig wasps—is giving way to a realization that at least some fig species are pollinated by more than one fig wasp species. A molecular survey of eight Panamanian fig species uncovered cryptic fig-pollinating wasp species in four fig species (Molbo et al. 2003). In yet other interactions, the plants are pollinated by a combination of pollinating floral parasites and unrelated pollinators that do not oviposit into the flowers. Some populations of the plant *Lithophragma parviflorum* rely heavily upon *Greya politella* for pollination, but other populations are pollinated more by bombyliid flies and some solitary bees (Thompson and Cunningham 2002). Given the millions of years over which these interactions have coevolved and diversified, it should not be surprising that geographic ranges of plants and pollinating floral parasites have expanded and contracted repeatedly, producing geographically variable networks in these mutualistic interactions.

ANT-TENDING MUTUALISMS

Mutualistic interactions between ants and plants can also span the continuum from symbiosis to interactions among free-living species, but in a different way from pollinating floral parasites. At the one extreme, some ant species live symbiotically on one or a few related plant taxa, colonizing new plants through infectious spread. These interactions show the complex phylogenetic and geographic structure characteristic of most mutualistic symbioses with horizontally transmitted symbionts. At the other extreme are more facultative interactions between ants and plants with extrafloral nectaries or elaiosome-bearing seeds, which have a long history of study in evolutionary ecology and are still providing new perspectives on how these interactions may increase plant fitness (Koptur, Rico-Gray, and Palacios-Rios 1998; Oliveira et al. 1999) and ant fitness (Morales and Heithaus 1998). These interactions have diversified into multispecific networks that attract groups of unrelated species. Experimental studies that have excluded ants, wasps, or both have shown that the evolution of extrafloral nectaries can even favor the development of networks of interaction with predatory Hymenoptera that extend beyond ants to include wasps (Cuautle and Rico-Gray 2003). In between the extremes of symbiosis in ant-plant mutualisms is a wide range of interactions that vary among species and habitats in their potential for sustained symbiotic relationships with individual plants (Alonso 1998).

Although multiple studies have characterized the costs and benefits of these ecologically important interactions (Bronstein 1998; Hoeksema and Bruna 2000; Heil and McKey 2003), none has yet characterized the community-wide network structure in a way that would allow analyses of the degree of nestedness or compartmentalization that may result from coevolutionary selection. A few studies, however, have begun a systematic evaluation of geographic variation in network structure that will provide the basis for these analyses. Rico-Gray et al. (1998) found 135 pairs of interaction among 13 ant species and 42 plant species at Zapotitlan valley within Mexico and 312 associations among 30 ant species and 102 plant species in the coastal Veracruz region. When undertaken, formal studies of nestedness and compartmentalization of these networks should provide new insights into how sociality, mutualism, and different degrees of symbiosis shape coevolving networks.

Comparison of ant-plant interactions with those between ants and lycaenid butterflies or homopterans should provide additional insights into the geographic mosaic of coevolution and the network structure of interactions involving eusocial species (Morales 2002; Pierce et al. 2002). For example, some ants simply tend lycaenid butterfly larvae on their host plants, milking the larvae for the secretions they provide. Others, however, herd individual larvae each day between the plants on which the larvae feed and corrals in which the ants keep the larvae safe when not feeding. There is some evidence of geographic mosaics in these interactions. Ovipositing females of the lycaenid butterfly *Jalmenus evagorus* preferentially lay eggs on plants in response to cues from their natal ant populations rather than on plants with ants of the same species from other populations (unpublished work of A. Fraser cited in Pierce et al. 2002). Currently needed are analyses of the local and geographic network structure of ant-lycaenid interactions and similar interactions that differ along the continuum of husbandry and degree of symbiosis.

Variable Outcomes and Cheating in Mutualisms

The network structure of mutualisms between free-living species almost guarantees that interactions between species will vary among years and across environments. Pairs of species are known to vary among habitats and regionally in their effectiveness as mutualists (Bronstein 1994a; Thompson and Cunningham 2002; Thomson 2003). Moreover, indirect interactions among species create conflicting or correlated selection pressures that are likely to vary geographically (Armbruster et al. 1997; Strauss 1997a; Strauss et al. 1999; Herrera et al. 2002; van Ommeren and Whitham 2002). Local populations of cheater species are one of the consequences. Cheating is a loaded term, but here it simply means any species that exploits a mutualism without providing a fitness gain to the other species. Exploiters of mutualism have been reported within some populations of most mutualisms that have been studied in detail (Addicott 1996; Taylor and Bruns 1997; Pellmyr 1999; Stanton et al. 1999; Marr, Brock, and Pellmyr 2001; Ferriere et al. 2002). They are part of the cost of mutualism and the accumulation of species into mutualistic networks.

Although the dissolution of mutualisms through cheating has long been considered a problem for the long-term maintenance of mutualism, part of the problem arises from lack of consideration of the geographic mosaic of coevolution. Cheaters are unlikely to invade all local populations throughout the geographic range of a mutualism, and their effects are likely to vary among populations. Among yucca moths, for example, cheater lineages have arisen repeatedly (Pellmyr and Leebens-Mack 2000). These antagonistic moth species lay their eggs in yucca fruits but do not pollinate the flowers (Addicott 1996; Pellmyr, Leebens-Mack, and Huth 1996; Bronstein and Ziv 1997; Marr, Brock, and Pellmyr 2001). Within the *T. yuccasella* group, two of the ten species are antagonistic (Pellmyr 1999). Both cheater species are highly derived taxa within the phylogeny and overlap with pollinator species in complex geographic patterns across North America (fig. 14.6). Geographic studies of the distribution of sister pairs, one a pollinator and one a cheater, suggest a mosaic pattern of species distributions that surely must have changed repeatedly as yuccas, mutualistic yucca moths, and parasitic yucca moths have coevolved over millions of years (Pellmyr and Leebens-Mack 2000). The five illustrated species are only a subset of the currently recognized seventeen yucca moth species that occur across North America, including Mexico, in various sympatric combinations. The overall distribution of yuccas and yucca moths includes multiple regions lacking cheater populations of the moths.

Local instances of cheating in populations therefore seem unlikely to result in global dissolution of a mutualism, although they could change both the local and global coevolutionary trajectories. Mathematical models of the geographic mosaic of coevolution show that, if a mutualist becomes a cheater (i.e., antagonist) in part of its geographic range, it can prevent fixation of mu-

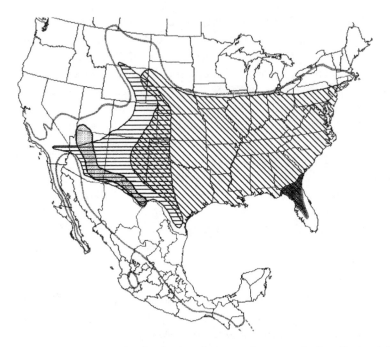

Fig. 14.6   The overall geographic distribution of yuccas and yucca moths (gray line) and the distribution of pollinating yucca moths and their sister cheater species within that range: pollinating *Tegeticula cassandra* (dark shading) and cheating *T. intermedia* (diagonal lines); pollinating *T. elatella* and *T. superficiella* (light gray shading) and cheating *T. corruptrix* (horizontal lines). Broken lines at the very bottom of figure indicate the southern extension of yucca in cultivation in the absence of its pollinators. From Pellmyr and Leebens-Mack 2000.

tualistic traits if there is gene flow among populations (Nuismer, Thompson, and Gomulkiewicz 1999, 2000; Gomulkiewicz, Nuismer, and Thompson 2003; Nuismer, Thompson, and Gomulkiewicz 2003). Viewed another way, however, these models also suggest that cheating by one species in part of the geographic range of an interaction will not necessarily lead to complete dissolution of the mutualism throughout its geographic range. The overall outcomes depend upon the relative strength of mutualistic and antagonistic selection across landscapes, the pattern and level of gene flow, and the mode of density regulation among populations (Gomulkiewicz, Nuismer, and Thompson 2003). Similarly, invasion of a pairwise mutualism by an antagonistic species or coevolution among three species in which one species has the potential to be antagonistic in some situations does not necessarily lead to dissolution of the mutualism in models that incorporate geographic structure

(Gomulkiewicz, Nuismer, and Thompson 2003; Wilson, Morris, and Bronstein 2003). Future geographic models with more complex network structure will help further resolve how variable mutualisms shape the dynamics of mutualistic networks.

Variable outcomes in mutualism between any particular pair of free-living species should leave a phylogenetically patchy imprint, with mutualism arising and dissipating repeatedly among populations and species. Mutualistic interactions persist among the interacting lineages, but contributions of individual populations and species come and go either locally or throughout their geographic ranges. Although the data are still few, long-term stability of mutualistic networks between free-living species appears to result from the stabilizing effects of network structure played out across complex landscapes, despite local failures of mutualism between particular pairs of species.

Future Studies of the Coevolutionary Dynamics of Mutualistic Networks

Overall, the coevolutionary dynamics of mutualistic networks are more intricately structured, both locally and geographically, than most treatments of mutualism have so far been able to confront. But we now have better ecological, molecular, mathematical, and statistical tools to evaluate how mutualistic networks coevolve and become stabilized over time. Ecological approaches to understanding the structure of local mutualistic selection within a community context have improved greatly in recent years. Phylogenetic analyses have made it possible to understand the overall patterns of divergence among taxa involved in mutualistic relationships. Phylogeographic methods are making it possible to evaluate patterns of secondary contact among taxa, showing how multiple closely related species have become involved in local mutualisms. Molecular and ecological approaches together have advanced to the point where it is now even possible to begin estimating in some cases the number of genes involved in switching from one set of mutualistic partners to another (Schemske and Bradshaw 1999; Bradshaw and Schemske 2003). Finally, evolutionary approaches to community ecology are beginning to help decipher how unrelated species converge on mutualisms in different ways in different communities.

The major goal of research on mutualism among free-living species should be to understand how coevolution shapes network structure across landscapes. Once we understand that, we will be in a much better position to understand how traits evolve within mutualisms and how mutualistic networks have persisted and spread over millions of years. Those results will also put us in a much better position to understand how to conserve the coevolutionary dynamics of mutualisms across landscapes that are dominated increasingly by introduced species. Many mutualistic networks readily accumulate new species, including invasive species. We therefore need to understand the consequences of species introductions and landscape fragmentation within the broader overall context of coevolutionary selection on mutualistic networks.

Conclusions

The multispecific structure of mutualisms between free-living species is not the result of some highly diffuse and inexplicable process. It is a direct result of coevolutionary selection to form mutualistic networks that specialize on a core set of mutualistic traits. The process involves both complementarity in traits important to fitness and convergence of unrelated species on those traits. Species composition and degree of specialization within any particular species are highly likely to vary across landscapes as a direct result of the coevolutionary process. Current work suggests that, through this process, coevolution within mutualistic networks of free-living species favors nested networks. Such trophic mutualistic networks may be more likely to be statistically nested than trophic antagonistic networks, because mutualistic networks are not subject to the coevolutionary alternation that may shape antagonistic networks.

As mutualistic networks accumulate species, coevolutionary selection favors asymmetries in specialization built around a core of generalist species. These networks create novel lifestyles that evolve as a direct result of the predictable local availability of multiple mutualistic partners. Coevolved mutualistic networks therefore create a wide range of lifestyles, including species that interact with many species within the network. The process itself creates the illusion over time of a loose, noncoevolved assemblage.

Mutualistic networks among free-living species are a crucial part of all major communities worldwide and can no longer be excluded from any scientific evaluation of the ecological and coevolutionary organization of biological communities. These networks connect taxa within and among communities and sometimes even among biogeographic regions. Many pollinators and frugivores use multiple plant species, and some migrating hummingbirds and frugivorous birds interact with plants each year across two continents. Eliminate pollination and seed dispersal by animals, and reproduction would simply cease in a large proportion of plant species. In turn, a high proportion of insects, birds, and mammals would no longer be able to survive in the absence of the nectar or fruit on which they depend. The loss of these mutualisms would collapse the trophic structure and organization of terrestrial communities worldwide.

Together, these three chapters on mutualism suggest that the continuum between symbiotic and nonsymbiotic interactions is crucial to our understanding of coevolving mutualisms. Selection on symbiotic mutualisms generally favors genetically restricted mutualistic interactions and sets of complementary mutualists, whereas selection on mutualisms among free-living species favors not only complementarity among mutualists but also convergence of unrelated species on the complementary traits. The varying importance of coevolutionary selection for complementarity and convergence along the continuum from symbiotic mutualisms to mutualisms among free-living species creates different patterns in the geographic mosaic of coevolution.

Coevolutionary complementarity and convergence also help us understand why mutualisms have become central to the organization of biodiversity. Through coevolutionary complementarity, natural selection has favored a wide range of obligate symbiotic associations, and many of these associations have persisted as the interacting lineages have diversified worldwide over millions of years. Coevolution has also favored the development of mutualistic networks among free-living species that remain stable as species diversity and accumulate species through coevolutionary convergence. The persistence and continued spread of mutualisms almost guarantees that most species, or at least most multicellular species, are part of one or more mutualistic interactions.

These chapters on mutualism also reveal just how much our understanding of the evolutionary processes shaping mutualisms has matured over the past decade. We now have working hypotheses on how coevolutionary selection varies along the continuum from antagonistic to mutualistic symbiosis and along the continuum from mutualistic symbioses to mutualisms between free-living species. We also have some predictions on how the structure of mutualism should vary across landscapes. These hypotheses and predictions can help organize coevolutionary research on mutualisms by reaching beyond cost-benefit analyses to address broader questions, including the conditions favoring high or low genetic diversity in mutualistic interactions, different forms of network structure, and the geographic mosaic of outcomes in coevolving interactions.

15 Coevolutionary Displacement

Coevolutionary displacement is a specific local outcome of the geographic mosaic of coevolution. Any geographic region of overlap between species that use the same habitat or resources is a potential coevolutionary hotspot that may result in displacement in morphological, physiological, behavioral, or life-history characters. Local character displacement, whether coevolutionary or focused on one species within a network, is only part of the geographic mosaic. If a guild of similar species overlaps geographically in multiple places and in different combinations, all or some species may become displaced in morphological traits in one region but become displaced in behavior or habitat use in another, depending upon the local community context in which the interaction takes place. Clines may develop in characters across landscapes as populations share genes across coevolutionary hotspots and coldspots. As in other forms of interaction, then, geographic selection mosaics, coevolutionary hotspots, and trait remixing across populations may all contribute to the overall geographic pattern of character displacement found among species that share similar ecological niches. Viewing character displacement in this way emphasizes that any local particular outcome of coevolutionary displacement is part of the broader geographic mosaic of coevolution among species.

Consideration of geographic selection mosaics and the comparative structure of coevolutionary hotspots will become increasingly important in studies of character displacement now that it is evident that modes of interaction other than competition may contribute to displacement in characters and habitat use among coexisting species. Species may compete directly in sympatry for limiting resources, favoring individuals with traits that allow them to specialize on the resources not used by the other species. Alternatively, sympatric populations may share predators or parasites, favoring divergent individuals that suffer lower rates of predation or parasitism because they have different traits or use different habitats. Even shared mutualists have the potential to drive displacement in sympatric species. Coevolutionary displacement has therefore turned out to be more than simple divergence of two or more closely related species competing directly for limiting resources.

In this chapter, I evaluate our current understanding of coevolutionary displacement among pairs or groups of species as a specific form of the geographic mosaic of coevolution. Throughout the chapter, I use the specific phrase "coevolutionary displacement" rather the more general phrase "ecological character displacement" whenever I am emphasizing the role of coevolution in displacement. The chapter focuses on four specific hypotheses:

- Displacement of competitors across coevolutionary hotspots
- Displacement through apparent competition across coevolutionary hotspots
- Replicated guild structure among coevolutionary hotspots
- Trait overdispersion in multispecific competitive networks

Characterizing Coevolutionary Displacement

When closely related species overlap in their geographic ranges, they sometimes differ in more traits in the region of overlap than in regions of allopatry. These differences can have three nonexclusive causes. Two species that are geographically separated for a long period of geological time will almost inevitably differ in some respects as they evolve in different environments. Where these species come into secondary contact, additional differences between them could arise simply because the physical environment in that region is different from the other environments in which these species occur. Alternatively, the species may diverge in sympatry through reproductive character displacement, if the interspecific hybrids have inherently lower survival or reproduction. Under these conditions, natural selection may favor individuals with traits that decrease the change of mating with the other species. The third potential cause is ecological character displacement in which the differences in sympatry are driven directly or indirectly by ecological (e.g., competitive) rather than reproductive interactions. At the extreme, only one of the species may diverge in traits in regions of overlap. Most well-studied cases of ecological character displacement, however, suggest that both species sometimes undergo selection for displacement where their ranges overlap, thereby leading to coevolutionary displacement.

Character displacement is evident in three observed patterns of traits found among sympatric species (Schluter 2000b), although the patterns themselves do not necessarily imply that coevolutionary displacement has produced them. Some pairs or groups of species show exaggerated divergence of traits in sympatry as compared with the same species in allopatry. Some larger groups of closely related species (e.g., *Anolis* lizards in the Caribbean) show replicated, but independently derived, guild structures throughout their geographic ranges. Observation of these patterns depends directly upon knowledge of the geographic structure of traits in the interacting species. In addition, even larger local assemblages of species show overdispersion of traits among species within a guild that shares common resources. Schluter's (2000a, 2000b) comprehensive analysis of examples of character displacement indicates that exaggerated divergence in sympatry is the most common kind of evidence for displacement. It is also, however, probably the most common kind of evidence evaluated in studies of character displacement, because it can be shown through displacement of only two species in a zone of sympatry when compared to neighboring zones of allopatry.

Schluter's (2000a) analysis evaluated published studies of character displacement based upon six criteria first proposed by Schluter and McPhail (1992). The criteria are a synthesis of the criteria proposed by various researchers over the previous twenty years: (1) phenotypic differences among populations and species have a genetic basis; (2) chance can be ruled out as an explanation of the pattern; (3) population and species differences are evolutionary shifts and not simply a result of species sorting; (4) shifts in resource use match changes in phenotypic traits associated with use of those resources; (5) environmental differences between sympatric and allopatric sites are not the cause of the pattern; and (6) independent evidence suggests that similar phenotypes compete for resources. By Schluter's analysis, there were, at end of the twentieth century, sixty-four examples of interactions between two or more congeneric species that met at least one of the six criteria. Most of the examples were of vertebrates, and additional examples continue to accumulate (e.g., Adams and Rohlf 2000; Melville 2002). Only a few examples are of plants, and most of those involve reproductive characters (e.g., Armbruster, Edwards, and Debevec 1994; Caruso 2000; Hansen, Armbruster, and Antonsen 2000).

Schluter and McPhail's criteria have become the gold standard for studies of ecological character displacement based upon interspecific competition, and they have helped impose increased rigor on studies of local divergence in coevolving competitors. Few of the published examples of character displacement come close to meeting all the criteria, and few have been studied through field experiments rather than observational or laboratory experiments. Meeting the criteria, however, is only the beginning rather than an end to the analysis of coevolutionary character displacement driven by competition. Local regions of coevolutionary character displacement are coevolutionary hotspots and are, as such, only one part of an overall analysis of the geographic mosaic of coevolution between coevolving species.

Exaggerated Divergence in Sympatry I:
Competitive Character Displacement across Coevolutionary Hotspots

There are currently two coevolutionary hypotheses for the ecological causes of exaggerated divergence of characters within local coevolutionary hotspots: one driven by direct competition among species for limiting resources, and the other, by shared interactions with predators, parasites, or mutualists. The current working hypothesis on competitive character displacement can be cast as follows within the framework of the geographic mosaic theory of coevolution:

> *The hypothesis of displacement of competitors across coevolutionary hotspots.*
>
> *Closely related species that compete for a potentially limiting resource coevolve through divergence in traits or habitat use when populations come into contact in regions where the resource is limiting to both species (i.e., coevolutionary hotspots), but the pattern of displacement may differ among hotspots.*

Closely related species are likely to be strong competitors for some resources, because they are often slightly variant forms of their common ancestor. These descendant species seldom have geographic ranges identical to those of their ancestor. The regions of allopatry are structural coevolutionary coldspots, which have been the historical "controls" for evaluating character divergence in sympatric populations of competitors. Where their ranges overlap, species may compete for resources in some environments but not in others. Only some environments may therefore become coevolutionary hotspots for these species. Each hotspot may favor different patterns of competitive character displacement, because species differ in their relative competitive ability among environments, as shown by a wide range of experiments (e.g., Goodnight and Craig 1996; Hedrick and King 1996; Joshi and Thompson 1996; Aarssen and Keogh 2002). Such experiments differ in the degree to which the measured traits reflect true relative fitness differences shaped by competition among environments (Aarssen and Keogh 2002), but they nevertheless often show that the relative direct or indirect ecological responses of species to interspecific interactions with competitors are not uniform across environments. Overall, then, interspecific competition should show strong evidence for geographic selection mosaics, coevolutionary hotspots, and coevolutionary coldspots that are either structural (i.e., one of the competitors is missing) or functional (i.e., the species are sympatric in regions in which resources are not limiting).

In recent years, mathematical models of character displacement along geographic gradients have added a broader perspective to the way in which geographic mosaics may be shaped by competitive divergence (Case and Taper 2000). In these models, interspecific competition lowers population densities in sympatry, which produces a large asymmetry in gene flow between regions of sympatry and allopatry. The result is a lower probability of local adaptation in some regions of sympatry, creating the possibility of local extinction in some populations and sharp borders between the species. Extrapolating from the models, coevolutionary hotspots should vary in the length of time they persist in different regions of overlap. Productivity gradients across the ranges of these interactions can modify the resulting geographic mosaic of coevolution in these models, as has also been shown in models of predator-prey coevolutionary hotspots (Hochberg and van Baalen 1998).

If competition is responsible for coevolutionary character displacement across landscapes, then the pattern of displacement within coevolutionary hotspots should reflect the pattern of competitive intensity. In highly mobile species with much gene flow among neighboring populations, there may be little internal pattern within the zone of sympatry. Displacement may be evident only as differences in means of characters between allopatric and sympatric populations. In less mobile species, however, the zone of sympatry may be a micromosaic of populations showing different degrees of character displacement. Displacement may then be visible as a cline in traits across the coevolutionary hotspot that matches the relative frequency of the competing species across the hotspot (fig. 15.1). Such clines, however, could also be

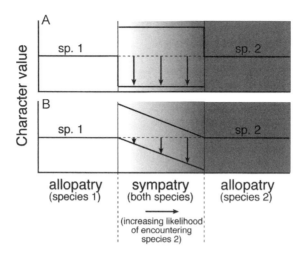

Fig. 15.1   Two possible internal structures to a competitive coevolutionary hotspot. (A) Co-evolutionary displacement throughout the hotspot with no internal structure. (B) Clinal displacement across the hotspot, matching differences in the relative frequency of the two species. A third possible internal structure is a micromosaic of displacement rather than a cline, if the environment does not vary continuously across the hotspot region. After Pfennig and Murphy 2002.

driven by other forms of interspecific interaction, such as predation, or by hybridization between species across a contact zone. If those causes can be eliminated, then competitive character displacement becomes a likely cause of the pattern.

Few studies have attempted to analyze the internal dynamics of competitive coevolutionary hotspots, but results for spadefoot toads in western North America suggest a pattern that may be common (Pfennig and Murphy 2000, 2002, 2003). Two species of spadefoot toad, *Spea bombifrons* and *S. multiplicata*, are sympatric over a broad area of southwestern United States and northern Mexico and occur in the same ponds. Allopatric populations of both species span equally broad regions (fig. 15.2). The species compete for food, and competition can be particularly high for access to fairy shrimp as prey. These toad species therefore show the classic interspecific geographic structure under which competitive character displacement should be possible.

Both species are capable of developing into two morphological forms as tadpoles. All the tadpoles are initially omnivorous, with small heads suited to feeding on detritus. But if they ingest fairy shrimp early in life, they can

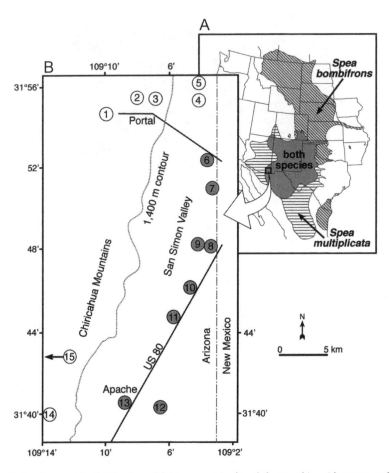

Fig. 15.2   Geographic distribution of (A) two species of spadefoot toad in midwestern and southwestern North America and (B) study sites within a region of sympatry. Gray circles represent ponds containing both species, whereas open circles represent ponds containing only *S. multiplicata*. After Pfennig and Murphy 2002.

develop into large-headed carnivores that specialize on these shrimp (fig. 15.3). Geographic analyses of traits and laboratory experiments on trait response under competition suggest that these toad species have undergone coevolutionary character displacement for expression of these morphs when in sympatry. Expression of the different morphs by the two species is more exaggerated in sympatric than in allopatric populations, indicating that the degree of plasticity in morph expression has been a focus of natural selection

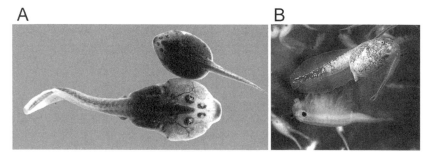

Fig. 15.3   Different morphs of the spadefoot toads (genus *Spea*). (A) The omnivore morph is the upper tadpole, and the carnivore morph is the lower tadpole. (B) The omnivore in the upper part of the photograph is eating a fairy shrimp. From Pfennig and Murphy 2002.

on these competitors. *Spea bombifrons* are more carnivorelike in sympatric populations, and *S. multiplicata* are more omnivorelike (Pfennig and Murphy 2002).

Competition can be intense among individuals of the same morph (Pfennig 1992). High levels of competition for the shrimp limit the chance that an individual will ingest shrimp early in life and develop into a carnivorous morph (Frankino and Pfennig 2001). Geographic differences in the genetic tendency to develop into a carnivore morph or an omnivore morph can therefore be used as an index of the pattern of competitive intensity across a zone of sympatry. It becomes a probe into the internal structure of a co-evolutionary hotspot.

Within southeastern Arizona, ponds at different elevations vary in presence of the two species. At elevations above fourteen hundred meters, only *S. multiplicata* occurs, but at lower elevations both species occur in various proportions. Omnivorous morphs of *Spea multiplicata* are more frequent in ponds in which the relative abundance of *S. bombifrons* is highest, suggesting that the presence of *S. bombifrons* decreases the chance that *S. multiplicata* tadpoles will encounter fairy shrimp early in life and transform into the carnivorous morph (Pfennig and Murphy 2002). In controlled laboratory experiments, *S. multiplicata* tadpoles become more omnivorelike in response to interspecific competition, and *S. bombifrons* tadpoles become more carnivorelike (fig. 15.4). These species therefore show frequency-dependent character displacement mediated by this phenotypically plastic response. Pfennig and Murphy (2002) refer to such cases as facultative character displacement.

Selection to minimize competition with *S. bombifrons* has generated her-

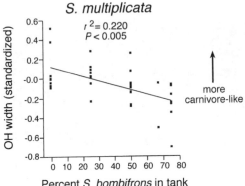

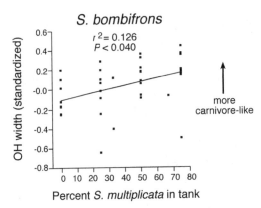

Fig. 15.4   Change in the mean orbitohyoideus muscle (OH) width in two competing spadefoot toads when placed in interspecific competition at various frequencies of the other species. Greater OH widths are associated with more carnivore-like jaw morphology. The standardized values of OH are the residuals of the regression of OH width on snout-vent length of the tadpoles to control for differences among tadpoles in overall body size. After Pfennig and Murphy 2002.

itable divergence in carnivore production between neighboring high- and low-elevation populations of *S. multiplicata*. At high elevations, where *S. bombifrons* are absent, *S. multiplicata* populations maintain plasticity to express the carnivore morphology. In contrast, at lower elevations, where *S. bombifrons* are present, *S. multiplicata* populations have become canalized and have evolved reduced propensities (and, in some cases, completely lost the ability) to express the carnivore morphology. Moreover, this pattern of divergence in morph production is replicated in different populations that experience similar selective pressures (D. W. Pfennig, personal communication). Thus, selection imposed by *S. bombifrons* has led repeatedly to the evolution of canalized populations of *S. multiplicata*, which express only the omnivore phenotype.

One other major study has evaluated frequency-dependent divergent selection across landscapes in competing species (Schluter 2003). Threespine

Fig. 15.5   Representative benthic and limnetic forms of threespine sticklebacks (*Gasterosteus aculeatus*) from Priest Lake, British Columbia. The fish have been stained to highlight bone. Black scaling bars: 5 mm. From Peichel et al. 2001.

stickleback fish (*Gasterosteus aculeatus*) in western Canada differ in feeding morphology and the food resources they use (fig. 15.5). Sympatric fish in small lakes always consist of a genetically larger benthic-feeding form that eats invertebrates on the sediment and a genetically smaller limnetic-feeding form that eats zooplankton. In contrast, solitary species are intermediate in morphology and diet (McPhail 1992; Schluter and McPhail 1992; Schluter 2000a). Each pair of species in small lakes in British Columbia appears to have originated near the end of the Pleistocene through two invasions of freshwater by the marine zooplankton-feeding species (Taylor and McPhail 1999, 2000). The first invasion favored intermediate forms, and the second invasion favored character displacement between coexisting species (fig. 15.6). A series of careful experimental studies has shown that threespine sticklebacks compete for food and that competition promotes divergence in feeding morphology in sympatric species (e.g., Schluter 1994, 1996b; Pritchard and Schluter 2001; Schluter 2003). When an intermediate form is challenged with either a limnetic form or a benthic form, it suffers reduced fitness. Moreover, competition is strongest between phenotypes that are most similar, and the intensity of selection on intermediate forms depends upon the frequency of the other forms. Hence, selection on character divergence varies with frequency-dependent competition (Schluter 2003).

Character displacement also appears to have occurred in stickleback species in eastern Canada. Brook sticklebacks (*Culaea inconstans*) that are sympatric with ninespine sticklebacks (*Pungitius pungitius*) have diverged in

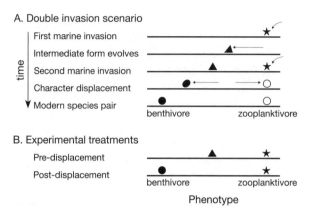

Fig. 15.6   (A) Hypothesized sequence of invasion of coastal lakes by threespine stickleback species and character displacement in lakes with coexisting species. Symbols (stars, triangles, and circles) indicate the mean position of populations along the morphological continuum. Curved arrows indicate invasion by a marine ancestor, and straight arrows indicate the direction of evolutionary change in morphology. (B) The structure of experimental treatments designed to test for stages in the hypothesized sequence before and after displacement. From Schluter 2000b.

morphology, and experimental studies indicate that interspecific competition may have driven the divergence (Gray and Robinson 2002). These studies of other stickleback species suggest that competition for food in sticklebacks is not simply a result of some particular characteristic of threespine sticklebacks. Competitive character displacement appears to be a possible outcome wherever stickleback species co-occur.

Exaggerated Divergence in Sympatry II: Coevolutionary Displacement Driven by Causes Other than Direct Competition

Exaggerated divergence of characters in sympatry need not always be driven by direct competition for limiting resources. The alternative hypothesis invoking interspecific interactions can be stated as followed, framed within the context of the geographic mosaic of coevolution:

The hypothesis of displacement through apparent competition across coevolutionary hotspots.

Species that share predators, parasites, or mutualists coevolve through divergence of traits, habitat use, or use of mutualists when populations come into contact in regions where their fitnesses decline as a result of these shared interspecific interactions. The pattern of displacement may differ among coevolutionary hotspots.

Species often differ in traits that affect not only their direct access to resources (e.g., food, different parts of trees) but also in traits that enhance avoidance of enemies (e.g., running speed, use of particular microhabitats) or access to mutualists. Some mathematical models of coevolution have suggested that two prey species that compete for the same resources and share a predator in sympatry could undergo character displacement driven largely by the shared predator (Abrams 2000b; Abrams and Chen 2002). The result could be a geographic selection mosaic in which competition, predation, and other modes of interaction differ in their influence on character displacement among regions of overlap.

No study has shown that predation alone drives character displacement in any coexisting species, but there are indications that predation at least contributes to geographic patterns in resource use in coexisting species. Three-spine sticklebacks not only compete in small lakes in British Columbia; they are also preyed upon by cutthroat trout, which is the only other fish species native to lakes containing both limnetic and benthic sticklebacks. In the absence of trout in experimental ponds, limnetic sticklebacks survive better than benthic sticklebacks. When cutthroat trout are introduced into the ponds, benthic fish survive better because they are less vulnerable to predation (Vamosi and Schluter 2002). These differences in vulnerability are reflected in differences in traits between the stickleback forms. In contrast to limnetic fish, benthic fish are relatively large, have poorly developed armor, and do not school with conspecifics. Hence, competition and predation may potentially interact to shape the patterns of character divergence observed across the geographic ranges of these sticklebacks (Vamosi and Schluter 2004).

Field studies on nesting bird assemblages have also suggested that both competition and predator avoidance may drive character displacement (Martin 1993). Artificial and real nests of songbirds that occur together within microhabitats in Arizona suffer a higher level of predation than nests that overlap less in placement within microhabitats. Moreover, census studies of these birds indicate that habitat partitioning for nest sites is greater than for foraging sites. Hence, as for sticklebacks, competition and predation may interact to shape character displacement among some birds.

Character displacement may also be driven by mutualism, and the clearest examples involve divergence of floral traits of congeneric plants to attract different pollinators, as has occurred in *Dalechampia* in multiple tropical regions (Armbruster 1997; Armbruster et al. 1997; Hansen, Armbruster, and Antonsen 2000) and triggerplants (*Stylidium* spp.) across multiple regions of overlap in southwestern Australia (Armbruster, Edwards, and Debevec 1994). Such divergence in sympatry could be driven by natural selection to compete for a limited number of efficient local pollinators, as has been suggested for *Dalechampia* and *Stylidium*. Alternatively, or additionally, such divergence among some plant species could be driven by reproductive character displacement, by favoring plants with floral traits that minimize the chance of deposition of heterospecific pollen or by indirectly favoring specialized pollinators that remain locally constant to that one plant species.

Symbiotic mutualisms that are horizontally transmitted also have the potential to favor character displacement, because these mutualisms sometimes favor reciprocal specificity (see chapter 12). Hosts that acquire the symbionts of a sympatric congener may suffer lower fitness, either because the symbiont is poorly adapted to that host or because the novel symbiont competes with the original symbiont. Under these conditions, natural selection could favor use of different habitats by hosts or some other form of character displacement that minimizes the potential to acquire the symbiont of the other host species. No studies have yet evaluated the potential for this kind of mutualist-driven character displacement.

The most obvious situation in which coevolutionary character displacement is mediated by symbionts—ancient, obligate symbionts in this case—is in competition for light among species that harbor photosynthetic mutualists. Competition for light in plants is actually competition for light required for the plant's ancient coevolved mutualism with chloroplasts. Interspecific competition for light should favor plant morphologies that optimize the contribution of chloroplasts. Similarly, different morphologies in coexisting corals and fungi that form lichens may also reflect character divergence to maximize use of the light required by these species for mutualistic photosynthetic symbioses. These are extreme examples, but they highlight the possibility that character displacement among species may be driven by selection to maximize the mutualistic effects of symbiotic mutualisms.

The Geographic Mosaic of Replicated Guild Structure

Increasingly, studies of coevolutionary character displacement are taking broad geographic approaches that draw interpretation from a combination of phylogenetic, phylogeographic, and ecological studies. Some of these studies suggest that interacting species coevolve in predictable ways when they come into contact. Individual species may differ regionally in their evolutionary responses, but the overall interaction takes on a predictable structure when species come into contact. The limnetic and benthic sticklebacks are one example, but other examples suggest multispecific character displacement within guilds of species. The current view can be summarized as follows:

The hypothesis of replicated guild structure among coevolutionary hotspots.

Multispecific assemblages of competing congeners undergo similar patterns of character displacement wherever they occur in sympatry, although individual species may be displaced in different ways in different geographic regions.

Anolis lizards in the Caribbean and Bahamas have been perhaps the most intensively studied species for geographic patterns of coexistence, interspecific competition, and repeated patterns of multispecific character displacement (Williams 1969; Losos 1994; Roughgarden 1995; Miles and Dunham 1996; Losos et al. 2003; Schoener, Spiller, and Losos 2003). Anoles are a diverse group of almost 400 species, of which approximately 140 occur in the Caribbean and 111 in the Greater Antilles (Cuba, Hispaniola, Jamaica, and Puerto Rico) (Jackman et al. 1999). The islands differ in the number of anole species they contain, and individual species differ in their probability of successful colonization of some islands (Losos and Spiller 1999). Extinction, driven by natural disasters, can occur on smaller islands, suggesting that assemblage structure has been dynamic on at least these smaller islands over geologic time (Spiller, Losos, and Schoener 1998). Nevertheless, species have specialized in habitat use in similar ways across all these islands, producing small-twig specialists, base-of-tree specialists, trunk specialists, arboreal specialists, tree-canopy giants, and grass specialists. Not all islands have all these specialists, but the specialists look similar on each island on which they occur.

Remarkably, each of these habitat specialists has arisen multiple times, yet assemblage structure among islands is highly convergent (Losos 1995; Losos et al. 1998). The Greater Antilles have six ecomorphs, and four of them are common to all the islands. Phylogenetic analyses have suggested that none of the four ecomorphs shared among islands is a monophyletic group. The ecomorphs have arisen repeatedly through at least seventeen transitions among ecomorph classes. Moreover, there has been very little niche conservatism during divergence of these species (Losos et al. 2003). Cuba harbors fifty-eight *Anolis* species, and eleven species co-occur in western Cuba. Phylogenetic reconstruction of the western Cuba assemblage suggests that closely related species are no more ecologically similar than chance would dictate. Coevolving interspecific interactions have been more important than phylogenetic conservatism in shaping competitive coevolution among these congeners.

The ecomorph structure of anole assemblages has apparently remained intact amid repeated environmental changes over tens of millions of years. Fossil tree-canopy specialists in amber from twenty million years ago look almost the same as extant tree-canopy ("trunk-crown") specialists (de Queiroz, Chu, and Losos 1998). Apparently, these assemblages have a very limited range of coevolutionary options wherever competing anoles come into contact. Nevertheless, the process of displacement is ongoing and may occur quickly within those bounds, as indicated by experimental introductions of lizards onto islands. Two species of *Anolis* were introduced onto small islands in the Bahamas in 1977 and 1981 and measured for traits in 1991 (Schoener and Schoener 1983; Losos et al. 2001). Both species had undergone morphological divergence, although the current results cannot eliminate phenotypic plasticity rather than genetic change as the cause (Losos, Warheit, and Schoener 1997; Losos et al. 2001).

Besides the evolution of discrete ecomorphs, anoles also undergo further differentiation within ecomorphs on large islands. Ecomorphs are based upon differences in microhabitat use (e.g., trunk-ground specialists), but variants of an ecomorph may be distributed across macrohabitats that differ in vegetation type, topography, and climate on an island. Within Hispaniola the *A. cybotes* species complex consists of trunk-ground specialists, but populations differ in morphology among macrohabitats (Glor et al. 2001). The overall results for anoles suggest coevolutionarily dynamic interactions that are conservative in some respects, such as number of potential ecomorphs, but subject to ongoing natural selection that has molded the same species in different ways on different islands and in different habitats within an island. Even the role of coevolutionary character displacement itself is variable, generating the usual patterns of guild structure in some regions but possibly not in others (Losos 1996).

Trait Overdispersion within Competitive Networks

As the number of potentially competing species increases beyond congeners, so do the number of ways in which coevolutionary displacement can shape interactions across landscapes. The exact same composition of local competitor species will almost never be replicated across most populations of any widespread species. For example, populations of the frog *Pseudacris crucifer* have the potential to interact competitively with only one or two anuran species in the northern part of their geographic range, but they may encounter up to ten or more other anuran species in the southern part of their range (Morin 2003). Moreover, environmental conditions vary among populations, adjusting relative competitive ability across landscapes.

How coevolution molds networks of potential competitors beyond congeners remains one of the great current challenges in coevolutionary biology. It is a specific version of the broader question of how the geographic mosaic of coevolution may contribute to the formation of predictable coevolved network structures of species across landscapes. In some mathematical models, competitive coevolution can result in fewer local interacting species than in model assemblages that allow invasion but no coevolution (Rummel and Roughgarden 1983). A long-standing hypothesis in evolutionary ecology is that local multispecific competitive networks may sometimes coevolve in predictable ways (Brown and Wilson 1956), resulting in constant size ratios among species for traits undergoing coevolutionary divergence (Hutchinson 1959) (fig. 15.7). As a coevolutionary hypothesis, trait overdispersion implies that all locally competing species in sympatry become displaced in traits along a single resource axis relative to their allopatric populations.

The hypothesis of trait overdispersion in multispecific competitive networks.

Competing species diverge within coevolutionary hotspots to form assemblages that differ by a constant ratio in the traits under coevolutionary selection. Individual species may be displaced in different ways in different geographic regions.

Dozens of studies have shown statistical overdispersion of traits in networks of coexisting potential competitors, and the ratios for the presumed traits under selection generally range between 1.03 and 1.98 (Schluter 2000a). Most known examples of character displacement through trait overdispersion are for vertebrate assemblages and involve traits presumed, or in some cases demonstrated, to govern use of food resources (Schluter 2000a). These traits include skull and tooth or beak dimensions. Examples for plants usually involve floral characters that affect pollinator access to resources, but it remains difficult in most of these studies to fully distinguish between ecological and reproductive character displacement.

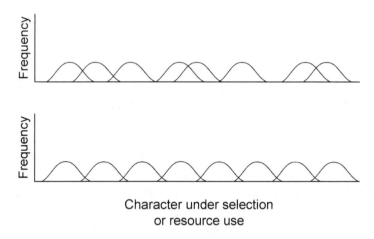

Fig. 15.7   Trait overdispersion along a logarithmically scaled axis of characters or resource use within an assemblage of competing species. The upper panel shows a random distribution of means of a trait among the species or a random distribution of the frequency of use of a resource. The lower panel shows a constant ratio among means along the same axis.

Nevertheless, the extent to which this hypothesis generally holds within local assemblages of species is still poorly understood. As Schluter (2000a) has argued, constant size ratios are neither a necessary nor even a likely general outcome of processes favoring character displacement. It should result only when competition occurs along a single resource gradient that can be subdivided in a symmetric way among species, with morphology and resource use varying linearly along the gradient (Taper and Case 1985). Part of the problem in evaluating constant size ratios is to determine the traits on which to construct the ratios. Species may show constant ratios for the trait under coevolutionary selection (e.g., tooth size or shape) but not for other traits (e.g., body size), and males and females need to be evaluated effectively as separate species for some analyses (Dayan and Simberloff 1994, 1998).

Moreover, two other potential complications make few analyses of size ratios straightforward. One is that some species undergo major ontogenetic shifts in resource use. Juveniles of one species may overlap in resource use with adults of another species. Successful persistence during coevolution may require an ontogenetic progression of morphologies that collectively minimize competition. Analyses of size ratios may therefore need to incorporate ontogenetic shifts in morphology and resource use within each species in the interacting assemblage. The other complication is that coevolution must be separated from nonevolutionary species sorting (Chase and Leibold 2003). Successful colonization of habitats may be most likely in species that are already different in resource use from other local species. Such biased colonization could create the same pattern as coevolutionary divergence toward a constant size ratio. Only ecological data linked to phylogeographic data can begin to sort out these two alternative ways of producing constant size ratios.

Even with increasingly broad geographic analyses and improved ecological analyses of competition using refined statistical approaches, few studies remain unchallenged (e.g., McDonald 2002). That is not surprising, given the breadth, scale, and diversity of analyses needed for these studies. Nevertheless, understanding the conditions favoring trait overdispersion remains important, because this hypothesis directly confronts the role of interspecific competition in shaping the geographic mosaic of multispecific coevolution.

Coevolution and Deterministic Assembly Rules within Competitive Networks

There is still much to learn about how local coevolving competitive networks derive from broader regional assemblages. Following Diamond's (1975) influential analysis of birds of the Bismarck Archipelago, most studies of community assembly rules have used ecological and morphological approaches to predict patterns of local species co-occurrence within regional assemblages of species (Weiher and Keddy 1999; Belwood et al. 2002; Gotelli and McCabe 2002). These studies have spawned an increasingly powerful set of null models of assembly structure from a variety of perspectives (e.g., Gotelli and Graves 1996; Hubbell 2001). Few of the analyses, however, have begun to incorporate evaluation of geographic selection mosaics and coevolutionary hotspots.

Future studies that include the geographic mosaic of coevolution and phylogeographic patterns will provide additional power for analyses of assembly rules. Some long-standing hypotheses on deterministic assemblage dynamics imply a shifting geographic mosaic of coevolution mediated by hotspots and coldspots. The hypothesis of taxon cycles was originally proposed to explain the shifting structure of assemblages, especially within archipelagos (Wilson 1959, 1961). The hypothesis proposed that species' ranges undergo sequential phases of expansion and contraction that are usually associated with shifts in ecological distributions and adaptations during the cycle. Various versions of the taxon cycle have invoked directional coevolutionary change through competition or a combination of competition and parasitism as the cause. Some colonizers initially may be better competitors than local species. These invaders may then evolve toward even greater usurpation of resources, thereby forcing evolution of the poorer competitors into a specialized niche with few resources, followed by extinction (Roughgarden and Pacala 1989; Roughgarden 1995). Alternatively, colonizers may initially be disproportionately free of parasites and predators and later decline in abundance as predators and parasites evolve to exploit them. Increased predation and parasitism would then favor individuals that use the subset of resources or habitats in which they are relatively free of enemies. These changes would contract the geographic range of a species and increase the chance of local extinction (Ricklefs and Cox 1972; Ricklefs and Bermingham 2002).

The hypothesis of favored states in assemblages is yet another view of how competitive networks may create a predictable structure through ecological assembly rules. The hypothesis assumes a higher likelihood of competitive exclusion within groups of morphologically and ecologically similar species than among dissimilar species. It argues that competitive assemblages initially develop by filling a region with one species from each "functional group" of unrelated species. Only then are more species added from each group (Fox 1987; Fox and Brown 1993). The implication is that any coevolutionary displacement that occurs will be concentrated in the latter stages of community assembly, and that there is likely to be regional variation in the importance of coevolution in current assemblage structure.

There is still much debate over deterministic sequences in community assembly and the role of coevolution in those sequences (e.g., Brown, Fox, and Kelt 2000; Stone, Dayan, and Simberloff 2000). It is, however, an important problem to solve for our understanding of how the geographic mosaic of coevolution continually reshapes local, regional, and broader geographic structure of potentially competing assemblages of species. The problem is also important because how species partition resources is an essential part of how communities are constrained across multiple spatial scales (Brown 1995). It may turn out that certain combinations of geographic selection mosaics, coevolutionary hotspots, and trait remixing create particular patterns of deterministic regional or broader geographic assembly structure. We are only now entering a phase of coevolutionary biology in which it is possible to address these questions. All these broader questions require a combined knowledge of phylogeography and the geographic mosaic of coevolving assemblages of species.

Character Release in Coevolutionary Coldspots

The flip side of character displacement is character release. Populations within assemblies that lack selection for character displacement should show evidence of the evolution of character release. Evidence for character release in coevolutionary coldspots can include evolution toward intermediate character states, increased variance in characters, and an overall broader use of resources or habitats. Of these three forms of evidence, the last is the most difficult to interpret with respect to coevolution without additional phylogenetic, geographic, and ecological data. Species sometimes respond rapidly to removal of a competitor by expanding their use of resources or habitats, and the response may be caused without any evolutionary change. Some other species, however, show little initial response to competitor removal. That lack of response can mean either that the species were not competitors after all or that coevolution has displaced the species onto adaptive peaks that they cannot easily abandon over the time scale of most local ecological studies.

The remarkable long-term studies of competition within rodent assemblages in the Chihuahuan desert of the southwestern United States over the past quarter century have shown the complexity of assemblage response to competitor release (Brown and Munger 1985; Brown et al. 2001; Ernest and Brown 2001). These rodent assemblages show intense competition and sort into guilds of species that use resources in different ways. In 1977 three species of kangaroo rat were removed from four experimental plots, and the response of the remaining rodent assemblage has been assessed relative to control plots every month since then. The remaining rodents in the community did not broaden their use of resources over the following eighteen years through 1995. During those years eight other rodent species colonized the plots, for up to a few years each, in higher densities than in the control plots. In 1996, however, a species of pocket mouse never previously found at the study site colonized the experimental plots, expanded in numbers, and began using virtually all the resources that had been used by the kangaroo rats prior to their removal (Ernest and Brown 2001).

This experiment far exceeds in number of years of study any comparable experiment on the dynamic structure of competitive assemblages. Direct comparisons with other studies of competition are therefore impossible, and for now we are left with extrapolation from this precious research. The pocket mice fit a largely vacant local niche left open for almost two decades after the removal of the kangaroo rats. The implication is that assemblages that have partitioned resources, possibly through competitive character displacement, cannot always respond instantaneously to loss of a member of the assemblage. These studies show that at least some of the subsequent structure of local community networks can arise, after a perturbation, from differential sorting among communities rather than coevolution. What remains unknown, for now, are the subsequent coevolutionary dynamics that follow such a colonization.

Evaluation of any longer-term response among extant species requires phylogenetic and phylogeographic analyses of competing taxa across coevolutionary hotspots and coldspots. Through use of these approaches, *Cnemidophorus* lizards on islands in the Sea of Cortez show evidence of character release relative to populations living on the mainland in Baja California (Radtkey, Fallon, and Case 1997). Multiple *Cnemidophorus* species occur throughout Mexico and the southwestern United States, and two species are broadly sympatric throughout Baja California. Islands in the adjacent Sea of Cortez harbor either one or both species. On five of the six islands with one species, lizards are intermediate in size, whereas islands with two species show displacement in size. Phylogenetic analyses based upon cytochrome *b* DNA sequences suggest that the islands have been colonized at least five times from the mainland and that character release has evolved at least twice. The genetic basis of these geographic differences has not yet been determined, but these results provide a template for future analyses of the genetic and ecological structure of character release in these taxa.

Coevolutionary approaches to the study of the fossil record are also providing new insights into the temporal structure of character release. A striking feature of coral reefs is the great diversity of morphological forms that coexist. The community structure of corals and coral forms varies among habitats and depth, creating patterns of co-occurrence that are evident among living species and fossil assemblages (Pandolfi and Jackson 2001). Studies of living species have shown that coral morphology affects resource use, and coral species commonly undergo intense competition for the space they need to provide light for their mutualistic photosynthetic dinoflagellates (Van Veghel, Cleary, and Back 1996). Hence, competition and mutualism are bound together in these coevolving interactions, and there is strong potential for character displacement among sympatric species.

The coral genus *Montastraea* consists of a group of coral species that differ considerably in growth form (Pandolfi, Lovelock, and Budd 2002). Within the shallow-water reef of Barbados, a widespread organ-pipe *Montastraea* dominated the zone between five and eight meters in water depth for

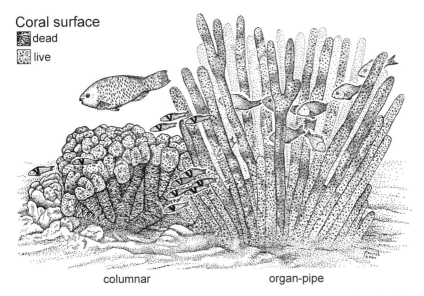

Coral surface
- dead
- live

columnar   organ-pipe

Fig. 15.8   Differences in morphology among coral species of the *Montastraea* "*annularis*" complex in a Pleistocene reef. The organ-pipe species had columns that maintained living corallites along the entire length of the thin columns and grew fastest in shallow water. The columnar lineage maintained living corallites only along the top and periphery of the columns and grew more slowly. From Pandolfi, Lovelock, and Budd 2002.

more than 500,000 years (fig. 15.8). Sometime between 82,000 and 3,000 years ago, it became extinct, and columnar species within the *M.* "*annularis*" complex diversified in morphology in its absence. That diversification has provided the evidence for character release. Morphological evidence indicates that the extinct organ-pipe species was truly distinct from living and fossil members of the complex. Another Pleistocene columnar species of *Montastraea* also inhabited the region and was probably distinct from the three living members of the complex.

Prior to extinction of the organ-pipe *Montastraea,* the columnar species *M. annularis* showed a significant shift away from organ-pipe morphology when in sympatry with the organ-pipe species. After extinction of the organ-pipe species, *M. annularis* sensu stricto underwent a shift in width of columns toward a morphology more similar to that of the organ-pipe species (Pandolfi, Lovelock, and Budd 2002). This change in morphology probably resulted from competitive release, because the direction of change was toward a morphology that allowed faster growth rates and, by inference, increased fitness. The extent to which this morphological shift was genetically based is unknown, but it occurred during a time at which little other environmental change occurred in these habitats, suggesting that the change probably had a genetic basis.

More Complex Dynamics within the Geographic Mosaic of Coevolution

As with all forms of coevolution outcome, variation in character displacement among coevolutionary hotspots and coldspots creates the potential for biased extinction of local demes across landscapes. Such biased extinction, if based on genetic differences among the populations, creates the possibility of higher-level selective units within coevolving interactions. Wade (2003) has argued that covariances among communities created by the geographic mosaic of interspecific interactions may provide the opportunity for hierarchical selection in interspecific interactions. Debate continues in evolutionary biology over how the genetic substructuring of species affects the hierarchical structure of natural selection (Wade and Goodnight 1998; Wade and Griesemer 1998; Coyne, Barton, and Turelli 2000; Wade 2002; Antonovics 2003; Chase and Knight 2003; Collins 2003; Wade 2003; Whitham et al. 2003; Wilson and Swenson 2003). There seems little question that local demes are likely to differ in persistence depending upon the traits under selection and the forms of selection molding coevolutionary change, as indicated by most mathematical models. The controversy is over the potential for interdemic selection to shape the direction and rate of evolutionary and coevolutionary change over and above traditional spatial models of evolutionary change.

Laboratory studies have shown that if competing species are divided into genetically differentiated sets of populations and allowed to coevolve without gene flow for multiple generations, the various populations sometimes coevolve toward different outcomes (Joshi and Thompson 1995; Goodnight and Craig 1996; Hedrick and King 1996). Some of these experiments have shown the potential for interdemic selection to shape coevolution by molding the pattern of variance across populations (Goodnight and Craig 1996). At the very least, we know from mathematical models that the geographic mosaic of coevolution creates dynamics different from those observed in local interacting populations. Hence, the effect of the geographic mosaic may often be more subtle than direct interdemic selection. As indicated by the models and examples throughout all these chapters, substructuring of interactions into locally coevolving populations linked by gene flow creates local, regional, and species-wide outcomes and trajectories of coevolution that are unlikely at the level of locally isolated populations. Future geographic models will help refine predictions of how different forms and hierarchical structures of selection mold variances and trajectories of coevolution among populations of interacting species.

Conclusions

Coevolutionary displacement is inherently a specific local outcome of the geographic mosaic of coevolution. It can produce three geographic patterns within coevolutionary hotspots relative to coldspots, including exaggerated divergence of characters in sympatry, repeated guild structure, and trait overdispersion within competitive networks. Local coevolutionary displacement may result directly from interspecific competition or through the combined effects of competition and other interspecific interactions. The challenge is to understand how the distribution of geographic selection mosaics, coevolutionary hotspots, and trait remixing mold the overall evolutionary trajectories of competing species.

There is only limited experimental evidence that coevolutionary displacement in geographically overlapping species is due directly to resource competition rather than to a combination of competition and other forms of interspecific interaction. The competition-based view of the earth resulted from a restricted focus on resources, competition, and predation that dominated the sciences of ecology for the first two-thirds of the twentieth century. As our understanding of the pervasive role of parasitism and mutualism has expanded in recent decades, competition is increasingly taking its place among these other forms of interaction. In fact, the lines are blurring as it becomes clearer that parasites and mutualists are often major agents of competitive interaction. It is not yet clear where interspecific competition will finally end up in the hierarchy of interactions shaping the overall coevolutionary structure of the earth. Nevertheless, results from recent years have shown convincingly that some species do compete and coevolve for limiting resources.

A full analysis of the geographic mosaic of coevolution—incorporating geographic selection mosaics, coevolutionary hotspots, and trait remixing—has yet to be determined for any pair or assemblage of interacting species undergoing coevolutionary displacement. Some analyses, such as those for *Ano-*

lis lizards and stickleback fish, are closing in on a full understanding as more and more populations and approaches are included in these well-studied taxa. With these and similar studies, it will become possible to evaluate better how coevolutionary character displacement continually reshapes species across geographic ranges and how it may sometimes determine species boundaries.

16 Applied Coevolutionary Biology

The future of human societies depends upon a better understanding of the geographic mosaic of coevolution. The more we learn about the coevolutionary process, the more apparent it is that ongoing coevolution among species permeates every aspect of our societies, including medicine, epidemiology, agriculture, forestry, aquaculture, and conservation efforts. And the more we study the role of coevolution in all these human endeavors, the more evident it becomes that coevolutionary theory absent attention to the structure and dynamics of the geographic mosaic will be of little practical aid. We increasingly require a science of applied coevolutionary biology that incorporates what we have learned about coevolution in natural populations and the differences between natural coevolution and human-induced coevolution (Thompson 1999c, 1999d).

Our developing knowledge of coevolutionary dynamics in natural landscapes can tell us much about what we still need to know as we follow the path of manipulating the coevolutionary process. There are, however, important differences between natural coevolution and some of the new coevolutionary processes we are developing. This chapter explores those similarities and differences within the framework of the geographic mosaic of coevolution. I also argue in these closing pages why, for the future of our societies, it is important for us as scientists to make nonscientists more aware of the pervasiveness of rapid evolution and coevolution, and why it is important for us to preserve long-coevolved interactions.

The Developing Opportunity for Applied Coevolutionary Biology

Realization of the need for applied coevolutionary biology is beginning to spread simultaneously across multiple disciplines. Reviews of immunogenetics and major diseases, for example, have increasingly called for a greater understanding of genetic differences in disease resistance among populations (Hill 1998; Miller et al. 2002). Hill (1998) noted in a review of the immunogenetics of human infectious diseases that the "extent of inter-population heterogeneity in infectious disease resistance alleles [is] only beginning to be addressed." That lack of knowledge is rapidly changing in this first decade of the twenty-first century. The increasing availability to researchers of fully sequenced genomes of parasites and hosts has created opportunities that we are only now starting to appreciate.

I participated in a symposium on genomics and coevolution in 2003—after the human genome, the *Anopheles* mosquito genome, and the *Plasmodium* genome had all been sequenced. The possibility for linking genomics and coevolution was now in place, but it was evident during the symposium just how much interdisciplinary collaboration is needed to translate coevolutionary theory of natural populations into applied coevolutionary theory. How do we manipulate the genetics of *Anopheles* species or deploy vaccines to our advantage for interactions like these that show clear indications of geographic selection mosaics, coevolutionary hotspots, and continual trait remixing across landscapes? Comparative genomics are only the first step toward a combined molecular and ecological understanding of how we can manipulate continually coevolving interactions.

The next great advances in our coevolutionary fight against parasites and pathogens will result at least as much from an increased understanding of evolutionary population biology and coevolutionary theory as from further refinements in molecular biology and biological engineering. We already know how to move genes between species, and the techniques are becoming more efficient every year. We also know much about the molecular biology of how these genes function to confer resistance against enemies. What we still lack is a solid understanding of how to deploy these genes (and surrogate genes such as vaccines) within and among populations in ways that slow the coevolved responses of parasites and pathogens. In a world in which we routinely move among continents and oceans, that understanding can come only from studies of the coevolutionary process. As with all ecological questions, those studies must be linked to other disciplines in innovative ways that accept the now pervasive influence of our societies on all ecological processes (Ehrlich 1997, 2001).

Human Consciousness and the New Coevolutionary Rules

We began the long process toward applied coevolutionary biology thousands of years ago. Three fundamental changes in the organization of life on earth occurred with the rise of human consciousness and created the possibility for new forms of rapid coevolutionary change. The first has been the increasing importance of human cultural inheritance in governing the fate of our own and other species. This form of transgenerational transmission has been called biocultural evolution by E. O. Wilson and metabiological evolution by Jonas Salk, and the units of transmission have been dubbed memes by Richard Dawkins. Unlike any other species, we have the ability as individuals or groups of individuals to draw on a transgenerational learned knowledge base greater than can be accumulated by a population of locally interacting individuals. The second major change has been our increasingly systematic accumulation, categorization, and dissemination of knowledge of the earth and the universe at large. We can rearrange information in endless ways and transmit it within seconds to others halfway around the world. The third major change is our ability to shape consciously the evolution of other species and to manipulate the rules by which our coevolution with other species proceeds. Like the first two, this third change affects every aspect of how our societies will continue to develop. It began with the development of agriculture and the domestication of animals. It has proceeded to a state that Daniel Janzen (1999) has called the gardenification of the earth, driven by a species he has called the most coevolutionary animal of them all (Janzen 1984).

Our ability to mold populations of other species by consciously selecting for particular traits and transmitting that knowledge to others living on different continents has made it possible for the first time for one species to direct the process of evolution and coevolution with another species toward a specific worldwide goal. This potential to direct evolution toward a specific *future global* goal is fundamentally different from the way in which natural selection has acted on species over the billions of years of life on earth. Natural selection adapts populations only to current local conditions, favoring those individuals best adapted to current physical and biotic environments. Most of evolution will continue to be governed by natural selection, but artificial selection toward a future—sometimes transcontinental—purpose is now reshaping all major biological communities on the earth. This third change has made it possible for us to direct how we and other species coevolve with one another.

We can view our increasing ability to co-opt the coevolutionary process as two key innovations. The first is our ability to preserve and spread naturally occurring genotypes of other species that benefit our societies. The second innovation is our recently acquired ability to create and disseminate novel genotypes through interbreeding of separate populations or through genetic engineering and control of their spread. Through these innovations, we can control the genetics and molecular biology of our acquired coevolutionary power, but we remain unsure of the ecological consequences of our actions.

Selective Breeding and Agricultural Coevolution

As we co-opt the coevolutionary process, we are rapidly creating communities of species designed through selective breeding. Agricultural coevolution is the process by which we select for novel genotypes in other species and distribute those genotypes almost instantaneously across landscapes in response to rapidly evolving parasites and pathogens. In a real sense, leaf-cutter ants practice agricultural coevolution by favoring particular genotypes of fungi within their colonies (Wirth et al. 2003). Human agriculture, however, pushes beyond the agricultural systems of ants by actively breeding for novel genotypes. Through this process, our societies create genotypes in plant and animal species that exceed the range of traits found within natural populations. Selective breeding in agriculture has been accompanied in recent centuries by a change that is equally profound for the future of the coevolutionary process: our attempts to directly control crop and livestock diseases.

Agriculture appears to have originated through artificial selection of a few species in a few places and spread later to other cultures. The exact number of separate origins of human agriculture is still under debate, but in the Fertile Crescent of southwest Asia, humans domesticated cereals, legumes, goats, and sheep, while in China, agriculture was based upon rice, millet, pigs, and chickens (Harris 1996; Diamond 1997; Salamini et al. 2002). In the Americas, maize and other crops became staples. The cultures of New Guinea cultivated taro and bananas (Denham et al. 2003).

These agricultural societies began the human transformation of species interactions that continues today, and some of the manipulations are multispecific. Human cultures started manipulating grapes and yeasts to make wine at least seven thousand years ago (McGovern et al. 1996). The oldest known wine jar, which comes from an excavation of a Neolithic site in Iran's northern Zagros Mountains, held a resinated wine. The resin, taken from terebinth trees (*Pistacia atlantica*), served to inhibit the growth of the bacteria that turn wine to vinegar and was a common way of preserving wine in the ancient Near East. What is remarkable about this wine is that it shows that as early as seven thousand years ago, humans were actively manipulating—both directly and indirectly—up to four other species or groups of species at a time (grapes, yeasts, terebinth, and bacteria).

Although selective breeding by humans began thousands of years ago, it has been only in the past two hundred years, and especially the past one hundred, that we have created extensive monocultures and bicultures. Our societies have become so skilled at transforming geographic and genetic landscapes that there are now places on earth where you can stand and see nothing but several crop genotypes as far as your vision allows. The uninterrupted wheat- and pea-blanketed hills of eastern Washington State and the corn and soybean expanses of the midwestern United States became possible and practical only with industrialization.

At one level these are awe-inspiring landscapes that highlight the power of our increasing ability to manipulate the genetic structure of large regions of the earth. The spread of aquaculture has the potential to similarly affect parts of the earth's oceans and lakes as genetically differentiated stocks of fish species are replaced with selected lines kept partially in pens. In some parts of the world entire lakes are maintained for selectively bred lines of fish.

These novel environments are fundamentally different from the environments in which species commonly coevolve. We may apply to them our understanding of the molecular genetics of coevolution, but we must do so with an understanding that the resulting dynamics of coevolution will be quite different because the population genetics, geographic structure, and coevolutionary pace are completely different. Up until now our understanding of the dynamics of agriculture coevolution has progressed mostly through selection on the structural genetics of defense and counterdefense and observations on the pace at which parasites and enemies respond to introduction of novel host genotypes within large monoculture. In fact, plant pathologists have for decades been at the forefront in applying coevolutionary concepts to selective breeding programs and agriculture. The major gene-for-gene paradigm in plant pathology, developed by Flor (1942, 1955, 1956) in the mid-twentieth century, is explicitly coevolutionary. The first mathematical model of coevolution, and the paper that coined the word (although not the concept), was of gene-for-gene coevolution between flax and flax rust (Mode 1958).

Plant pathologists have used the genetic architecture of coevolution between plants and pathogens as a tool for selective breeding, but achieving what is commonly called durable resistance has been difficult. Success requires an understanding of how pathogens evolve when faced with multiple host genotypes distributed across landscapes (Gould 1998; Rausher 2001). Intercropping, planting of trap crops to divert pests from more important crops, and limitations on proportions of novel resistant crop genotypes allowed within regions are all versions of our attempt to manipulate the pace of agricultural coevolution.

That is why the long-term studies of the coevolutionary dynamics of Australian wild flax and flax by Jeremy Burdon and his colleagues are so profoundly important (Burdon and Thrall 2000; Burdon, Thrall, and Lawrence 2002). They are the only studies that have shown how gene-for-gene coevolutionary dynamics occur across natural landscapes over the time scale of decades. These studies have demonstrated that ongoing gene-for-gene coevolution is an inherently geographic process. The process is maintained by metapopulation structure, which is in turn embedded in a broader geographic structure. This structure is exactly what large-scale monocultures lack. We are not going to replace modern agribusiness with small, genetically diverse populations of plants. We can, however, use the knowledge gained from the few precious coevolutionary studies now in hand to begin designing new strategies on how to develop durable resistance in crops.

The current attempt to understand virulence in corn smut within the framework of the geographic mosaic of coevolution is an example of what may be possible for other interactions between crops and pests. Corn smut (*Ustilago maydis*) is a naturally occurring pathogen of teosinte and maize. It has coevolved with maize throughout its domestication and spread as a crop of worldwide importance. The pedigree of all maize varieties grown in North America is known, which provides a firm phylogeographic template for understanding geographic patterns of resistance against corn smut (Neuhauser et al. 2003). The interaction between cultivated maize and corn smut is unusual in that selective breeding for resistance against smut in the early 1900s was successful and has remained durable through much of the past century. Neuhauser et al. (2003) have developed a research program to understand the geographic stability of resistance against smut. The goal is to understand why a pathogen with high potential for evolving increased virulence has failed to do so. By using a combination of phylogeographic, genetic, and ecological approaches, their research program is asking how the geographic history of founder effects and the genetic structure of maize and corn smut may have created a huge coevolutionary coldspot.

New molecular tools are continuing to help fill in the sketchy historical

Fig. 16.1   Herbarium specimen collected in the Royal Botanic Gardens, Dublin, Ireland, in 1846 of a potato plant infected with the pathogen *Phytophthora infestans*. From Ristaino, Groves, and Parra 2001.

evidence of how novel genotypes in crops and pathogens have spread across landscapes and how the genetic diversity of pathogens has shaped epidemics. Evaluation of the DNA of historical specimens in herbaria, museums, and agronomic collections will become an increasingly crucial source for these studies, if these valuable collections can continue to be preserved for scientific use. DNA analyses of the Irish potato famine of 1845 and the potato-blight epidemics in Europe and North America during the late nineteenth century and early twentieth centuries show the utility of these historical specimens. These epidemics were caused by the oomycete plant pathogen *Phytophthora infestans*, which causes late blight. Although there are four known modern haplotypes, only one clonal lineage, the 1b mitochondrial DNA haplotype (or US-1 genotype), dominated *P. infestans* populations worldwide prior to the 1980s. This is the genotype thought to have caused the epidemics. Fortunately, infected potato leaves were preserved on herbarium samples from Britain, France, and Ireland collected from 1845 through 1847 (fig. 16.1), making it possible to amplify the DNA of the pathogens on those specimens. Amplification of a region of the mitochondrial DNA found only in the US-1 genotype showed that the lesions on the mid-1800s specimens were not caused by this genotype of the pathogen (Ristaino, Groves, and Parra 2001). That means that the epidemic was caused either by one of the other three known genotypes or by another genotype now unknown.

At smaller geographic scales, analyses of genetic differentiation in pest species are showing how local adaptation can develop within regions as pest species interact either with a combination of agricultural crops or with a combination of crops and native plant species. In North America, pea aphids on alfalfa and clover crops within a particular region not only suffer different levels of successful attack by parasitic wasps, they also show evidence of strong divergent selection to form host races (Via 1994, 1999; Via, Bouck, and Skillman 2000; Via and Hawthorne 2002). In some parts of Europe, the European corn borer (*Ostrinia nubilalis*) feeds on both mugwort (*Artemisia vulgaris*) and maize, but in France it includes mugwort only in the northern part of the country, and those populations are genetically differentiated from populations found on maize (Martel et al. 2003). Hence, there is a clear geographic mosaic to the interaction between this insect and its agricultural and native hosts.

Similar efforts are needed to evaluate how polyploidy shapes the genetic and ecological structure of interactions between plants and their herbivores and pollinators. Much of modern agriculture is build upon polyploid plants, including forage and crop plants as diverse as alfalfa, wheat, oats, coffee, potatoes, sugarcane, cotton, peanuts, bananas, tobacco, and strawberries. Moreover, we now know that polyploidy is much more geographically dynamic within plant taxa than previously thought (Soltis and Soltis 1995, 1999). Some studies are beginning to evaluate how the evolution of polyploidy among natural plant populations shapes the evolution of interactions with parasites and mutualists across multiple spatial scales (Thompson et al. 1997; Segraves and Thompson 1999; Husband 2000; Nuismer and Thompson 2001; Thompson, Nuismer, and Merg 2004). These studies provide clues on how selective breeding for polyploidy may reshape interactions with pests or mutualists, but they need to be expanded to a broader range of natural and managed landscapes.

Collectively these studies are in the vanguard of approaches that are attempting to assess how the geographic mosaic of agricultural coevolution and that of natural coevolution may differ. They are an attempt to go beyond local intercropping and trap-crop strategies as means of slowing the evolution of pests that overcome resistance built into selected varieties. Starting with broad geographic patterns of resistance, virulence, or host specialization, these studies will help us understand how we can maintain coevolutionary coldspots between plants and pests across regions as we continue to mainipulate crop species.

Genetic Modification (GM) of Foods and Diseases

Genetic modification (GM) of foods and diseases has become the next step in manipulation of the coevolutionary process. We are attempting to escalate the rate of the evolution of defenses in crops and domesticated animals against parasites and pathogens, and we are attempting to do it in a way that minimizes the rate at which these enemies evolve counterdefenses. Research programs for genetic modification of plants against pathogens now include all major crops. Proposals to genetically modify and release a wide range of species have increased since the beginning of the new millennium. A report from the Royal Society in Britain in 2001 called for increased research on genetic modification of livestock for resistance to foot-and-mouth disease and sleeping sickness in order to help developing countries (Adam 2001). Calls for genetically modifying the vectors of human diseases have also increased. High on that list has been a discussion of the genetic modification of mosquitoes that transmit diseases such as malaria. The goal of these efforts would be to release into natural populations genetically modified individuals unable to harbor the disease or transmit it to humans (Scott et al. 2002). As these mosquitoes spread throughout local populations, the number and proportion of mosquitoes capable of transmitting disease would decrease.

The potential stumbling blocks to these efforts have been well rehearsed in recent years as epidemiologists, molecular biologists, population biologists, and evolutionary ecologists have compared ideas and concerns. Much of the uncertainty about what would work, and work safely, stems from an increasing realization that we are only now starting to understand the complexity of the geographic structure and dynamics of the coevolutionary process. Development of a science of the release of genetically modified organisms therefore must be an outgrowth of what we have learned in the past decade and are still learning about the geographic mosaic of coevolution. Such attempts are increasing. The special issue on *Anopheles gambiae* mosquitoes and malaria in the journal *Science* in 2002 included articles calling for a better understanding of how the relationship between mosquito density and human infection varies across time and geographic locations (e.g., Scott et al.

2002). A major workshop on the release of genetically manipulated insects in the control of vector-borne diseases noted that the issues associated with release programs will be specific to local populations (Alphey et al. 2002).

Given the astonishingly fast rates at which many diseases and disease vectors evolve in response to our attempts to control their populations, we can anticipate that any such attempts at genetically manipulating vector populations will be an ongoing process. That is the past history of all such attempts. Drug therapy for malaria led to the evolution of chloroquine resistance in Southeast Asian and South American *Plasmodium* populations in the 1950s, Papua New Guinea populations in the 1960s, and East African populations in the 1970s (Wellems 2002). Insecticide-treated bed nets have been highly effective at reducing the number of bites from *Anopheles* mosquitoes in Africa, but some populations are evolving resistance to these pesticides, as happened with earlier pesticide efforts (Miller and Greenwood 2002).

The debate over genetic modification of other species is a complex scientific, social, and political problem partly because our societies are entering uncharted coevolutionary territory. In the 1980s and 1990s, the initial concerns over genetically modified species were over the problems of whether GM species would pose a hazard to humans, spread invasively into surrounding habitats, or hybridize with wild relatives and transform those species. Experiments on transgenic crops have started to provide results on the patterns of local spread and local hybridization with other species or genotypes (Crawley et al. 2001). Related studies of genetically modified herbicide-tolerant crops have begun to show how maintenance of landscapes as even stricter monocultures affects the local biodiversity of nontarget species (Firbank 2003). Ultimately, all these studies will require evaluation within a broader framework of ongoing coevolution across regions and continents.

Surrogate Coevolution

The discovery and medical use of antibiotics and the development of vaccines against pathogens during the twentieth century created an entirely new form of interaction that mimics coevolutionary change. Through antibiotics and vaccines, we transform genetically susceptible individuals into resistant individuals. We change the apparent, but not the real, genetic structure of the host population. Antibiotics and vaccines therefore function as surrogate genes, and they generate a process that can be called surrogate coevolution (Thompson 1999c).

We can think of surrogate coevolution as the means by which we manipulate our interactions with other species through rapid deployment of surrogate genes such as antibiotics or vaccines. The process of surrogate coevolution differs from both natural and agricultural coevolution. Antibiotics and vaccines function as induced traits, but they are not subject to the rules of population genetics. Surrogate genes can be distributed almost instantaneously within populations across all age groups, thereby rapidly achieving high frequencies. They can spread rapidly among populations, following spatial patterns quite different from that found in the spread of real genes. They can be deployed simultaneously worldwide, or at least at multiple foci worldwide. And they are partially independent of the genotype and environment in which they occur. Regardless of the individual's inherited genotype, the administration of an antibiotic or vaccine can prevent or decrease attack by the pathogen in a broad range of environments. To be sure, antibiotics and vaccines have different effects on different individuals, whether people or livestock, but their remarkable effectiveness across so many genetic backgrounds and environments is part of the power of these important tools in human society.

That effectiveness has generated the inevitable evolutionary consequences. Over the past half century we have created an arsenal of antibiotics in our coevolutionary war against pathogenic bacteria, and new antibiotic-resistant populations have appeared with tremendous evolutionary speed. At least some of that increase has come from the increasingly widespread use of antibiotics within human populations, placing strong selection for resistance on pathogen populations. The problem, however, is more complex than just the evolution of resistance in target pathogens. Administration of antibiotics can also favor the evolution of pathogenic strains of otherwise commensal symbionts. Moreover, the transmission dynamics may be more complicated than in simple coevolutionary dynamics, because antibiotic resistance in organisms such as bacteria is often transmitted through plasmids. The evolutionary dynamics are therefore sometimes governed by the rate of exchange of transposable elements. Hence, this form of surrogate evolution may often be a problem in multispecific surrogate coevolution involving complex genetic dynamics.

Two of the best data sets showing the link between the volume of antibiotics administered to populations and antibiotic resistance in commensal bacteria that can become pathogenic are for β-lactamase-producing strains of *Moraxella catarrhalis* in Finland and penicillin-resistant strains of *Streptococcus pneumoniae* in Iceland (see fig. 16.2A and B). Both countries have kept

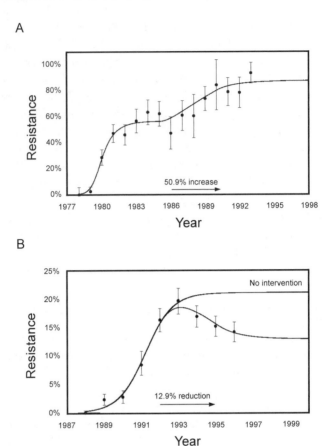

Fig. 16.2   Increase in the resistance of pathogenic forms of two bacteria commonly found as commensals in humans. (A) β-lactamase-producing strains of *Moraxella catarrhalis* from children less than six years old in Finland. (B) Penicillin-resistant strains of *Streptococcus pneumoniae* from children less than seven years old in Iceland. The line showing no intervention is the predicted frequency of resistance if no intervention program had been initiated. The error bars show 95 percent confidence intervals. The lines are from an epidemiological model fitted to the data. From Austin, Kristinsson, and Anderson 1999.

detailed records over several decades of antibiotic distribution and cases of antibiotic resistance for these diseases. In Finland, *M. catarrhalis* increased during the late 1970s and early 1980s as antibiotic volume remained fairly constant. Increased antibiotic use between 1985 and 1990 led to a further increase in the frequency of resistant strains (Austin, Kristinsson, and Anderson 1999). Similarly, penicillin-resistant strains of pneumococci increased in

Iceland during a period of high usage of antibiotics. Concerted measures to reduce antibiotic usage between 1992 and 1995, however, led to a reduction in the frequency of resistant strains (Austin, Kristinsson, and Anderson 1999). In effect, changes in the volume of antibiotics used in populations are altering the proportion of humans that impose selection for resistance on these bacteria.

Only since the 1990s have mathematical models been developed to explore how community-wide use of antibiotics shapes the rate of evolution of antibiotic resistance in pathogens and commensals (e.g., Bonhoeffer, Lipsitch, and Levin 1997; Castillo-Chavez and Feng 1997; Stewart et al. 1998; Austin, Kristinsson, and Anderson 1999). The models and associated epidemiological data suggest that drug volume is one of the major drivers of the rate of surrogate coevolution through antibiotic use. Increasing volume often increases the rate of development of antibiotic resistance. Some population-genetic models suggest that any of three equilibrium states are possible, depending upon the volume of antibiotic consumption and rate of plasmid transfer. These include failure of the resistant strain to become established, replacement of the sensitive strain by the resistant strain, or coexistence of the two forms. As in all coevolving polymorphisms, spatial structure is likely to have important effects on the rate of evolution and the equilibrium states.

More generally, despite the rapid pace of advancement in evolutionary and coevolutionary theory, it is only since the 1980s, and especially since the 1990s, that medical researchers have begun to incorporate evolutionary and coevolutionary dynamics into serious discussions of antibiotic and vaccine deployment protocols. When Robert May and Roy Anderson first started publishing their influential series of mathematical models on the evolutionary and coevolutionary aspects of epidemiology (Anderson and May 1979, 1982; May and Anderson 1983, 1990; Anderson and May 1991), their work focused the attention of evolutionary biologists on the high evolutionary rates of many parasites. Their models and the refinements in subsequent modeling approaches increasingly have had a powerful influence on discussions about protocols in the management of disease dynamics and immunity in human populations (Anderson, Donnelly, and Gupta 1997; Levin, Lipsitch, and Bonhoeffer 1999; Levin and Andreasen 1999; Koella and Restif 2001; Earn, Dushoff, and Levin 2002; Iwasa, Michor, and Nowak 2003). As with all questions related to the geographic mosaic of coevolution, one of the key issues has become the problem of how processes at one spatial or temporal scale shape processes at other scales (Levin et al. 1997).

Medical research is also beginning to use the natural process of coevolutionary dynamics to search for vaccines. By identifying specific allelic variants and proteins associated with our coevolution with parasites and pathogens, epidemiological research is moving closer to managing directly the coevolutionary process. Given what we know about the coevolutionary process, such research requires a strongly geographic perspective. At least five thousand candidate proteins have been identified that show some activity with pharmaceutical potential to combat malaria, and many of these have multiple variants (Richie and Saul 2002). At a Keystone Symposium on malaria in 2002, Louis Miller of the National Institutes for Health argued for research identifying unique allelic variants in human populations living in malarial regions as a way of identifying parasite proteins that could be targeted by new vaccines (Long and Hoffman 2002).

There is also increasing discussion about the ways in which different vaccination strategies may generate reciprocal responses in parasite and pathogen populations. Varying degrees of incomplete immunity, differences among regions in the proportion of a population immunized, and global movement of people among population centers will all contribute to different patterns in coevolutionary dynamics. Regions of moderate vaccine-induced immunity, for example, could become regions of surrogate coevolutionary hotspots, because selection acts strongly on the parasite populations. Regions of low vaccine-induced immunity may show traditional coevolutionary dynamics, and regions of high vaccine-induced immunity may act as coevolutionary coldspots. The result could be a strongly delineated geographic mosaic in the structure and dynamics of coevolution, with the added twist of the need to sort out how natural and surrogate coevolution differ in their effects on the dynamics.

There is no question that surrogate coevolution mediated by antibiotics and vaccines will become an increasingly important part of human culture. The impact of parasites and pathogens on human populations is so pervasive that we will continue to deploy as wide an arsenal of defenses as our societies can muster. Tropical protozoan diseases alone—such as malaria, leishmaniasis, and Chagas' disease—afflict three billion people worldwide (Gelb and Hol 2002). One recent estimate is that between twenty and thirty new drugs will be needed to bring tropical protozoan diseases under control (Gelb and Hol 2002). Because there will always be logistical difficulties in organizing such efforts worldwide, and because the effectiveness of these efforts will often differ among environments, the surrogate coevolutionary dynamics that result will always have a marked geographic structure.

As different combinations of drugs and vaccines are deployed in different countries, the inevitable rapid evolution of drug and vaccine resistance will continue to show strong geographic patterns. The geographic mosaic of surrogate coevolution will be further fueled by ongoing coevolution between many of our parasites and the vectors that transmit them among human individuals. Malaria, sleeping sickness, leishmaniasis, and Chagas' disease, for example, are all transmitted by insect vectors, and the specific insect vectors often vary geographically. Imposing strong selection on our parasites through surrogate coevolution will create ripple effects on the geographic mosaic of coevolution between these parasites and their vectors.

Some of the proposed cures for human diseases involve manipulating coevolutionary processes even more by disrupting coevolved interactions in disease vectors. Use of the antagonistic *Wolbachia* bacteria in arthropods and use of antibiotics against mutualistic *Wolbachia* in nematodes are among the proposed manipulations (Dobson 2003; Rasgon, Styer, and Scott 2003). Diseases linked to nematodes affect millions of people worldwide. Because nematodes generally rely upon their mutualistic *Wolbachia* bacteria, potential epidemiological solutions include disruption of that mutualism through the use of *Wolbachia*-specific antibiotics. Resistance to antibiotics has been reported in some *Wolbachia* and other symbiotic bacteria (Stevens, Giordano, and Fialho 2001), and it seems likely that any attempt to use antibiotics to disrupt the mutualism between these bacteria and their nematode hosts will create a geographic mosaic of resistance driven by coevolution between these species to thwart the effect of the antibiotics.

Evolutionary approaches to epidemiology and medicine have tremendous potential for aiding our societies (Stearns 1999; Ewald 2000). If we are to be successful, we will need a much more refined theory on the differences between surrogate coevolution and natural coevolution. We will also need to understand how surrogate coevolution between humans and parasites reshapes natural coevolution between those parasites and their vectors.

Coevolution amid Rapidly Changing Landscapes

Our new ability to reshape the coevolutionary process—acquired during a blink of geologic time—has been accompanied by five related human-imposed changes that are altering the earth's natural landscapes and the natural coevolutionary process (see table 16.1). Concern over how these changes may be affecting overall patterns of adaptation, speciation, and extinction of taxa is increasing (Bazzaz 2001; Ehrlich 2001; Mooney and Cleland 2001;

Table 16.1   Recent alterations of the earth's landscapes that are changing the global
dynamics of coevolution

Rapidly changing geographic structure of species
Alteration of the interaction structure of ecoregions
Worldwide redistribution of species
Increased hybridization among formerly separated species
Urbanization

Myers and Knoll 2001; Templeton et al. 2001; Western 2001; Woodruff 2001;
Thomas et al. 2004). Each of these changes has a great potential for reshaping the geographic mosaic of coevolution and the global dynamics of coevolutionary change.

RAPIDLY CHANGING GEOGRAPHIC STRUCTURE OF SPECIES

The geographic structure of many species is changing quickly from opportunities for colonization created by human modification of landscapes, active
movement of species among regions, and climate change (Parmesan et al.
1999; Warren et al. 2001; Hill, Thomas, and Huntley 2003). Species distributions are always changing, but the current changes are happening at remarkably fast rates, creating rapid ecological and evolutionary changes in species
at the edges of their ranges (Thomas et al. 2001). These changes are likely to
create nonequilibrium dynamics in novel coevolving interactions that require more direct study if we are to manage populations amid such rapid
change.

The earth's ever-changing mosaic of biological communities has until recently resulted from a combination of physical gradients and species interactions. We are now, however, fragmenting landscapes in nonecological dimensions. These changes are altering the dynamic interaction structure of
communities and restricting the movement of individuals among populations of some species (Cordeiro and Howe 2003; Ferraz et al. 2003). Fragmentation decreases the contribution of ongoing gene flow to the geographic
mosaic of coevolution. Current mathematical models of coevolution all suggest that the absolute and relative rates of gene flow among parasite and host
populations shape local patterns of host and parasite adaptation. Under some
conditions, decreased gene flow among host populations could result in
higher densities of coevolving parasites (Carlsson-Graner and Thrall 2002;
Nuismer and Kirkpatrick 2003).

We have fragmented landscapes in fundamentally novel ways. Imposition of large-scale monocultures distributed as patchworks with knife-edge ecotones, annual replacement of one crop community by another over large areas, and networks of irrigated green islands in desert landscapes create patterns very different from those found within natural environments. These environments have the potential to force the formation of local populations at very different scales—either larger or smaller—than would normally occur across regions where natural selection would often favor local adaptation. We can guess that these human-produced mosaics have essentially the same effects on regional and continental scale coevolutionary dynamics as those imposed by natural mosaics, but we simply do not know.

These are the realities of the earth's new structure, and even major efforts cannot restore the geographic mosaic structure of long-coevolved networks across most landscapes once they are lost. Even where we have attempted to conserve and restore native assemblages of species, the chosen genotypes have sometimes come from populations or mixes of populations distant from the restoration site. Movement of genotypes over great distances creates a dynamic genetic structure in coevolving interactions very different from that produced by the natural geographic mosaic of coevolution. The challenge now is to develop a science of biogeography of fragmented environments (Daily, Ehrlich, and Sánchez-Azofeifa 2001; Ricketts, Daily, and Ehrlich 2001) and couple it with an understanding of how the geographic mosaic of coevolution shapes interactions across these new kinds of landscapes. The goal of this effort must be the preservation of the complexity of ecological, evolutionary, and coevolutionary processes.

ALTERATION OF THE INTERACTION STRUCTURE OF ECOREGIONS

Within many ecoregions worldwide, we have changed species composition nonrandomly. While elevating the overall background rate of extinction (Dirzo and Raven 2003), we have especially eliminated or greatly reduced the numbers of large predators. In some regions, the changes have occurred as a consequence of creating population fragments too small to allow large species to persist (Terborgh et al. 2001). In other regions, active hunting of large mammals has produced ripple effects on other interspecific interactions (Estes et al. 1998).

There is plenty of informed discussion about how these effects have reshaped the ecological structures of ecoregions (Power et al. 1996; Jackson et al. 2001), but little of it so far has included consideration of how these changes may have led to rapid coevolutionary change in the remaining species. Given what we now know of the speed at which populations may evolve and interactions may coevolve in response to altered conditions, it would be surprising if some of the observed changes within and among ecoregions are not the result of altered coevolutionary dynamics among the surviving species.

WORLDWIDE REDISTRIBUTION OF SPECIES

We have moved thousands of species among continents and oceans during the past few centuries. During this short period of time, new and emerging diseases appear to have developed in marine environments as a result of climate change and the introduction of diseases into environments where they did not occur previously (Harvell et al. 1999). On land, the biodiversity of entire regions has been transformed through movement of hundreds of species among continents. In California alone, there are now more than a thousand introduced plant species distributed among that region's almost five thousand native plant species (Hickman 1993; Bossard, Randall, and Hoshovsky 2000). This worldwide redistribution of the earth's species is rapidly creating new geographic mosaics in species interactions.

Within some ecoregions, novel species and interactions are expanding so quickly that they are becoming the dominant characteristics of these ecoregions. In the steppe communities of the North American intermountain West and the oak woodlands of California, alien grasses now often form the matrix within which the remaining species of plants and animals interact. Across these landscapes, alien species remain relatively uncommon only in harsh environments such as those created by serpentine soils (Harrison 1999). Some highly invasive alien species may be at least temporarily decoupled from most enemies and mutualists in their new environments (Mitchell and Power 2003; Torchin et al. 2003). For example, a compilation of fungi and viruses attacking 473 plant species in the United States naturalized from Europe found 84 percent fewer fungal species and 24 percent few viral species in the naturalized ranges than in the native ranges (Mitchell and Power 2003). Other studies, however, have shown that some introduced species may rapidly accumulate new enemies or mutualists in their new environments (Parker 1997; Agrawal and Kotanen 2003).

As introductions of species continue worldwide, old interaction networks may even form again in new settings. European honeybees commonly visit

European plants on continents where both the bees and the plants have been introduced (Goulson 2003). In some cases, re-forming of the network may involve rapid evolution of the interacting species. The anther smut *Microbotryum violaceum* attacks both *Silene alba* and *Silene vulgaris* in Europe. Both plant species were introduced into eastern North America, but, until recently, the anther smut attacked only one of these two species, *S. alba* (Thrall, Biere, and Antonovics 1993; Alexander and Antonovics 1995; Alexander et al. 1996). In 1998, however, a diseased population of *S. vulgaris* was discovered in Virginia (Antonovics, Hood, and Partain 2002). Disease expression of the pathogen on this plant population is often abnormal, but this may be the early stages of rapid evolution as this anther smut and its host plants re-establish some version of their previous network structure.

We now know that some interactions with introduced species can evolve quickly, and there are dozens of well-documented examples of rapid evolution of interspecific interactions. The examples span a range of traits, including defensive chemistry, host preference in parasites, and morphology. Introduced wild parsnip populations in the midwestern United States have coevolved with native parsnip webworm populations across highly disturbed landscapes since the 1800s, favoring populational differences in chemical defenses in the plants and counterdefenses in the insects (Berenbaum and Zangerl 1998). Swallowtail butterflies in the central valley of California, where roadsides are dominated by introduced fennel, exhibit a genetically based higher preference for fennel than do west coast populations that feed on native plant species (Thompson 1998b). And soapberry bugs in some populations in the southeastern United States have evolved novel mouthpart lengths in two different regions, allowing them to successfully attack introduced sapindaceous plant species (Carroll and Boyd 1992; Carroll et al. 2001).

Some invasive species, and invasive species assemblages, have become cosmopolitan in distribution. As these species coevolve with each other and with native species, they are creating global interspecific networks in which they are increasingly among the core species. In effect, our societies are homogenizing some interactions over large scales. From what we now know of the geographic mosaic of coevolution, we may be creating a situation in which some invasive species are increasingly stabilized because they are establishing the geographic mosaic structure that is being lost by more geographically restricted native species. Over time, that imbalance could lead to further homogenization of species networks.

INCREASED HYBRIDIZATION AMONG FORMERLY SEPARATED SPECIES

Our global redistribution of species may be creating opportunities for the formation of new and invasive species. Hybridization and polyploidization are two of the most common ways by which new plant species are formed in natural assemblages of species (Rieseberg, Van Fossen, and Desrochers 1995; Rieseberg et al. 2003). Multiple new plant species have arisen since the nineteenth century by a combination of hybridization and polyploidy (allopolyploidy) mediated by human activities. One of the clearest examples of the origin of an invasive species through this process is the documented origin and spread of a new salt-marsh species, *Spartina anglica,* which is now a serious invader in salt marshes on multiple continents. The species was formed when the North American salt-marsh grass *Spartina alternifolia* was introduced to southern England and western France in the late nineteenth century. It hybridized with the native *Spartina maritime* and gave rise by chromosomal doubling to the allopolyploid species that is now called *S. anglica* (Ayres and Strong 2001).

The formation of new species through allopolyploidy in the plant genus *Tragopogon* adds another twist. In these cases, three introduced species hybridized among themselves to produce new allopolyploid species. Three original species were introduced from Europe into eastern Washington State and northern Idaho around 1950. Since then these species have repeatedly formed two named polyploid species that have spread throughout the region (Soltis et al. 1995; Cook et al. 1998). These new species are easily differentiated from their parental species and are now common plants in southeastern Washington and adjacent Idaho.

Other instances of new allopolyploid species continue to appear (Abbott and Forbes 2002), but little is known about the ways in which any of these new species interact with native herbivores, pathogens, and pollinators. We already know, however, that both hybridization (Whitham, Morrow, and Potts 1994; Fritz et al. 1996; Whitham et al. 1999) and polyploidy (Thompson, Nuismer, and Merg 2004) can strongly affect interactions with enemies and mutualists among native species. We should therefore expect the formation of new allopolyploid species to result in rapid coevolutionary change.

URBANIZATION

The expansive cities that dot the globe have created fundamentally new geographic mosaics in a wide range of interspecific interactions. Concentrations of tens of millions of people connected by daily air flights create complicated patterns of gene movement in hosts and vectors. Strains of human influenza are spread globally, requiring an increasingly worldwide effort to predict which strains need to be in each year's set of vaccines (Abbott 2001). The global alert issued by the World Health Organization in response to severe acute respiratory syndrome (SARS) in 2002 emphasized the developing global dynamics of infectious diseases. The sheer speed at which new genotypes of parasites and pathogens can be spread among major urban centers in different continents is creating patterns of rapid evolution in pathogens that we are only starting to understand.

For some parasites or their vectors, these huge urban environments are also creating novel stable environments, which are, in turn, linked to surrounding populations that show more genetic differentiation. For example, dengue fever is transmitted from one human to another by *Aedes* mosquitoes and has expanded in tropical and subtropical regions worldwide in regions of dense occupation. Part of the spread has resulted from the spread of the disease organism's highly effective vector, *A. aegypti*, which is now pervasive in some urban environments. In Vietnam, it was first reported in 1913 and was present in 89 percent of houses in Saigon by 1967 (Huber et al. 2002). In Ho Chi Minh City and the surrounding countryside it has already developed a distinctive geographic structure. Mosquitoes within the city have stable breeding populations, the result of year-round availability of larval breeding sites and a human population of more than four million. The city-dwelling *A. aegypti* population appears, on the basis of analysis of microsatellite markers, to be almost panmictic. In contrast, considerable microsatellite differentiation occurs among populations in the outskirts of the city (Huber et al. 2002). In this case the geographic differentiation may result mostly from random genetic drift in the populations found in the outskirts rather than from local adaptation in those environments. The overall pattern, however, highlights the potential effects of urbanization on the geographic mosaic of coevolution among species, not only for humans and our parasites and vectors but also for the thousands of other species that occur within and among urban environments.

CONSEQUENCES FOR THE GEOGRAPHIC MOSAIC OF COEVOLUTION

These five alterations of natural landscapes (see table 16.1) undoubtedly have different consequences for different interactions. The current challenge is to take what we are learning about the geographic mosaic of coevolution and use it in ways that preserve biodiversity and the coevolutionary process for future generations. Our rearrangement of the earth's biodiversity is surely resulting in the loss of geographic structure in some interactions, shifting the dynamics to local rather than geographic scales for some species in the absence of gene flow. For these interactions, coevolution may proceed, for example, toward a rapid loss of genetic diversity in defenses against enemies. In contrast, our worldwide movement of genes and species may be creating more global coevolutionary dynamics in other interactions. That seems to be so in some emerging diseases of humans, crops, and livestock.

At the very least, these changes are likely to shift the coevolutionary dynamics of many interactions back toward earlier stages of nonequilibrium dynamics. Some infectious diseases exhibit traveling waves of infection across novel landscapes (Russell et al. 2004), and most mathematical models indicate that interspecific interactions show different rates and patterns of change in the early stages of coevolution than in later stages. Although, in an ever-changing world, few interactions are probably ever at genetic equilibrium, we may be creating an earth that keeps many interactions at the earlier, highly dynamic stages of coevolutionary change. We may therefore be altering coevolutionary dynamics in ways that make it increasingly difficult to predict the consequences of our actions as we create new interaction networks.

Potential Impediments to the Science of Applied Coevolutionary Biology

LACK OF APPRECIATION OF THE PERVASIVENESS OF ONGOING EVOLUTION AND COEVOLUTION

We can confront these ecological and epidemiological challenges through the development of applied coevolutionary biology. One major impediment, though, is the lack of knowledge and appreciation in our society of the pervasiveness of rapid evolution and coevolution. We are still struggling to explain to explain to society at large the uses of evolutionary biology (Futuyma 1995). We can change that lack of appreciation only once we begin to make the possibility of ongoing evolution and coevolution an explicit central working hypothesis in all studies of ecological and epidemiological dynamics. Our failure to do so helps reinforce the impression that evolution is something we cannot observe in action. It makes evolutionary biology appear to be an esoteric science of long-term change rather than a discipline that affects people's daily lives. When examples such as the rapid evolution of Darwin's finches and of pathogens are reported, to most people they seem like odd exceptions. Yet they should not.

We scientists and science writers make the situation worse when we fail to use the word "evolution" when writing about rapid evolution. A whole vocabulary of phrases has developed to provide surrogates for the word. Entire articles on rapid evolution routinely appear in newspapers and the science reporting section of major science journals without a single mention of the word "evolution" or "evolve," substituting instead such phrases as "acquired new traits," "resistant strains spread," "could . . . ultimately adjust," and "could . . . ultimately breach." Phrases such as "acquired resistance" and "developed resistance" have become automatic substitutes for "rapid evolution of pathogens" in much popular science writing. Although biologists translate these phrases to mean "evolution," nonbiologists often do not, because they simply do not know that populations generally "develop resistance" by evolving through the process of natural selection.

Some of this problem with vocabulary is historical artifact. Evolution was once the development hypothesis, coined by Haller in 1744 as a name for preformationism, in which future generations were already present in compact form within the bodies of the current generation, like Russian dolls (Bowler 1975; Gould 1977). In the 1700s and 1800s, others used the word as a synonym for the ontogeny of organisms. Similarly, phrases such as "acquired resistance" are historical artifacts of wording associated with Lamarck's theory of the inheritance of acquired characteristics. All these words intermingle across time, even as the meaning of the word "evolution" has changed.

With just a few changes of words to say what we actually mean, we have at our disposal a powerful tool for educating nonscientists about the pervasiveness of ongoing evolution and coevolution in our lives. Each time a popular science article on rapid evolution fails to use the word "evolution," we lose another opportunity to educate nonscientists about rapid evolutionary change. If our societies are to understand the ongoing roles of evolution and coevolution, then we must do a better job as scientists in using the words that make the process causing the change absolutely clear.

LOSS OF THE FREE R&D OF LONG-COEVOLVED INTERACTIONS AND WILDERNESS

The other impediment to the development of a science of applied coevolutionary biology is the continuing loss of wilderness. As we continue to transform the earth through the alteration of landscapes, agricultural coevolution, and surrogate coevolution, we are simultaneously creating opportunities for studying the dynamics of the early stages of coevolutionary change. It is impossible for us, however, to use these results alone to predict the long-term consequences of our actions. That is one of the main reasons the last remaining large wilderness areas are so vitally important to our future. These environments retain remnants of interactions that have coevolved over millions of years. The structure of these coevolved relationships can tell us much about how coevolution shapes species across space and time in complex landscapes.

By wilderness I do not mean landscapes completely unmodified by human activities, because such landscapes are now mostly gone. All the remaining wilderness regions are becoming—or must become—in some sense gardens, as Janzen (1999) has argued. Humans now dominate almost all ecosystems worldwide (Vitousek et al. 1997), and we are a pervasive evolutionary force (Palumbi 2001). By wilderness I mean the small percentage of landscapes that are relatively unmodified by human activities (see Mittermeier et al. 2003) and in which the dominant form of interaction is among species that have coevolved for many millennia. These are the landscapes in which we can still evaluate the genetic, populational, and regional processes of coevolution that keep players in the evolutionary game for long periods of time.

Each of these long-coevolved interactions is irreplaceable free research and development for our societies. Each of them is a product of thousands or even millions of years of evolution and coevolution. These species have geographically structured genomes that have been honed sharply by natural selection across unforgiving landscapes, and their gene sequences carry crucial information on what works when coevolving with parasites, predators, and mutualists in particular kinds of environments. No amount of money could ever replace the information found within these natural populations on how to combine genes in ways that are effective at thwarting enemies and attracting mutualisms in particular environments. It is simply impossible to set up an experiment and ask how genomes in truly complex, multispecific environments will evolve over millions of years.

Wilderness and long-coevolved interactions are therefore among humanity's most undervalued assets. Like the other undervalued assets, such as clean water, we have begun to appreciate them only as they have become rare (Balmford et al. 2002). Increasing the value of natural assets in general requires what Gretchen Daily and Katherine Ellison (Daily and Ellison 2002) have called a "new economy of nature" that makes maintenance of the assets profitable for societies and individuals. Part of that profitability is coming from an increased appreciation of the potential medicines and engineering designs that we can develop through careful study of the "wild solutions" already attained within natural communities (Beattie and Ehrlich 2001). Another part of the profitability comes from an increasing appreciation of ecosystem services and of the cost to societies when those services break down (Daily 1997). We can, of course, obtain some of the information and some of those services without maintaining wildernesses in which long-coevolved interactions continue to coevolve. In the process, however, we lose the context in which those solutions have worked, because we lose the population-level and regional-level dynamics that make some solutions more stable than others.

Each time we lose a long-coevolved interaction, we therefore lose some of that free research and development that natural selection has already done for us. At great expense, we can clean up polluted water, but even with unlimited money we cannot replace long-coevolved interactions and explore them for societal solutions they may have harbored. Each time we lose one of the few remaining wilderness areas, we lose one of our touchstones for understanding the long-term outcomes of coevolution in particular environments. At a time when agriculture, forestry, and aquaculture are struggling with the problem of breeding for durable resistance against parasites and pathogens, species interactions in wilderness areas can tell us about the number of defense alleles that are maintained within local natural populations over long periods of time. Moreover, these interactions can tell us about the combination of defenses that are effective in particular kinds of environments.

These interactions can show us the kinds of population and genetic dynamics that result from ongoing coevolution between species beyond the early decades after an interaction first forms. That kind of information is crucial if we are to have a scientific theory of conservation that conserves the coevolutionary process. We are creating hundreds of thousands of new interactions as we move species globally. Most of these species interact with multiple other species and differ geographically in the particular species with which they interact most. Our scientific knowledge of how introduced species interact with other species is restricted to a small number of species, each studied for a few generations. Consequently, we simply have no idea how these interactions will coevolve and stabilize over the long term.

Long-coevolved interactions can therefore give us the clues we need to make more informed decisions. We will, though, have to piece together the clues across multiple fragmented areas. Through those attempts we can develop a more complete understanding of the geographic mosaic of coevolution over the short, intermediate, and long-term. We can, for example, compare an interaction in post-Pleistocene environments in northern North America and Europe with the same interaction in older populations from nonglaciated areas. We can compare interactions across gradients to see how species coevolve in different ways in different environments. The possibilities are endless and the scientific knowledge to be gained is immense, if we retain these remaining touchstone wilderness environments and use the information they contain.

As we co-opt the coevolutionary process for our own ends, we therefore should be able to maintain the coevolutionary process in ways that reduce human suffering while preserving biodiversity for future generations. But we have to truly understand the process. The few remaining relatively undisturbed communities worldwide are among the most precious resources we can offer future generations. They show us how species coevolve as geographic mosaics across complex landscapes, and they tell us how species can remain in the evolutionary game for millions of years despite the onslaught of rapidly evolving enemies. The rich diversity of life is a result of the ever-changing geographic mosaic of coevolution, and the future of the earth depends upon our maintaining the coevolutionary process. That is the essential message of Darwin's entangled bank.

Conclusions

Our developing understanding of the geographic mosaic of coevolution provides a hopeful framework both for conserving the coevolutionary process in rapidly changing landscapes and for developing a science of applied coevolutionary biology. The task is daunting, because we are rapidly changing the spatial and temporal scales at which coevolution operates and even the process of coevolution itself. Almost instantaneous transport of genes and species over wide geographic scales differs from anything that has occurred historically in most taxa. Perhaps most important, we are developing surrogate genes such as antibiotics and vaccines that are creating a completely new form of coevolutionary dynamics. We are, in fact, actively attempting to manipulate coevolutionary trajectories at a time when we have only begun to understand how selection mosaics, coevolutionary hotspots, and trait remixing drive coevolution. Our chances for success will be greatest if we combine explicit studies of coevolution in these novel interactions with studies that use the free R&D found in long-coevolved interactions.

Appendix: *Major Hypotheses on Coevolution*

Overall Framework: The Geographic Mosaic Theory of Coevolution

The major components of the geographic mosaic theory of coevolution (chapters 1–6) can be encapsulated as follows.

MAJOR ASSUMPTIONS

Species are groups of genetically differentiated populations, and most interacting species do not have identical geographic ranges.

Species are phylogenetically conservative in their interactions, and that conservatism often holds interspecific relationships together for long periods of time.

Most local populations specialize their interactions on a few other species.

The ecological outcomes of these interspecific interactions differ among communities.

Species often become locally adapted to local populations of other species and continue to evolve rapidly.

EVOLUTIONARY HYPOTHESIS

Geographic selection mosaics. Natural selection on interspecific interactions varies among populations partly because there are geographic differences in how fitness in one species depends upon the distribution of genotypes in another species. That is, there is often a genotype-by-genotype-by-environment interaction in fitnesses of interacting species.

Coevolutionary hotspots. Interactions are subject to reciprocal selection only within some local communities. These coevolutionary hotspots are embedded in a broader matrix of coevolutionary coldspots, where local selection is nonreciprocal.

Trait remixing. The genetic structure of coevolving species also changes through new mutations, gene flow across landscapes, random genetic drift, and extinction of local populations. These processes contribute to the shifting geographic mosaic of coevolution by continually altering the spatial distributions of potentially coevolving alleles and traits. Even if an interaction is antagonistic across all populations, local populations may differ at any moment in time in the range of defense and counter-defense alleles on which selection can act.

ECOLOGICAL PREDICTIONS

Populations differ in the traits shaped by an interaction.

Traits of interacting species are well matched only in communities.

Few coevolved traits spread across all populations to become fixed traits within species, because few coevolved traits will be favored across all populations.

Chapters 7–15 explore how the major hypotheses in coevolutionary biology fit within this framework. Chapter 7 evaluates hypotheses on lineage diversification, whereas chapters 8–15 evaluate how the seven major classes of local coevolutionary dynamics fit within the broader geographic mosaic of coevolution.

Coevolution and Speciation

The geographic mosaic of coevolution shapes speciation through the general process of diversifying coevolution. The hypothesis of escape-and-radiate coevolution is the only current specific hypothesis of how diversifying coevolution shapes the overall diversification of lineages.

Diversifying coevolution. Through the geographic mosaic of coevolutionary change, some populations of interacting species diverge into new species. The genetic feedback resulting from local coevolutionary adaptation may allow more rapid speciation of populations than speciation that results from adaptation to physical environments.

Escape-and-radiate coevolution. Interacting parasites and hosts may diverge through a multistep process that creates reciprocal starbursts of speciation. The process develops as a new mutation spreads throughout a host population, allowing the mutant individuals to avoid attack by its parasites. The mutant population enters a new adaptive zone, allowing it to spread and diversify across landscapes in the absence of the interaction with their previous enemies. Host diversification therefore occurs primarily during time periods in which the interaction is not occurring. Subsequently, a mutant parasite population overcomes these new defenses and radiates in species onto the now-diversified mutant host clade.

Coevolution and Adaptation

In addition, the geographic mosaic shapes the seven major trajectories of local coevolutionary dynamics. The specific hypotheses are summarized here, in some cases in slightly more expanded form than in the chapters to summarize the discussion of the hypothesis in those chapters.

COEVOLVING POLYMORPHISMS

The geographic mosaic of oscillating polymorphisms. Coevolving polymorphisms are shaped by natural selection on parasites to adapt to the most common local host genotypes and by selection on hosts to recognize and resist attack by parasites. Selection on hosts favors individuals with genes that allow detection of multiple parasite species or genotypes, but it also favors rare genotypes to which the local parasites are not adapted. Local persistence of a small number of coevolving polymorphisms is maintained by multiple forms of selection, but maintenance of high allelic diversity of coevolving polymorphisms within species across millennia requires geographic selection mosaics, coevolutionary hotspots, and trait remixing.

Optimal allelic diversification in enemy-recognition systems. Coevolution between hosts and multiple parasites proceeds through selection on hosts to accumulate alleles that allow recognition of parasite attack. Simultaneously, selection acts on each parasite species to be sufficiently different from other parasites (e.g., in its outer membrane proteins) that individuals escape detection by host alleles evolved to detect other parasite species. Because parasite assemblages differ geographically, host populations will differ in the recognition alleles they accumulate, and parasite populations will differ in how they diverge from co-occurring parasites to escape host detection. Host recognition alleles will diversify within local populations to encompass the range of detectably different local parasites and may reach an upper limit, beyond which alleles may interfere with each other in parasite detection.

Red Queen mosaics. Local adaptation in parasite assemblages and selection favoring locally rare host genotypes together create continually shifting geographic selection mosaics and coevolutionary hotspots that are further modified through ongoing trait remixing among populations. Sexual individuals are more likely than asexual individuals to produce offspring with higher than average fitness amid constantly changing coevolutionary selection across landscapes. Sexual reproduction is therefore maintained by the geographic mosaic of coevolution.

Red Queen captured. Asexual reproduction in some species results from coevolution between parasites and host, such that selection on hosts favoring sexual reproduction is countered by selection on parasites favoring asexual reproduction in their hosts.

COEVOLUTIONARY ALTERNATION

Coevolutionary alternation. Natural selection favors predators (or grazers or parasites) that preferentially attack prey populations with relatively low levels of defense, and it favors increased levels of defense in those prey populations. Meanwhile, selection favors the evolution of reduced defenses in unattacked prey populations, because defenses impose fitness costs in the absence of predation. As relative levels of defense change among prey species over time, selection favors those predators that preferentially attack prey species that are currently least defended. The hypothesis is most likely to apply to predators, grazers, and those parasites that preferentially choose among multiple potential victims.

Coevolutionary alternation with escalation. Coevolutionary escalation results from coevolutionary alternation when predator, grazer, or parasite populations evolve their relative preferences among victim species faster than previously preferred, but now unattacked, victim populations completely lose their defenses. Under these conditions, the overall levels of defense and counterdefense ratchet upward over time. Relentless escalation is prevented by occasional evolution of qualitatively new (but potentially less costly) defenses and counterdefenses within local populations, by gene flow among coevolutionary hotspots and coldspots across landscapes, and by adaptive compromises imposed on populations within multispecific networks.

ATTENUATED ANTAGONISM

Coevolution with symbionts toward attenuated antagonism. Coevolution between parasites and hosts leads to attenuated antagonism when local symbiont genetic diversity is low, the rate of infectious spread of symbionts is low, and the geographic structure of species allows high rates of local adaptation in hosts and parasites. Which combination of these factors is most important in coevolution toward attenuated antagonism is becoming the subject of much current research, as new models and empirical studies evaluate coevolutionary change rather than simply the evolution of virulence in parasites attacking host populations that are prevented from undergoing a coevolutionary response.

COEVOLUTIONARY COMPLEMENTARITY AND CONVERGENCE

Coevolved monocultures and complementary symbionts. Symbiotic mutualisms have evolved from the subset of symbiotic interactions in which the interacting species have complementary traits that confer reciprocal fitness benefits and in which hosts are able to restrict the interaction locally to single symbiont genotypes or complementary sets of noncompeting symbionts. The major mechanisms enforcing mutualism between symbionts and hosts include vertical transmission of symbionts and complex genetic handshaking with horizontally transmitted symbionts.

Symbiont partitioning. Large and modular organisms maintain complex assemblages of symbiotic mutualists by partitioning them among body parts, maintaining "controller" symbionts, selectively eliminating modules, or altering assemblage structure at different ages and in different environments.

Convergence within mutualistic networks. Reciprocal selection on mutualisms between free-living species favors genetically variable, multispecific networks in which species converge and specialize on the core traits of the mutualism rather than directly on other species. Mutualistic networks should therefore coevolve toward highly malleable geographic mosaics of interacting species. Current results suggest that these networks should show a more nested structure than antagonistic networks.

COEVOLUTIONARY DISPLACEMENT

Displacement of competitors across coevolutionary hotspots. Closely related species that compete for a potentially limiting resource coevolve through divergence in traits or habitat use when populations come into contact in regions where the resource is limiting to both species (i.e., coevolutionary hotspots) but the pattern of displacement may differ among hotspots.

Displacement through apparent competition across coevolutionary hotspots. Species that share predators, parasites, or mutualists coevolve through divergence of traits, habitat use, or mutualists when populations come into contact in regions where their fitnesses decline as a result of these shared interspecific interactions. The pattern of displacement may differ among coevolutionary hotspots.

Replicated guild structure among coevolutionary hotspots. Multispecific assemblages of competing congeners undergo similar patterns of character displacement wherever they occur in sympatry, although individual species may be displaced in different ways in different geographic regions.

Trait overdispersion in multispecific competitive networks. Competing species diverge within coevolutionary hotspots to form assemblages that differ by a constant ratio in the traits under coevolutionary selection. Individual species may be displaced in different ways in different geographic regions.

Literature Cited

Aanen, D. K., P. Eggleston, C. Rouland-Lefevre, T. Guldberg-Frøslev, S. Rosendahl, and J. J. Boomsma. 2002. The evolution of fungus-growing termites and their mutualistic fungal symbionts. *Proceedings of the National Academy of Sciences USA* 99:14887–14892.

Aarssen, L. W., and T. Keogh. 2002. Conundrums of competitive ability in plants: What to measure? *Oikos* 96:531–542.

Abbott, A. 2001. The flu HQ. *Nature* 414:10–11.

Abbott, P., and N. A. Moran. 2002. Extremely low levels of genetic polymorphism in endosymbionts (*Buchnera*) of aphids (*Pemphigus*). *Molecular Ecology* 11:2649–2660.

Abbott, R. J., and D. G. Forbes. 2002. Extinction of the Edinburgh lineage of the allopolyploid neospecies, *Senecio cambrensis* Rosser (Asteraceae). *Heredity* 88:267–269.

Abrahamson, W. G., and A. E. Weis. 1997. *Evolutionary ecology across three trophic levels: Goldenrods, gallmakers, and natural enemies.* Princeton University Press, Princeton, New Jersey.

Abrams, P. A. 1999. The adaptive dynamics of consumer choice. *American Naturalist* 153:83–97.

———. 2000a. The evolution of predator-prey interactions: Theory and evidence. *Annual Review of Ecology and Systematics* 31:79–105.

———. 2000b. Character shifts of species that share predators. *American Naturalist* 156:S45–S61.

———. 2001. Predator-prey interactions. Pages 277–289 in C. W. Fox, D. A. Roff, and D. J. Fairbairn, eds., *Evolutionary ecology: Concepts and case studies.* Oxford University Press, Oxford.

Abrams, P. A., and X. Chen. 2002. The evolution of traits affecting resource acquisition and predator vulnerability: Character displacement under real and apparent competition. *American Naturalist* 160:692–704.

Abrams, P. A., and H. Matsuda. 1996. Fitness minimization and dynamic instability as a consequence of predator-prey coevolution. *Evolutionary Ecology* 11:1–20.

Adam, D. 2001. Plans for GM livestock fail the poor. *Nature* 411:403.

Adams, D. C., and F. J. Rohlf. 2000. Ecological character displacement in *Plethodon*: Biomechanical differences found from a geometric morphometric study. *Proceedings of the National Academy of Sciences USA* 97:4106–4111.

Adams, R. M. M., U. G. Mueller, A. K. Holloway, A. M. Green, and J. Narozniak. 2001. Garden sharing and garden stealing in fungus-growing ants. *Naturwissenschaften* 87:491–493.

Addicott, J. F. 1996. Cheaters in yucca/moth mutualism. *Nature* 380:114–115.

————. 1998. Regulation of mutualism between yuccas and yucca moths: Population level processes. *Oikos* 81:119–129.

Agrawal, A. A. 2000. Host-range evolution: Adaptation and trade-offs in fitness of mites on alternative hosts. *Ecology* 81:500–508.

————. 2001. Phenotypic plasticity in the interactions and evolution of species. *Science* 294:321–326.

Agrawal, A., and P. M. Kotanen. 2003. Herbivores and success of exotic plants: A phylogenetically-controlled experiment. *Ecology Letters* 6:712–715.

Agrawal, A. F, and C. M. Lively. 2002. Infection genetics: Gene-for-gene versus matching-alleles models and all points in between. *Evolutionary Ecology Research* 4:79–90.

————. 2003. Modelling infection as a two-step process combining gene-for-gene and matching-allele genetics. *Proceedings of the Royal Society of London Biological Sciences Series B* 270:323–334.

Akino, T., J. J. Knapp, J. A. Thomas, and G. W. Elmes. 1999. Chemical mimicry and host specificity in the butterfly *Maculinea rebeli*, a social parasite of *Myrmica* ant colonies. *Proceedings of the Royal Society of London Biological Sciences Series B* 266:1419–1426.

Alexander, H. M., and J. Antonovics. 1995. Spread of anther-smut disease (*Ustilago violacea*) and character correlations in a genetically variable experimental population of *Silene alba*. *Journal of Ecology* 83:783–794.

Alexander, H. M., P. H. Thrall, J. Antonovics, A. M. Jarosz, and P. V. Oudemans. 1996. Population dynamics and genetics of plant disease: A case study of anther-smut disease. *Ecology* 77:990–996.

Allen, M. F., W. Swenson, J. I. Querejeta, L. M. Egerton-Warburton, and K. K. Treseder. 2003. Ecology of mycorrhizae: A conceptual framework for complex interactions among plants and fungi. *Annual Review of Phytopathology* 41:271–303.

Alonso, L. E. 1998. Spatial and temporal variation in the ant occupants of a facultative ant-plant. *Biotropica* 30:201–213.

Alphey, L., C. B. Beard, P. Billingsley, M. Coetzee, A. Crisanti, C. Curtis, P. Eggleston et al. 2002. Malaria control with genetically manipulated insect vectors. *Science* 298:119–121.

Als, T. D., D. R. Nash, and J. J. Boomsma. 2002. Geographical variation in host-ant specificity of the parasitic butterfly *Maculinea alcon* in Denmark. *Ecological Entomology* 27:403–414.

Althoff, D. M., and J. N. Thompson. 1999. Comparative geographic structures of two parasitoid-host interactions. *Evolution* 53:818–825.

————. 2001. Geographic structure in the searching behaviour of a specialist parasitoid: Combining molecular and behavioural approaches. *Journal of Evolutionary Biology* 14:406–417.

Altshuler, D. L., and A. G. Clark. 2003. Darwin's hummingbirds. *Science* 300:588–589.

Anderson, R. A., B. J. G. Knols, and J. C. Koella. 2000. *Plasmodium falciparum* sporozoites increase feeding-associated mortality of their mosquito hosts, *Anopheles gambiae s. l. Parasitology* 120:329–333.

Anderson, R. M., C. A. Donnelly, and S. Gupta. 1997. Vaccine design, evaluation, and community-based use for antigenically variable infectious agents. *Lancet* 350:1466–1470.

Anderson, R. M., and R. M. May. 1979. Population biology of infectious disease. Part 1. *Nature* 280:361–367.

————. 1982. Coevolution of hosts and parasites. *Parasitology* 85:411–426.

————. 1991. *Infectious diseases of humans: Dynamics and control.* Oxford University Press, Oxford.

Antonovics, J. 2003. Toward community genomics? *Ecology* 84:598–601.

Antonovics, J., M. Hood, and J. Partain. 2002. The ecology and genetics of a host shift: *Microbotryum* as a model system. *American Naturalist* 160:S40–S53.

Antonovics, J., and P. H. Thrall. 1994. The cost of resistance and the maintenance of genetic polymorphism in host-pathogen systems. *Proceedings of the Royal Society of London Biological Sciences Series B* 257:105–110.

Antonovics, J., P. H. Thrall, and A. M. Jarosz. 1998. Genetics and the spatial ecology of species interactions: The *Silene-Ustilago* system. Pages 158–80 in P. Kareiva, ed., *Spatial ecology: The role of space in population dynamics and interspecific interactions.* Princeton University Press, Princeton, New Jersey.

Antonovics, J., P. Thrall, A. Jarosz, and D. Stratton. 1994. Ecological genetics of metapopulations: The *Silene-Ustilago* plant-pathogen system. Pages 146–170 in L. A. Real, ed., *Ecological Genetics.* Princeton University Press, Princeton, New Jersey.

Archibald, J. M., M. B. Rogers, M. Toop, K. Ishida, and P. J. Keeling. 2003. Lateral gene transfer and the evolution of plastid-targeted proteins in the secondary plastid-containing alga *Bigelowiella natans. Proceedings of the National Academy of Sciences USA* 100:7678–7683.

Arkhipova, I., and M. Meselson. 2000. Transposable elements in sexual and ancient asexual taxa. *Proceedings of the National Academy of Sciences USA* 97:14473–14477.

Armbruster, W. S. 1997. Exaptations link evolution of plant-herbivore and plant-pollinator interactions: A phylogenetic inquiry. *Ecology* 78:1661–1672.

Armbruster, W. S., and B. G. Baldwin. 1998. Switch from specialized to generalized pollination. *Nature* 394:632.

Armbruster, W. S., M. E. Edwards, and E. M. Debevec. 1994. Floral character displacement generates assemblage structure of western Australian triggerplants (*Stylidium*). *Ecology* 75:315–329.

Armbruster, W. S., J. J. Howard, T. P. Clausen, E. M. Debevec, J. C. Loquvam, M. Matsuki, B. Cerendolo et al. 1997. Do biochemical exaptations link evolution of plant defense and pollination systems? Historical hypotheses and experimental tests with *Dalechampia* vines. *American Naturalist* 149:461–484.

Arnold, A. E., Z. Maynard, G. S. Gilbert, P. D. Coley, and T. A. Kursar. 2000. Are tropical endophytes hyperdiverse? *Ecology Letters* 3:267–274.

Arnold, A. E., L. C. Mejía, D. Kyllo, E. I. Rojas, Z. Maynard, N. Robbins, and E. A. Herre. 2003. Fungal endophytes limit pathogen damage in a tropical tree. *Proceedings of the National Academy of Sciences USA* 100:15649–16654.

Arnold, S. J. 1981. Behavioral variation in natural populations. II. The inheritance of a feeding response in crosses between geographic races of the garter snake, *Thamnophis elegans. Evolution* 35:510–515.

Atmar, W., and B. D. Patterson. 1993. The measure of order and disorder in the distribution of species in fragmented habitat. *Oecologia* 96:373–382.

Austin, D. J., K. G. Kristinsson, and R. M. Anderson. 1999. The relationship between the volume of antimicrobial consumption in human communities and the frequency of resistance. *Proceedings of the National Academy of Sciences USA* 96:1152–1156.

Avise, J. C. 1994. *Molecular markers, natural history and evolution.* Chapman and Hall, New York.

———. 2000. *Phylogeography: The history and formation of species.* Harvard University Press, Cambridge, Massachusetts.

Ayres, D. R., and D. R. Strong. 2001. Origin and genetic diversity of *Spartina anglica* (Poaceae) using nuclear DNA markers. *American Journal of Botany* 88:1863–1867.

Baker, A. C. 2003. Flexibility and specificity in coral-algal symbiosis: Diversity, ecology, and biogeography of *Symbiodinium*. *Annual Review of Ecology and Systematics* 34:661–689.

Ball-Ilosera, N. S., J. L. Garcia-Marin, and C. Pla. 2002. Managing fish populations under mosaic relationships: The case of brown trout (*Salmo trutta*) in peripheral Mediterranean populations. *Conservation Genetics* 3:385–400.

Balmford, A., A. Bruner, P. Cooper, R. Costanza, S. Farber, R. E. Green, M. Jenkins et al. 2002. Economic reasons for conserving wild nature. *Science* 297:950–953.

Bandi, C., T. J. C. Anderson, C. Genchi, and M. L. Blaxter. 1998. Phylogeny of *Wolbachia* bacteria in filarial nematodes. *Proceedings of the Royal Society of London Biological Sciences Series B* 265:2407–2413.

Bandi, C., M. Sironi, G. Damiani, L. Magrassi, C. A. Nalepa, U. Laudani, and L. Sacchi. 1995. The establishment of intracellular symbiosis in an ancestor of cockroaches and termites. *Proceedings of the Royal Society of London Biological Sciences Series B* 259:293–299.

Bascompte, J., P. Jordano, C. J. Melián, and J. M. Olesen. 2003. The nested assembly of plant-animal mutualistic networks. *Proceedings of the National Academy of Sciences USA* 100: 9383–9387.

Baumann, P., C. Y. Lai, L. Baumann, D. Rouhbakhsh, N. A. Moran, and M. A. Clark. 1995. Mutualistic associations of aphids and prokaryotes: Biology of the genus *Buchnera*. *Applied and Environmental Microbiology* 61:1–7.

Bazzaz, F. A. 1996. *Plants in changing environments: Linking physiological, population, and community ecology.* Cambridge University Press, Cambridge.

———. 2001. Plant biology in the future. *Proceedings of the National Academy of Sciences USA* 98:5441–5445.

Beattie, A. J., and P. R. Ehrlich. 2001. *Wild solutions: How biodiversity is money in the bank.* Yale University Press, New Haven, Connecticut.

Becerra, J. X. 1997. Insects on plants: Macroevolutionary chemical trends in host use. *Science* 276:253–256.

Becerra, J. X., and D. L. Venable. 1999. Macroevolution of insect-plant associations: The relevance of host biogeography to host affiliation. *Proceedings of the National Academy of Sciences USA* 12626–12631.

Bekkevold, D., J. Frydenberg, and J. J. Boomsma. 1999. Multiple mating and facultative polygony in the Panamanian leaf-cutter ant *Acromyrmex echinatior*. *Behavioral Ecology and Sociobiology* 46:103–109.

Beldade, P., and P. M. Brakefield. 2002. The genetics and evo-devo of butterfly wing patterns. *Nature Reviews Genetics* 3:442–452.

Beldade, P., K. Koops, and P. M. Brakefield. 2002. Developmental constraints *versus* flexibility in morphological evolution. *Nature* 416:844–847.

Bell, G. 1982. *The masterpiece of nature: The evolution and genetics of sexuality.* University of California Press, Berkeley, California.

———. 1988. *Sex and death in protozoa: The history of an obsession.* Cambridge University Press, Cambridge.

Belwood, D. R., P. C. Wainwright, C. J. Fulton, and A. Hoey. 2002. Assembly rules and functional groups at global and biogeographic scales. *Functional Ecology* 16:557–562.

Benassi, V., F. Frey, and Y. Carton. 1998. A new specific gene for wasp cellular immune resistance in *Drosophila*. *Heredity* 80:347–352.

Benkman, C. W. 1987. Crossbill foraging behavior, bill structure, and patterns of food profitability. *Wilson Bulletin* 99:351–368.

———. 1989. On the evolution and ecology of island populations of crossbills. *Evolution* 43:1324–1330.

———. 1993. Adaptation to single resources and the evolution of crossbill (*Loxia*) diversity. *Ecological Monographs* 63:305–325.

———. 1999. The selection mosaic and diversifying coevolution between crossbills and lodgepole pine. *American Naturalist* 153:S75–S91.

———. 2003. Divergent selection drives the adaptive radiation of crossbills. *Evolution* 57:1176–1181.

Benkman, C. W., W. C. Holimon, and J. W. Smith. 2001. The influence of a competitor on the geographic mosaic of coevolution between crossbills and lodgepole pine. *Evolution* 55:282–294.

Benkman, C. W., T. L. Parchman, A. Favis, and A. M. Siepielski. 2003. Reciprocal selection causes a coevolutionary arms race between crossbills and lodgepole pine. *American Naturalist* 162:182–194.

Benzie, J. A. H. 1999. Major genetic differences between crown-of-thorns starfish (*Acanthaster planci*) populations in the Indian and Pacific oceans. *Evolution* 53:1782–1795.

Berenbaum, M. R. 1995. Chemistry and oligophagy in the Papilionidae. Pages 27–38 in J. M. Scriber, Y. Tsubaki, and R. C. Lederhouse, eds., *Swallowtail Butterflies: Their Ecology and Evolutionary Biology.* Scientific Publishers, Gainesville, Florida.

Berenbaum, M. R., and A. R. Zangerl. 1992. Genetics of physiological and behavioral resistance to host furanocoumarins in the parsnip webworm. *Evolution* 46:1373–1384.

———. 1998. Chemical phenotype matching between a plant and its insect herbivore. *Proceedings of the National Academy of Sciences USA* 95:13743–13748.

Berenbaum, M. R., A. R. Zangerl, and J. K. Nitao. 1986. Constraints on chemical coevolution: Wild parsnip and the parsnip webworm. *Evolution* 40:1215–1228.

Bergelson, J., G. Dwyer, and J. J. Emerson. 2001. Models and data on plant-enemy coevolution. *Annual Review of Genetics* 35:469–499.

Bergelson, J., and C. B. Purrington. 1996. Surveying patterns in the cost of resistance in plants. *American Naturalist* 148:536–558.

Berlocher, S. H. 2000. Radiation and divergence in the *Rhagoletis pomonella* species group: Inferences from allozymes. *Evolution* 2000:543–557.

Berlocher, S. H., and J. L. Feder. 2002. Sympatric speciation in phytophagous insects: Moving beyond controversy? *Annual Review of Entomology* 47:773–815.

Bermingham, E., and J. C. Avise. 1986. Molecular zoogeography of freshwater fishes in the southeastern United States. *Genetics* 113:939–965.

Bernardi, G. 2000. Barriers to gene flow in *Embiotoca jacksoni,* a marine fish lacking a pelagic larval stage. *Evolution* 54:226–237.

Bernatchez, L., and C. C. Wilson. 1998. Comparative phylogeography of nearctic and palearctic fishes. *Molecular Ecology* 7:431–452.

Bever, J. D. 1999. Dynamics within mutualism and the maintenance of diversity: Inference from a model of interguild frequency dependence. *Ecology Letters* 2:52–61.

Bever, J. D., P. A. Schultz, A. Pringle, and J. B. Morton. 2001. Arbuscular mycorrhizal fungi: More diverse than meets the eye, and the ecological tale of why. *Bioscience* 51:923–932.

Bidartondo, M. I., and T. D. Bruns. 2001. Extreme specificity in epiparasitic Monotropoideae (Ericaceae): Widespread phylogenetic and geographic structure. *Molecular Ecology* 10:2285–2295.

————. 2002. Fine-level mycorrhizal specificity in the Monotropoideae (Ericaceae): Specificity for fungal species groups. *Molecular Ecology* 11:557–569.

Blackwell, M. 2000. Terrestrial life—fungal life from the start? *Science* 289:1884–1885.

Blanford, S., M. B. Thomas, C. Pugh, and J. K. Pell. 2003. Temperature checks the Red Queen? Resistance and virulence in a fluctuating environment. *Ecology Letters* 6:2–5.

Blum, M. J. 2002. Rapid movement of a *Heliconius* hybrid zone: Evidence for phase III of Wright's shifting balance theory? *Evolution* 56:1992–1998.

Boag, P. T., and P. R. Grant. 1981. Intense natural selection in a population of Darwin's finches (Geospizinae) in the Galápagos. *Science* 214:82–85.

Bohannan, B. J. M., and R. E. Lenski. 2000a. Linking genetic change to community evolution: Insights from studies of bacteria and bacteriophage. *Ecology Letters* 3:362–377.

————. 2000b. The relative importance of competition and predation varies with productivity in a model community. *American Naturalist* 156:329–340.

Bohannan, B. J. M., M. Travisano, and R. E. Lenski. 1999. Epistatic interactions can lower the cost of resistance to multiple consumers. *Evolution* 53:292–295.

Bohonak, A. J. 1999. Dispersal, gene flow, and population subdivision. *Quarterly Review of Biology* 74:21–45.

Bonhoeffer, S., M. Lipsitch, and B. R. Levin. 1997. Evaluating treatment protocols to prevent antibiotic resistance. *Proceedings of the National Academy of Sciences USA* 94:12106–12111.

Bonhoeffer, S., and P. Sniegowski. 2002. The importance of being erroneous. *Nature* 420:367–368.

Boots, M., P. J. Hudson, and A. Sasaki. 2004. Large shifts in pathogen virulence related to host population structure. *Science* 303:842–844.

Borghans, J. A. M., J. B. Beltman, and R. J. De Boer. 2004. MHC polymorphism under host-parasite coevolution. *Immunogenetics* 55:732–739.

Bossard, C. C., J. M. Randall, and M. C. Hoshovsky, eds. 2000. *Invasive plants of California's wildlands.* University of California Press, Berkeley, California

Bot, A. N. M., S. A. Rehner, and J. J. Boomsma. 2001. Partial incompatibility between ants and symbiotic fungi in two sympatric species of *Acromyrmex* leaf-cutting ants. *Evolution* 55:1980–1991.

Bouchon, D., T. Rigaud, and P. Juchault. 1998. Evidence for widespread *Wolbachia* infection in isopod crustaceans: Molecular identification and host feminization. *Proceedings of the Royal Society of London Biological Sciences Series B* 265:1081–1090.

Bowler, P. J. 1975. The changing meaning of "evolution." *Journal of the History of Ideas* 36:95–114.

Boyes, D. C., J. Nam, and J. L. Dangl. 1998. The *Arabidopsis thaliana RPM1* disease resistance gene product is a peripheral plasma membrane protein that is degraded coincident with the hypersensitive response. *Proceedings of the National Academy of Sciences USA* 95:15849–15854.

Bradshaw, H. D., Jr., and D. W. Schemske. 2003. Allele substitution at a flower colour locus produces a pollinator shift in monkeyflowers. *Nature* 426:176–178.

Brakefield, P. M. 2001. Structure of a character and the evolution of butterfly eyespot patterns. *Journal of Experimental Zoology* 291:93–104.

————. 2003. Artificial selection and the development of ecologically relevant phenotypes. *Ecology* 84:1661–1671.

Brakefield, P. M., V. French, and B. J. Zwaan. 2003. Development and the genetics of evolutionary change within insect species. *Annual Review of Ecology and Systematics* 34:633–660.

Brakefield, P. M., and T. G. Liebert. 2000. Evolutionary dynamics of declining melanism in the peppered moth in the Netherlands. *Proceedings of the Royal Society of London Biological Sciences Series B* 267:1953–1957.

Bremermann, H. J. 1980. Sex and polymorphism as strategies in host-pathogen interactions. *Journal of Theoretical Biology* 87:671–702.

Breznak, J. A. 2002. Phylogenetic diversity and physiology of termite gut spirochetes. *Integrative and Comparative Biology* 42:313–318.

Brodie, E. D., Jr. 1968. Investigations on the skin toxin of the adult rough-skinned newt, *Taricha granulosa. Copeia* 1968:307–313.

Brodie, E. D., Jr., J. L. Hensel, Jr., and J. A. Johnson. 1974. Toxicity of the urodele amphibians *Taricha, Notophthalmus, Cynops* and *Paramesotriton* (Salamandridae). *Copeia* 1974: 506–511.

Brodie, E. D., Jr., B. J. Ridenhour, and E. D. Brodie, III. 2002. The evolutionary response of predators to dangerous prey: Hotspots and coldspots in the geographic mosaic of coevolution between newts and snakes. *Evolution* 56:2067–2082.

Brodie, E. D., III, and E. D. Brodie, Jr. 1990. Tetrodotoxin resistance in garter snakes: An evolutionary response of predators to dangerous prey. *Evolution* 44:651–659.

———. 1999a. Costs of exploiting poisonous prey: Evolutionary trade-offs in a predator-prey arms race. *Evolution* 53:626–631.

———. 1999b. Predator-prey arms races. *Bioscience* 49:557–568.

Brodie, E. D., III, and B. J. Ridenhour. 2004. Reciprocal selection at the phenotypic interface of coevolution. *Integrative and Comparative Biology* (in press).

Brodo, I. M., S. D. Sharnoff, and S. Sharnoff. 2001. *Lichens of North America.* Yale University Press, New Haven, Connecticut.

Brody, A. K. 1997. Effects of pollinators, herbivores, and seed predators on flowering phenology. *Ecology* 78:1624–1631.

Brody, A. K., and N. M. Waser. 1995. Oviposition patterns and larval success of a pre-dispersal seed predator attacking two confamilial host plants. *Oikos* 74:447–452.

Bronstein, J. L. 1994a. Conditional outcomes in mutualistic interactions. *Trends in Ecology and Evolution* 9:214–217.

———. 1994b. Our current understanding of mutualism. *Quarterly Review of Biology* 69: 31–51.

———. 1998. The contribution of ant-plant protection studies to our understanding of mutualism. *Biotropica* 30:150–161.

Bronstein, J. L., and M. Hossaert-McKey. 1996. Variation in reproductive success within a subtropical fig/pollinator mutualism. *Journal of Biogeography* 23:433–446.

Bronstein, J. L., W. G. Wilson, and W. F. Morris. 2003. Ecological dynamics of mutualist/ antagonist communities. *American Naturalist* 162:S24–S39.

Bronstein, J. L., and Y. Ziv. 1997. Costs of two non-mutualistic species in a yucca/yucca moth mutualism. *Oecologia* 112:379–385.

Brooks, D. R., and D. A. McLennan. 1991. *Phylogeny, ecology, and behavior.* University of Chicago Press, Chicago.

———. 2002. *The nature of diversity: An evolutionary voyage of discovery.* University of Chicago Press, Chicago.

Brouat, C., N. Garcia, C. Andary, and D. McKey. 2001. Plant lock and ant key: Pairwise coevolution of an exclusion filter in an ant-plant mutualism. *Proceedings of the Royal Society of London Biological Sciences Series B* 268:2131–2141.

Broughton, W. J., S. Jabbouri, and X. Perret. 2000. Keys to symbiotic harmony. *Journal of Bacteriology* 182:5641–5652.

Brown, J. H. 1995. *Macroecology.* University of Chicago Press, Chicago.

Brown, J. H., B. J. Fox, and D. A. Kelt. 2000. Assembly rules: Desert rodent communities are structured at scales from local to continental. *American Naturalist* 156:314–321.

Brown, J. H., and J. C. Munger. 1985. Experimental manipulation of a desert rodent community: Food addition and species removal. *Ecology* 66:1545–1563.

Brown, J. H., T. G. Whitham, S. K. Morgan Ernest, and C. A. Gehring. 2001. Complex species interactions and the dynamics of ecological systems: Long-term experiments. *Science* 293:643–650.

Brown, J. M., J. H. Leebens-Mack, J. N. Thompson, O. Pellmyr, and R. G. Harrison. 1997. Phylogeography and host association in a pollinating seed parasite *Greya politella* (Lepidoptera: Prodoxidae). *Molecular Ecology* 6:215–224.

Brown, W. L., Jr., and E. O. Wilson. 1956. Character displacement. *Systematic Zoology* 5: 49–64.

Bruns, T. D. 1995. Thoughts on the processes that maintain local species diversity of ectomycorrhizal fungi. *Plant and Soil* 170:63–73.

Bruns, T. D., M. I. Bidartondo, and D. L. Taylor. 2002. Host specificity in ectomycorrhizal communities: What do the exceptions tell us? *Integrative and Comparative Biology* 42:352–359.

Bruns, T. D., T. M. Szaro, M. Gardes, K. W. Cullings, J. J. Pan, D. L. Taylor, T. R. Horton, A. Kretzer, M. Garbelotto, and Y. Li. 1998. A sequence database for the identification of ectomycorrhizal basidiomycetes by phylogenetic analysis. *Molecular Ecology* 7:257–272.

Brunsfeld, S. J., J. Sullivan, D. E. Soltis, and P. S. Soltis. 2002. Comparative phylogeography of north-western North America: A synthesis. Pages 319–339 in J. Silvertown, ed., *Integrating ecology and evolution in a spatial context.* Cambridge University Press, Cambridge.

Buckling, A., and P. B. Rainey. 2002. The role of parasites in sympatric and allopatric host diversification. *Nature* 420:496–499.

Budiansky, S. 2002. Creatures of our own making. *Science* 298:80–86.

Bull, J. J. 1994. Virulence. *Evolution* 48:1423–1437.

Burdon, J. J. 1994. The distribution and origin of genes for race-specific resistance to *Melampsora lini* in *Linum marginale. Evolution* 48:1564–1575.

———. 1997. The evolution of gene-for-gene interactions in natural pathosystems. Pages 245–262 in I. R. Crute, E. B. Holub, and J. J. Burdon, eds., *The gene-for-gene relationship in plant-parasite interactions.* CAB International, New York.

Burdon, J. J., L. Ericson, and W. J. Muller. 1995. Temporal and spatial changes in a metapopulation of the rust pathogen *Triphragmium ulmariae* and its host, *Filipendula ulmaria. Journal of Ecology* 83:979–989.

Burdon, J. J., A. H. Gibson, S. D. Searle, M. J. Woods, and J. Brockwell. 1999. Variation in the effectiveness of symbiotic associations between native rhizobia and temperate Australian *Acacia:* Within-species interactions. *Journal of Applied Ecology* 36:398–408.

Burdon, J. J., and J. K. Roberts. 1995. The population genetic structure of the rust fungus *Melampsora lini* as revealed by pathogenicity, isozyme and RFLP markers. *Plant Pathology* 44:270–278.

Burdon, J. J., and J. N. Thompson. 1995. Changed patterns of resistance in a population of *Linum marginale* attacked by the rust pathogen *Melampsora lini. Journal of Ecology* 83: 199–206.

Burdon, J. J., and P. H. Thrall. 1999. Spatial and temporal patterns in coevolving plant and pathogen associations. *American Naturalist* 153:S15–S33.

————. 2000. Coevolution at multiple spatial scales: *Linum marginale–Melampsora lini*— from the individual to the species. *Evolutionary Ecology* 14:261–281.

————. 2003. The fitness costs to plants of resistance to pathogens. *Genome Biology* 4:227.

Burdon, J. J., P. H. Thrall, and A. H. D. Brown. 1999. Resistance and virulence structure in two *Linum marginale–Melampsora lini* host-pathogen metapopulations with different mating systems. *Evolution* 53:704–716.

Burdon, J. J., P. H. Thrall, and G. J. Lawrence. 2002. Coevolutionary patterns in the *Linum marginale–Melampsora lini* association at a continental scale. *Canadian Journal of Botany* 80: 288–296.

Burton, R. S. 1998. Intraspecific phylogeography across the Point Conception biogeographic boundary. *Evolution* 52:734–745.

Bush, L. P., H. H. Wilkinson, and C. L. Schardl. 1997. Bioprotective alkaloids of grass-fungal endophyte symbioses. *Plant Physiology* 114:1–7.

Butlin, R. K., I. Schöen, and K. Martens. 1999. Origin, age and diversity of clones. *Journal of Evolutionary Biology* 12:1020–1022.

Caldarelli, G., P. G. Higgs, and A. J. McKane. 1998. Modelling coevolution in multispecies communities. *Journal of Theoretical Biology* 193:345–358.

Calsbeek, R., J. N. Thompson, and J. E. Richardson. 2003. Patterns of molecular evolution and diversification in a biodiversity hotspot: The California Floristic Province. *Molecular Ecology* 12:1021–1029.

Campbell, D. R., M. Crawford, A. K. Brody, and T. A. Forbis. 2002. Resistance to pre-dispersal seed predators in a natural hybrid zone. *Oecologia* 131:436–443.

Campbell, D. R., N. M. Waser, and E. J. Melendez-Ackerman. 1997. Analyzing pollinator-mediated selection in a plant hybrid zone: Hummingbird visitation patterns on three spatial scales. *American Naturalist* 149:295–315.

Cariveau, D., R. E. Irwin, A. K. Brody, L. S. Garcia-Mayeya, and A. von der Ohe. 2004. Direct and indirect effects of pollinators and seed predators to selection on plant and floral traits. *Oikos* 104:15–26.

Carlsson-Graner, U., and P. H. Thrall. 2002. The spatial distribution of plant populations, disease dynamics and evolution of resistance. *Oikos* 97:97–110.

Carroll, S. P., and C. Boyd. 1992. Host race radiation in the soapberry bug: Natural history with the history. *Evolution* 46:1052–1069.

Carroll, S. P., H. Dingle, T. R. Famula, and C. W. Fox. 2001. Genetic architecture of adaptive differentiation in evolving host races of the soapberry bug, *Jadera haemotoloma*. *Genetica* 112–113:257–272.

Carroll, S. P., H. Dingle, and S. P. Klassen. 1997. Genetic differentiation of fitness-associated traits among rapidly evolving populations of the soapberry bug. *Evolution* 51:1182–1188.

Carroll, S. P., S. P. Klassen, and H. Dingle. 1998. Rapidly evolving adaptations to host ecology and nutrition in the soapberry bug. *Evolutionary Ecology* 12:955–968.

Carton, Y., and A. J. Nappi. 1997. *Drosophila* cellular immunity against parasitoids. *Parasitology Today* 13:218–227.

Caruso, C. M. 2000. Competition for pollination influences selection on floral traits of *Ipomopsis aggregata*. *Evolution* 54:1546–1557.

Case, T. J., and M. L. Taper. 2000. Interspecific competition, environmental gradients, gene flow, and the coevolution of species' borders. *American Naturalist* 155:583–605.

Casiraghi, M., T. J. C. Anderson, C. Bandi, C. Bazzocchi, and C. Genchi. 2001. A phylogenetic analysis of filarial nematodes: Comparison with the phylogeny of *Wolbachia* endosymbionts. *Parasitology* 122:93–103.

Castellanos, M. C., P. Wilson, and J. D. Thomson. 2003. Pollen transfer by hummingbirds and bumblebees, and the divergence of pollination modes in *Penstamon*. *Evolution* 57:2742–2752.

Castillo-Chavez, C., and Z. Feng. 1997. To treat or not to treat: The case of tuberculosis. *Journal of Mathematical Biology* 35:629–656.

Cattin, M.-F., L.-F. Bersier, C. Banasek-Richter, R. Baltensperger, and J.-P. Gabriel. 2004. Phylogenetic constraints and adaptation explain food-web structure. *Nature* 427:835–839.

Cavender-Bares, J., and A. Wilczek. 2003. Integrating micro- and macroevolutionary processes in community ecology. *Ecology* 84:592–597.

Chapela, I. H., S. A. Rehner, T. R. Schultz, and U. G. Mueller. 1994. Evolutionary history of the symbiosis between fungus-growing ants and their fungi. *Science* 266:1691–1694.

Charlat, S., G. D. G. Hurst, and H. Merçot. 2003. Evolutionary consequences of *Wolbachia* infections. *Trends in Genetics* 19:217–223.

Charleston, M. A., and S. L. Perkins. 2003. Lizards, malaria, and jungles in the Caribbean. Pages 65–92 in R. D. M. Page, ed., *Tangled trees: Phylogeny, cospeciation, and coevolution.* University of Chicago Press, Chicago.

Charlesworth, B., and D. Charlesworth. 1983. The population dynamics of transposable elements. *Genetical Research* 42:1–28.

Charlesworth, B., D. Charlesworth, and N. H. Barton. 2003. The effects of genetic and geographic structure on neutral variation. *Annual Review of Ecology and Systematics* 34:99–125.

Charlesworth, B., P. Sniegowski, and W. Stephan. 1994. The evolutionary dynamics of repetitive DNA in eukaryotes. *Nature* 371:215–220.

Chase, J. M., and T. M. Knight. 2003. Community genetics: Toward a synthesis. *Ecology* 84:580–582.

Chase, J. M., and M. A. Leibold. 2003. *Ecological niches: Linking classical and contemporary approaches.* University of Chicago Press, Chicago.

Chavez, F. P., J. Ryan, S. E. Lluch-Cota, and M. Ñiquen C. 2003. From anchovies to sardines and back: Multidecadal changes in the Pacific Ocean. *Science* 299:217–221.

Chen, D. Q., C. B. Montllor, and A. H. Purcell. 2000. Fitness effects of two facultative endosymbiotic bacteria on the pea aphid, *Acyrthosiphon pisum,* and the blue alfalfa aphid, *A. kondoi. Entomologia Experimentalis et Applicata* 95:315–323.

Chen, X. A., S. Li, and S. Aksoy. 1999. Concordant evolution of a symbiont with its host insect species: Molecular phylogeny of the genus *Glossina* and its bacteriome-associated endosymbiont, *Wigglesworthia glossinidia. Journal of Molecular Evolution* 48:49–58.

Chenuil, A., and D. B. McKey. 1996. Molecular phylogenetic study of a myrmecophyte symbiosis: Did *Leonardoxa*/ant associations diversity via cospeciation? *Molecular Phylogenetics and Evolution* 6:270–286.

Christensen, K. M., T. G. Whitham, and P. Keim. 1995. Herbivory and tree mortality across a pinyon pine hybrid zone. *Oecologia* 101:29–36.

Christophides, G. K., E. Zdobnov, C. Barillas-Mury, E. Birney, S. Blandin, C. Blass, P. T. Brey et al. 2002. Immunity-related genes and gene families in *Anopheles gambiae. Science* 298:159–165.

Clark, J. W., S. Hossain, C. A. Burnside, and S. Kambhampati. 2001. Coevolution between a cockroach and its bacterial endosymbiont: A biogeographic perspective. *Proceedings of the Royal Society of London Biological Sciences Series B* 268:393–398.

Clark, M. A., N. A. Moran, P. Baumann, and J. J. Wernegreen. 2000. Cospeciation between bacterial endosymbionts (*Buchnera*) and a recent radiation of aphids (*Uroleucon*) and pitfalls of testing for phylogenetic congruence. *Evolution* 2000:517–525.

Clay, K. 1998. Fungal endophyte infection and the population biology of grasses. Pages 255–285 in G. P. Cheplick, ed., *The population biology of grasses.* Cambridge University Press, Cambridge.

Clay, K., and J. Holah. 1999. Fungal endophyte symbiosis and plant diversity in successional fields. *Science* 285:1742–1744.

Clay, K., and P. Kover. 1996. Evolution and stasis in plant-pathogen associations. *Ecology* 77:997–1003.

Clay, K., and C. Schardl. 2002. Evolutionary origins and ecological consequences of endophyte symbiosis with grasses. *American Naturalist* 160:S99–S127.

Clayton, D. H., S. E. Bush, B. M. Goates, and K. P. Johnson. 2003. Host defense reinforces host-parasite cospeciation. *Proceedings of the National Academy of Sciences USA* 100:15694–15699.

Cobb, N. S., S. Mopper, C. A. Gehring, M. Caouette, K. M. Christensen, and T. G. Whitham. 1997. Increased moth herbivory associated with environmental stress of pinyon pine at local and regional levels. *Oecologia* 109:389–397.

Coleman, A. W. 2002. Microbial eukaryote species. *Science* 297:337.

Coley, P. D., and J. A. Barone. 1996. Herbivory and plant defenses in tropical forests. *Annual Review of Ecology and Systematics* 27:305–335.

Collins, J. P. 2003. What can we learn from community genetics? *Ecology* 84:574–577.

Colpaert, J. V., P. Vandenkoornhuyse, K. Adriaensen, and J. Vangronsveld. 2000. Genetic variation and heavy metal tolerance in the ectomycorrhizal basidiomycete *Suillus luteus. New Phytologist* 147:367–379.

Conner, J. K. 2001. How strong is natural selection? *Trends in Ecology and Evolution* 16:215–217.

———. 2003. Artificial selection: A powerful tool for ecologists. *Ecology* 84:1650–1660.

Cook, J. M., D. Bean, S. A. Power, and D. J. Dixon. 2004. Evolution of a complex coevolved trait: Active pollination in a genus of fig wasps. *Journal of Evolutionary Biology* 17:238–246.

Cook, L. M., and B. S. Grant. 2000. Frequency of *insularia* during the decline in melanics in the peppered moth *Biston betularia* in Britain. *Heredity* 85:580–585.

Cook, L. M., P. S. Soltis, S. J. Brunsfeld, and D. E. Soltis. 1998. Multiple independent formations of *Tragopogon* tetraploids (Asteraceae): Evidence from RAPD markers. *Molecular Ecology* 7:1293–1302.

Cooper, V. S., and R. E. Lenski. 2000. The population genetics of ecological specialization in evolving *Escherichia coli* populations. *Nature* 407:736–739.

Cordeiro, N. J., and H. F. Howe. 2003. Forest fragmentation severs mutualism between seed dispersers and an endemic African tree. *Proceedings of the National Academy of Sciences USA* 100:14052–14056.

Cornell, H. V., and B. A. Hawkins. 2003. Herbivore responses to plant secondary compounds: A test of phytochemical coevolution theory. *American Naturalist* 161:507–522.

Coyne, J. A., N. H. Barton, and M. Turelli. 2000. Is Wright's shifting balance process important in evolution? *Evolution* 54:306–317.

Craig, T. P., J. D. Horner, and J. K. Itami. 1997. Hybridization studies on the host races of *Eurosta solidaginis:* Implications for sympatric speciation. *Evolution* 51:1552–1560.

Craig, T. P., J. K. Itami, C. Shantz, W. G. Abrahamson, J. D. Horner, and J. V. Craig. 2000. The influence of host plant variation and intraspecific competition on oviposition preference and offspring performance in the host races of *Eurosta solidaginis*. *Ecological Entomology* 25:7–18.

Crawley, M. J., S. L. Brown, R. S. Hails, D. D. Kohn, and M. Rees. 2001. Biotechnology: Transgenic crops in natural habitats. *Nature* 409:682–683.

Crespi, B. J. 2000. The evolution of maladaptation. *Heredity* 84:623–629.

Crow, J. F., and M. Kimura. 1965. Evolution in sexual and asexual populations. *American Naturalist* 99:439–450.

Crute, I. R., E. B. Holub, and J. J. Burdon, eds. 1997. *The gene-for-gene relationship in plant-parasite interactions.* CAB International, New York.

Cuautle, M., and V. Rico-Gray. 2003. The effects of wasps and ants on the reproductive success of the extrafloral nectaried plant *Turnera ulmifolia* (Turneraceae). *Functional Ecology* 17:417–423.

Currie, C. R. 2001a. A community of ants, fungi, and bacteria: A multilateral approach to studying symbiosis. *Annual Review of Microbiology* 55:357–380.

———. 2001b. Prevalence and impact of a virulent parasite on a tripartite mutualism. *Oecologia* 128:99–106.

Currie, C. R., A. N. M. Bot, and J. J. Boomsma. 2003. Experimental evidence of a tripartite mutualism: Bacteria protect ant fungus gardens from specialized parasites. *Oikos* 101:91–102.

Currie, C. R., U. G. Mueller, and D. Malloch. 1999. The agricultural pathology of ant fungus gardens. *Proceedings of the National Academy of Sciences USA* 96:7998–8002.

Currie, C. R., J. A. Scott, R. C. Summerbell, and D. Malloch. 1999. Fungus-growing ants use antibiotic-producing bacteria to control garden parasites. *Nature* 398:701–704.

Currie, C. R., B. Wong, A. E. Stuart, T. R. Schultz, S. A. Rehner, U. G. Mueller, G.-H. Sung, J. W. Spatafora, and N. A. Straus. 2003. Ancient tripartite coevolution in the attine ant-microbe symbiosis. *Science* 299:386–388.

Cushman, J. H., and A. J. Beattie. 1991. Mutualisms: Assessing the benefits to hosts and visitors. *Trends in Ecology and Evolution* 6:193–195.

Daily, G. C., ed. 1997. *Nature's services: Societal dependence on natural ecosystems.* Island Press, Washington, D.C.

Daily, G. C., P. R. Ehrlich, and G. A. Sánchez-Azofeifa. 2001. Countryside biogeography: Use of human-dominated habitats by the avifauna of southern Costa Rica. *Ecological Applications* 11:1–13.

Daily, G. C., and K. Ellison. 2002. *The new economy of nature: The quest to make conservation profitable.* Island Press, Washington, D.C.

Dale, C., G. R. Plague, B. Wang, H. Ochman, and N. A. Moran. 2002. Type III secretion systems and the evolution of mutualistic endosymbiosis. *Proceedings of the National Academy of Sciences USA* 99:12397–12402.

Daltry, J. C., G. Ponnudurai, C. K. Shin, N.-H. Tan, R. S. Thorpe, and W. Wüster. 1996. Electrophoretic profiles and biological activities: Intraspecific variation in the venom of the Malaysian pit viper (*Calloselasma rhodostoma*). *Toxicon* 34:67–79.

Daltry, J. C., W. Wüster, and R. S. Thorpe. 1996. Diet and snake venom evolution. *Nature* 379:537–540.

Davidson, S. K., and M. G. Haygood. 1999. Identification of sibling species of the bryozoan *Bugula neritina* that produce different anticancer bryostatins and harbor distinct strains of the bacterial symbiont "*Candidatus* Endobugula sertula." *Biological Bulletin* 196:273–280.

Davies, N. B., and M. De L. Brooke. 1989a. An experimental study of co-evolution between the cuckoo, *Cuculus canorus*, and its hosts. I. Host egg discrimination. *Journal of Animal Ecology* 58:207–224.

———. 1989b. An experimental study of co-evolution between the cuckoo, *Cuculus canorus*, and its hosts. II. Host egg markings, chick discrimination and general discussion. *Journal of Animal Ecology* 58:225–236.

Davies, S. J., S. K. Y. Lum, R. Chan, and L. K. Wang. 2001. Evolution of myrmecophytism in western Malesian *Macaranga* (Euphorbiaceae). *Evolution* 55:1542–1559.

Davis, D. R., O. Pellmyr, and J. N. Thompson. 1992. Biology and systematics of *Greya* Busck and *Tetragma*, new genus (Lepidoptera: Prodoxidae). *Smithsonian Contributions to Zoology* 524:1–88.

Davis, M. B., and R. G. Shaw. 2001. Range shifts and adaptive responses to quaternary climate change. *Science* 292:673–679.

Day, T., and J. G. Burns. 2003. A consideration of patterns of virulence arising from host-parasite coevolution. *Evolution* 57:671–676.

Dayan, T., and D. Simberloff. 1994. Character displacement, sexual dimorphism, and morphological variation among British and Irish mustelids. *Ecology* 75:1063–1073.

———. 1998. Size patterns among competitors: Ecological character displacement and character release in mammals, with special reference to island populations. *Mammal Review* 28:99–124.

De Boer, R. J., and A. S. Perelson. 1993. How diverse should the immune system be? *Proceedings of the Royal Society of London Biological Sciences Series B* 252:171–175.

de Jong, P. W., H. O. Frandsen, L. Rasmussen, and J. K. Nielsen. 2000. Genetics of resistance against defences of the host plant *Barbarea vulgaris* in a Danish flea beetle population. *Proceedings of the Royal Society of London Biological Sciences Series B* 267:1663–1670.

de Jong, P. W., and J. K. Nielsen. 1999. Polymorphism in a flea beetle for the ability to use an atypical host plant. *Proceedings of the Royal Society of London Biological Sciences Series B* 266:103–111.

———. 2002. Host plant use of *Phyllotreta nemorum:* Do coadapted gene complexes play a role? *Entomologia Experimentalis et Applicata* 104:207–215.

de Queiroz, K., L.-R. Chu, and J. B. Losos. 1998. A second Anolis lizard in Dominican amber and the systematics and ecological morphology of Dominican amber anoles. *American Museum Novitates* 3249:1–23.

Dedeine, F., F. Vavre, F. Fleury, B. Loppin, M. E. Hochberg, and M. Boulétreau. 2001. Removing symbiotic *Wolbachia* bacteria specifically inhibits oogenesis in a parasitic wasp. *Proceedings of the National Academy of Sciences USA* 98:6247–6252.

Denham, T. P., S. G. Haberle, C. Lentfer, R. Fullagar, J. Field, M. Therin, N. Porch, and B. Winsborough. 2003. Origins of agriculture at Kuk Swamp in highlands of New Guinea. *Science* 301:189–193.

Denison, R. F. 2000. Legume sanctions and the evolution of symbiotic cooperation by rhizobia. *American Naturalist* 156:567–576.

Diamond, J. M. 1975. Assembly of species communities. Pages 342–444 in J. M. Diamond, ed., *Ecology and evolution of communities.* Harvard University Press, Cambridge, Massachusetts.

———. 1997. *Guns, germs, and steel: The fates of human societies.* Norton, New York.

Dieckmann, U., and R. Law. 1996. The dynamical theory of coevolution: A derivation from stochastic ecological processes. *Journal of Mathematical Biology* 34:579–612.

Dieckmann, U., P. Marrow, and R. Law. 1995. Evolutionary cycling in predator-prey interactions: Population dynamics and the Red Queen. *Journal of Theoretical Biology* 176:91–102.

Dietl, G. P. 2003. Interaction strength between a predator and dangerous prey: *Sinistrofulgur* predation on *Mercenaria*. *Journal of Experimental Marine Biology and Ecology* 289:287–301.

Dietl, G. P., and P. H. Kelley. 2002. The fossil record of predator-prey arms races: Coevolution and escalation hypotheses. Pages 353–374 in M. Kowalewski and P. H. Kelley, eds., *The fossil record of predation*. Paleontological Society, New Haven, Connecticut.

Dingle, H. 1996. Migration: The biology of life on the move. Oxford University Press, Oxford.

Dirzo, R., and P. H. Raven. 2003. Global state of biodiversity and loss. *Annual Review of Environment and Resources* 28:137–167.

Ditchfield, A. D. 2000. The comparative phylogeography of neotropical mammals: Patterns of intraspecific mitochondrial DNA variation among bats contrasted to nonvolant small mammals. *Molecular Ecology* 9:1307–1318.

Dobson, S. L. 2003. Reversing *Wolbachia*-based population replacement. *Trends in Parasitology* 19:128–133.

Doebeli, M. 1997. Genetic variation and the persistence of predator-prey interactions in the Nicholson-Bailey model. *Journal of Theoretical Biology* 188:109–120.

Doebeli, M., and N. Knowlton. 1998. The evolution of interspecific mutualism. *Proceedings of the National Academy of Sciences USA* 95:8676–8680.

Doherty, P. F., Jr., G. Sorci, J. A. Royle, J. E. Hines, J. D. Nichols, and T. Boulinier. 2003. Sexual selection affects local extinction and turnover in bird communities. *Proceedings of the National Academy of Sciences USA* 100:5858–5862.

Doolittle, W. F., Y. Boucher, C. L. Nesbø, C. J. Douady, J. O. Andersson, and A. J. Roger. 2002. How big is the iceberg of which organellar genes in nuclear genomes are but the tip? *Philosophical Transactions of the Royal Society of London Biological Sciences Series B* 358:39–58.

Dopman, E. B., G. A. Sword, and D. M. Hillis. 2002. The importance of the ontogenetic niche in resource-associated divergence: Evidence from a generalist grasshopper. *Evolution* 56:731–740.

Douglas, A. E. 1998. Nutritional interactions in insect-microbial symbioses: Aphids and their symbiotic bacteria *Buchnera*. *Annual Review of Entomology* 43:17–37.

———. 1999. Mutualisms, ecology and ecologists. *Bulletin of the British Ecological Society* 30:14–15.

Douglas, A. E., and J. A. Raven. 2003. Genomes at the interface between bacteria and organelles. *Proceedings of the Royal Society of London Biological Sciences Series B* 358:5–18.

Drossel, B., P. G. Higgs, and A. J. McKane. 2001. The influence of predator-prey population dynamics on the long-term evolution of food web structure. *Journal of Theoretical Biology* 208:91–107.

Duda, R. F., Jr., A. J. Kohn, and S. R. Palumbi. 2001. Origins of diverse feeding ecologies within *Conus*, a genus of venomous marine gastropods. *Biological Journal of the Linnean Society* 73:391–409.

Duda, T. R., Jr., and S. R. Palumbi. 1999. Molecular genetics of ecological diversification: Duplication and rapid evolution of toxin genes of the venomous gastropod *Conus*. *Proceedings of the National Academy of Sciences USA* 96:6820–6823.

Duffy, J. E. 1996. Resource-associated population subdivision in a symbiotic coral-reef shrimp. *Evolution* 50:360–373.

Dupas, S., F. Frey, and Y. Carton. 1998. A single parasitoid segregating factor controls immune suppression in *Drosophila*. *Journal of Heredity* 89:306–311.

Dupas, S., Y. Carton, and M. Poirié. 2003. Genetic dimension of the coevolution of virulence-resistance in *Drosophila*–parasitoid wasp relationships. *Heredity* 90:84–89.

Dyall, S. D., M. T. Brown, and P. J. Johnson. 2004. Ancient invasions: From endosymbionts to organelles. *Science* 304:253–257.

Dybdahl, M. F., and C. M. Lively. 1995. Host-parasite interactions: Infection of common clones in natural populations of a freshwater snail (*Potamopyrgus antipodarum*). *Proceedings of the Royal Society of London Biological Sciences Series B* 260:99–103.

———. 1996. The geography of coevolution: Comparative population structures for a snail and its trematode parasite. *Evolution* 50:2264–2275.

———. 1998. Host-parasite coevolution: Evidence for rare advantage and time-lagged selection in a natural population. *Evolution* 52:1057–1066.

Earn, D. J. D., J. Dushoff, and S. A. Levin. 2002. Ecology and evolution of the flu. *Trends in Ecology and Evolution* 17:334–340.

Ebert, D. 1994. Virulence and local adaptation of a horizontally transmitted parasite. *Science* 265:1084–1086.

———. 1998. Experimental evolution of parasites. *Science* 282:1432–1435.

Ebert, D., and W. D. Hamilton. 1996. Sex against virulence: The coevolution of parasitic diseases. *Trends in Ecology and Evolution* 11:79–82.

Ebert, D., C. D. Zschokke-Rohringer, and H. J. Carius. 1998. Within- and between-population variation for resistance of *Daphnia magna* to the bacterial endoparasite *Pasteuria ramosa*. *Proceedings of the Royal Society of London Biological Sciences Series B* 265:2127–2134.

Ehrlich, P. R. 1997. *A world of wounds: Ecologists and the human dilemma*. Ecology Institute, Oldendorf/Luhe.

———. 2001. Intervening in evolution: Ethics and actions. *Proceedings of the National Academy of Sciences USA* 98:5477–5480.

Ehrlich, P. R., and P. H. Raven. 1964. Butterflies and plants: A study in coevolution. *Evolution* 18:586–608.

Eickbush, T. H., and A. V. Furano. 2002. Fruit flies and humans respond differently to retrotransposons. *Current Opinion in Genetics and Development* 12:669–674.

Elliott, P. F. 1974. Evolutionary responses of plants to seed-eaters: Pine squirrel predation on lodgepole pine. *Evolution* 28:221–231.

Ellner, S. P., N. G. Hairston, Jr., C. M. Kearns, and D. Kabaï. 1999. The roles of fluctuating selection and long-term diapause in microevolution of diapause timing in a freshwater copepod. *Evolution* 53:111–122.

Endler, J. A. 1980. Natural selection on color patterns in *Poecilia reticulata*. *Evolution* 34:76–91.

———. 1986. *Natural selection in the wild*. Princeton University Press, Princeton, New Jersey.

———. 1995. Multiple-trait coevolution and environmental gradients in guppies. *Trends in Ecology and Evolution* 10:22–29.

Eom, A.-H., D. C. Hartnett, and G. W. T. Wilson. 2000. Host plant species effects on arbuscular mycorrhizal fungal communities in tallgrass prairie. *Oecologia* 122:435–444.

Epperson, B. K. 2003. *Geographical genetics*. Princeton University Press, Princeton, New Jersey.

Ericson, L., J. J. Burdon, and W. J. Müller. 2002. The rust pathogen *Triphragmium ulmariae* as a selective force affecting its host, *Filipendula ulmaria*. *Journal of Ecology* 90:167–178.

Ernest, S. K., and J. H. Brown. 2001. Delayed compensation for missing keystone species by colonization. *Science* 292:101–104.

Escalante, A. A., E. Barrio, and F. J. Ayala. 1995. Evolutionary origin of the human and primate malarias: Evidence from the circumsporozoite protein gene. *Molecular Biology and Evolution* 12:616–626.

Estes, J. A., M. T. Tinker, T. M. Williams, and D. F. Doak. 1998. Killer whale predation on sea otters linking oceanic and nearshore ecosystems. *Science* 282:473–476.

Evans, A. G., and T. E. Wellems. 2002. Coevolutionary genetics of *Plasmodium* malaria parasites and their human hosts. *Integrative and Comparative Biology* 42:401–407.

Ewald, P. W. 1994. *Evolution of infectious disease*. Oxford University Press, Oxford.

———. 2000. *Plague time: How stealth infections cause cancers, heart disease, and other deadly ailments*. Free Press, New York.

Faegri, K., and L. van der Pijl. 1979. *The principles of pollination ecology*. Pergamon Press, Oxford.

Faeth, S. H. 2002. Are endophytic fungi defensive plant mutualists? *Oikos* 98:25–36.

Faeth, S. H., and W. F. Fagan. 2002. Fungal endophytes: Common host plant symbionts but uncommon mutualists. *Integrative and Comparative Biology* 42:360–368.

Faeth, S. H., and T. J. Sullivan. 2003. Mutualistic asexual endophytes in a native grass are usually parasitic. *American Naturalist* 161:310–325.

Falush, D., T. Wirth, B. Linz, J. K. Pritchard, M. Stephens, M. Kidd, M. J. Blaser et al. 2003. Traces of human migrations in *Helicobacter pylori* populations. *Science* 299:1582–1585.

Farrell, B. D. 1998. "Inordinate fondness" explained: Why are there so many beetles? *Science* 281:555–559.

———. 2001. Evolutionary assembly of the milkweed fauna: Cytochrome oxidase I and the age of *Tetraopes* beetles. *Molecular Phylogenetics and Evolution* 18:467–478.

Farrell, B. D., and C. Mitter. 1998. The timing of insect/plant diversification: Might *Tetraopes* (Coleoptera: Cerambycidae) and *Asclepias* (Asclepiadaceae) have co-evolved? *Biological Journal of the Linnean Society* 63:553–577.

Feder, J. L., J. B. Roethele, B. Wlazlo, and S. H. Berlocher. 1997. Selective maintenance of allozyme differences among sympatric host races of the apple maggot fly. *Proceedings of the National Academy of Sciences USA* 94:11417–11421.

Feldhaar, H., B. Fiala, J. Gadau, M. Mohamed, and U. Maschwitz. 2003. Molecular phylogeny of *Crematogaster* subgenus *Decacrema* ants (Hymenoptera: Formicidae) and the colonization of *Macaranga* (Euphorbiaceae) trees. *Molecular Phylogenetics and Evolution* 27:441–452.

Fellowes, M. D. E., and H. C. J. Godfray. 2000. The evolutionary ecology of resistance to parasitoids by *Drosophila*. *Heredity* 84:1–8.

Fellowes, M. D. E., A. R. Kraaijeveld, and H. C. J. Godfray. 1999. Cross-resistance following artificial selection for increased defense against parasitoids in *Drosophila melanogaster*. *Evolution* 53:966–972.

Felsenstein, J. 1974. The evolutionary advantage of recombination. *Genetics* 78:737–756.

Fenner, F., and P. J. Kerr. 1994. Evolution of the poxviruses, including the coevolution of virus and host in myxomatosis. Pages 273–292 in S. S. Morse, ed., *The evolutionary biology of viruses*. Raven Press, New York.

Ferguson, N. M., A. P. Galvani, and R. M. Bush. 2003. Ecological and immunological determinants of influenza evolution. *Nature* 422:428–433.

Ferraz, G., G. J. Russell, P. C. Stouffer, R. O. J. Bierregaard, Jr., S. L. Pimm, and T. E. Lovejoy. 2003. Rates of species loss from Amazonian forest fragments. *Proceedings of the National Academy of Sciences USA* 100:14069–14073.

Ferriere, R., J. L. Bronstein, S. Rinaldi, R. Law, and M. Gauduchon. 2002. Cheating and the evolutionary stability of mutualisms. *Proceedings of the Royal Society of London Biological Sciences Series B* 269:773–780.

Fiedler, K. 1996. Host-plant relationships of lycaenid butterflies: Large-scale patterns, interactions with plant chemistry, and mutualism with ants. *Entomologia Experimentalis et Applicata* 80:259–267.

Fiedler, K., and C. Saam. 1995. Ants benefit from attending facultatively myrmecophilous Lycaenidae caterpillars: Evidence from a survival study. *Oecologia* 104:316–322.

Finlay, B. J. 2002. Global dispersal of free-living microbial eukaryote species. *Science* 296:1061–1063.

Finlay, B. J., and T. Fenchel. 2002. Microbial eukaryote species. *Science* 297:337.

Firbank, L. G. 2003. Introduction: The farm scale evaluations of spring-sown genetically modified crops. *Philosophical Transactions of the Royal Society of London Biological Sciences Series B* 358:1777–1778.

Fleming, T. H., and J. N. Holland. 1998. The evolution of obligate pollination mutualisms: Senita cactus and senita moth. *Oecologia* 114:368–375.

Floate, K. D., and T. G. Whitham. 1993. The "Hybrid Bridge" hypothesis: Host shifting via plant hybrid swarms. *American Naturalist* 141:651–662.

Floate, K. D., G. D. Martinsen, and T. G. Whitham. 1997. Cottonwood hybrid zones as centres of abundance for gall aphids in western North America: Importance of relative habitat size. *Journal of Animal Ecology* 66:179–188.

Flor, H. H. 1942. Inheritance of pathogenicity in *Melampsora lini*. *Phytopathology* 32:653–669.

———. 1955. Host-parasite interaction in flax rust—its genetics and other implications. *Phytopathology* 45:680–685.

———. 1956. The complementary genic systems of flax and flax rust. *Advances in Genetics* 8:29–54.

Foitzik, S., B. Fischer, and J. Heinze. 2003. Arms races between social parasites and their hosts: Geographic patterns of manipulation and resistance. *Behavioral Ecology* 14:80–88.

Fox, B. J. 1987. Species assembly and the evolution of community structure. *Evolutionary Ecology* 1:201–213.

Fox, B. J., and J. H. Brown. 1993. Assembly rules for functional groups in North American desert rodent communities. *Oikos* 67:358–370.

Fox, C. W. 1993. A quantitative genetic analysis of oviposition preference and larval performance on two hosts in the bruchid beetle, *Callosobruchus maculatus*. *Evolution* 47:166–175.

Fox, C. W., K. J. Waddell, and T. A. Mousseau. 1994. Host-associated fitness variation in a seed beetle (Coleoptera: Bruchidae): Evidence for local adaptation to a poor quality host. *Oecologia* 99:329–336.

Frank, S. A. 1994a. Genetics of mutualisms: The evolution of altruism between species. *Journal of Theoretical Biology* 170:393–400.

———. 1994b. Recognition and polymorphism in host-parasite genetics. *Philosophical Transactions of the Royal Society of London Biological Sciences Series B* 346:283–293.

———. 1994c. Coevolutionary genetics of hosts and parasites with quantitative inheritance. *Evolutionary Ecology* 8:74–94.

———. 1995. The origin of synergistic symbiosis. *Journal of Theoretical Biology* 176:403–410.

———. 1996a. Host-symbiont conflict over the mixing of symbiotic lineages. *Proceedings of the Royal Society of London Biological Sciences Series B* 263:339–344.

———. 1996b. Models of parasite virulence. *Quarterly Review of Biology* 71:37–78.

———. 2000. Specific and non-specific defense against parasitic attack. *Journal of Theoretical Biology* 202:283–304.

———. 2002. *Immunology and evolution of infectious disease.* Princeton University Press, Princeton, New Jersey.

Frankino, W. A., and D. W. Pfennig. 2001. Condition-dependent expression of trophic polyphenism: Effects of individual size and competitive ability. *Evolutionary Ecology Research* 3:939–951.

Freeland, W. J., and W. J. Boulton. 1992. Coevolution of food-webs—parasites, predators and plant secondary compounds. *Biotropica* 24:309–327.

Fritz, R. S. 1999. Resistance of hybrid plants to herbivores: Genes, environment, or both? *Ecology* 80:382–391.

Fritz, R. S., C. G. Hochwender, S. J. Brunsfeld, and B. M. Roche. 2003. Genetic architecture of susceptibility to herbivores in hybrid willows. *Journal of Evolutionary Biology* 16:1115–1126.

Fritz, R. S., C. M. Nichols-Orians, and S. J. Brunsfeld. 1994. Interspecific hybridization of plants and resistance to herbivores: Hypotheses, genetics, and variable responses in a diverse herbivore community. *Oecologia* 97:106–117.

Fritz, R. S., B. M. Roche, S. J. Brunsfeld, and C. M. Orians. 1996. Interspecific and temporal variation in herbivore responses to hybrid willows. *Oecologia* 108:121–129.

Frost, S. D. W., M.-J. Dumaurier, S. Wain-Hobson, and A. J. L. Brown. 2001. Genetic drift and within-host metapopulation dynamics of HIV-1 infection. *Proceedings of the National Academy of Sciences USA* 98:6975–6980.

Fukatsu, T., and T. Hosokawa. 2002. Capsule-transmitted gut symbiotic bacterium of the Japanese common plataspid stinkbug, *Megacopta punctatissima. Applied and Environmental Microbiology* 68:389–396.

Funk, D. J., L. Helbling, J. J. Wernegreen, and N. A. Moran. 2000. Intraspecific phylogenetic congruence among multiple symbiont genomes. *Proceedings of the Royal Society of London Biological Sciences Series B* 267:2517–2521.

Funk, D. J., J. J. Wernegreen, and N. A. Moran. 2001. Intraspecific variation in symbiont genomes: Bottlenecks and the aphid-Buchnera association. *Genetics* 157:477–489.

Futuyma, D. J. 1995. The uses of evolutionary biology. *Science* 267:41–42.

———. 2000. Some current approaches to the evolution of plant-herbivore interactions. *Plant Species Biology* 15:1–9.

Futuyma, D. J., M. C. Keese, and D. J. Funk. 1995. Genetic constraints on macroevolution: The evolution of host affiliation in the leaf beetle genus *Ophraella. Evolution* 49:797–809.

Futuyma, D. J., M. C. Keese, and S. J. Scheffer. 1993. Genetic constraints and the phylogeny of insect-plant associations: Responses of *Ophraella communa* (Coleoptera: Chrysomelidae) to host plants of its congeners. *Evolution* 47:888–905.

Futuyma, D. J., and C. Mitter. 1996. Insect-plant interactions: The evolution of component communities. *Philosophical Transactions of the Royal Society of London Biological Sciences Series B* 351:1361–1366.

Galen, C. 1996. Rates of floral evolution: Adaptation to bumblebee pollination in an alpine wildflower, *Polemonium viscosum. Evolution* 50:120–125.

Galvani, A. P., R. M. Coleman, and N. M. Ferguson. 2003. The maintenance of sex in parasites. *Proceedings of the Royal Society of London Biological Sciences Series B* 270:19–28.

Gandon, S. 2002a. Local adaptation and the geometry of host-parasite coevolution. *Ecology Letters* 5:246–256.

———. 2002b. Coevolution between parasite virulence and host life-history traits. *American Naturalist* 160:374–388.

Gandon, S., Y. Capowiez, Y. Dubios, Y. Michalakis, and I. Olivieri. 1996. Local adaptation and gene-for-gene coevolution in a metapopulation model. *Proceedings of the Royal Society of London B* 263:1003–1009.

Gandon, S., D. Ebert, I. Olivieri, and Y. Michalakis. 1998. Differential adaptation in spatially heterogeneous environments and host-parasite coevolution. Pages 325–342 in S. Mopper and S. Y. Strauss, eds., *Genetic structure and local adaptation in natural insect populations: Effects of ecology, life history, and behavior.* Chapman and Hall, New York.

Gandon, S., and Y. Michalakis. 2002. Local adaptation, evolutionary potential and host-parasite coevolution: Interactions between migration, mutation, population size and generation time. *Journal of Evolutionary Biology* 15:451–462.

Ganusov, V. V., and R. Antia. 2003. Trade-offs and the evolution of virulence of microparasites: Do details matter? *Theoretical Population Biology* 64:211–220.

García-Ramos, G., and D. Rodríguez. 2002. Evolutionary speed of species invasions. *Evolution* 56:661–668.

Garrett, L. 1994. *The coming plague: Newly emerging diseases in a world out of balance.* Farrar, Straus and Giroux, New York.

Garrido, J. L., P. J. Rey, X. Cerdá, and C. M. Herrera. 2002. Geographical variation in diaspore traits of an ant-dispersed plant (*Hellborus foetidus*): Are ant community composition and diaspore traits correlated? *Journal of Ecology* 90:446–455.

Gavrilets, S. 1997. Coevolutionary chase in exploiter-victim systems with polygenic characters. *Journal of Theoretical Biology* 186:527–534.

———. 2003. Models of speciation: What have we learned in 40 years? *Evolution* 57:2197–2215.

Gavrilets, S., and A. Hastings. 1998. Coevolutionary chase in two-species systems with applications to mimicry. *Journal of Theoretical Biology* 191:415–427.

Geffeney, S., E. D. J. Brodie, P. C. Ruben, and E. D. I. Brodie. 2002. Mechanisms of adaptation in a predator-prey arms race: TTX-resistant sodium channels. *Science* 297:1336–1339.

Gehring, C. A., N. S. Cobb, and T. G. Whitham. 1997. Three-way interactions among ectomycorrhizal mutualists, scale insects, and resistant and susceptible pinyon pines. *American Naturalist* 149:824–841.

Gehring, C. A., and T. G. Whitham. 1991. Herbivore-driven mycorrhizal mutualism in insect-susceptible pinyon pine. *Nature* 353:556–557.

———. 1994. Interactions between aboveground herbivores and the mycorrhizal mutualists of plants. *Trends in Ecology and Evolution* 9:251–255.

———. 1995. Duration of herbivore removal and environmental stress affect the ectomycorrhizae of pinyon pines. *Ecology* 76:2118–2123.

Gelb, M. H., and W. G. J. Hol. 2002. Drugs to combat tropical protozoan parasites. *Science* 297:343–344.

Gemmill, A. W., and A. F. Read. 1998. Counting the cost of disease resistance. *Trends in Ecology and Evolution* 13:8–9.

Gemmill, A. W., M. E. Viney, and A. F. Read. 1997. Host immune status determines sexuality in a parasite nematode. *Evolution* 51:393–401.

Gibbs, H. L., M. D. Sorenson, K. Marchetti, M. De. L. Brooke, N. B. Davies, and H. Nakamura. 2000. Genetic evidence for female host–specific races of the common cuckoo. *Nature* 407:183–186.

Gil, R., B. Sabater-Muñoz, A. Latorre, F. J. Silva, and A. Moya. 2002. Extreme genome reduction in *Buchnera* spp.: Toward the minimal genome needed for symbiotic life. *Proceedings of the National Academy of Sciences USA* 99:4454–4458.

Gil, R., F. J. Silva, E. Zientz, F. Delmotte, F. González-Candelas, A. Latorre, C. Rausell et al. 2003. The genome sequence of *Blochmannia floridanus:* Comparative analysis of reduced genomes. *Proceedings of the National Academy of Sciences USA* 100:9388–9393.

Gilbert, L. E. 1980. Food web organization and the conservation of neotropical diversity.

Pages 11–33 in M. E. Soulé and B. A. Wilcox, eds., *Conservation biology: An evolutionary-ecological perspective.* Sinauer Associates, Sunderland, Massachusetts.

———. 1983. Coevolution and mimicry. Pages 263–281 in D. J. Futuyma and M. Slatkin, eds., *Coevolution.* Sinauer Associates, Sunderland, Massachusetts.

———. 2003. Adaptive novelty through introgression in *Heliconius* wing patterns: Evidence for a shared genetic "toolbox" from synthetic hybrid zones and a theory of diversification Pages 281–318 in C. L. Boggs, W. B. Watt, and P. R. Ehrlich, eds., *Butterflies: Ecology and evolution taking flight.* University of Chicago Press, Chicago.

Gill, D. E. 1986. Individual plants as genetic mosaics: Ecological organisms versus evolutionary individuals. Pages 321–343 in M. J. Crawley, ed., *Plant ecology.* Blackwell, Oxford.

Gill, D. E., L. Chao, S. L. Perkins, and J. B. Wolf. 1995. Genetic mosaicism in plants and clonal animals. *Annual Review of Ecology and Systematics* 26:423–444.

Gillespie, J. P., M. R. Kanost, and T. Trenczek. 1997. Biological mediators of insect immunity. *Annual Review of Entomology* 42:611–643.

Gillespie, R. 2004. Community assembly through adaptive radiation in Hawaiian spiders. *Science* 303:356–359.

Giovannetti, M., D. Azzolini, and A. S. Citernesi. 1999. Anastomosis formation and nuclear and protoplasmic exchange in arbuscular mycorrhizal fungi. *Applied and Environmental Microbiology* 65:5571–5575.

Giovannetti, M., P. Fortuna, A. S. Citernesi, S. Morini, and M. P. Nuti. 2001. The occurrence of anastomosis formation and nuclear exchange in intact arbuscular mycorrhizal networks. *New Phytologist* 151:717–724.

Glesener, R. R., and D. Tilman. 1978. Sexuality and the components of environmental uncertainty: Clues from geographical parthenogenesis in terrestrial animals. *American Naturalist* 112:659–673.

Glor, R. E., J. J. Kolbe, R. Powell, A. Larson, and J. B. Losos. 2001. Phylogenetic analysis of ecological and morphological diversification in Hispaniolan trunk-ground anoles (*Anolis cybotes* group). *Evolution* 57:2383–2397.

Goebel, W., and R. Gross. 2001. Intracellular survival strategies of mutualistic and parasitic prokaryotes. *Trends in Microbiology* 9:267–273.

Goetze, E. 2003. Cryptic speciation on the high seas; global phylogenetics of the copepod family Eucalanidae. *Proceedings of the Royal Society of London Biological Sciences Series B* 270:2321–2331.

Goldin, A. L. 2001. Resurgence of sodium channel research. *Annual Review of Physiology* 63:871–894.

Gomulkiewicz, R., S. L. Nuismer, and J. N. Thompson. 2003. Coevolution in variable mutualisms. *American Naturalist* 162:S80–S93.

Gomulkiewicz, R., J. N. Thompson, R. D. Holt, S. L. Nuismer, and M. E. Hochberg. 2000. Hot spots, cold spots, and the geographic mosaic theory of coevolution. *American Naturalist* 156:156–174.

Good, J. M., J. R. Demboski, D. W. Nagorsen, and J. Sullivan. 2003. Phylogeography and introgressive hybridization: Chipmunks (genus *Tamias*) in the northern Rocky Mountains. *Evolution* 57:1900–1916.

Goodnight, C. J., and D. M. Craig. 1996. The effect of coexistence on competitive outcome in *Tribolium castaneum* and *Tribolium confusum*. *Evolution* 50:1241–1250.

Gotelli, N. J., and G. R. Graves. 1996. *Null models in ecology.* Smithsonian Institution Press, Washington, D.C.

Gotelli, N. J., and D. J. McCabe. 2002. Species co-occurrence: A meta-analysis of J. M. Diamond's assembly rules model. *Ecology* 83:2091–2096.

Gould, F. 1998. Sustainability of transgenic insecticidal cultivars: Integrating pest genetics and ecology. *Annual Review of Entomology* 43:701–726.

Gould, S. J. 1977. *Ontogeny and phylogeny.* Harvard University Press, Cambridge, Massachusetts.

Gould, S. J., and R. C. Lewontin. 1979. The spandrels of San Marco and the Panglossian paradigm: A critique of the adaptationist programme. *Proceedings of the Royal Society of London Biological Sciences Series B* 205:581–598.

Goulson, D. 2003. Effects of introduced bees on native ecosystems. *Annual Review of Ecology and Systematics* 34:1–26.

Grant, B. S., and L. L. Wiseman. 2002. Recent history of melanism in American peppered moths. *Journal of Heredity* 93:86–90.

Grant, P. R., and B. R. Grant. 2002. Unpredictable evolution in a 30-year study of Darwin's finches. *Science* 296:707–711.

Gray, S. M., and B. W. Robinson. 2002. Experimental evidence that competition between stickleback species favours adaptive character divergence. *Ecology Letters* 5:264–272.

Green, A. M., U. G. Mueller, and R. M. M. Adams. 2002. Extensive exchange of fungal cultivars between sympatric species of fungus-growing ants. *Molecular Ecology* 11:191–195.

Greene, E. 1996. Effect of light quality and larval diet on morph induction in the polymorphic caterpillar *Nemoria arizonaria* (Lepidoptera: Geometridae). *Biological Journal of the Linnean Society* 58:277–285.

Groth, J. G. 1993. Evolutionary differentiation in morphology, vocalizations, and allozymes among nomadic sibling species in the North American red crossbill (*Loxia curvirostra*) complex. *University of California Publications in Zoology,* number 27, Berkeley, California.

Guillet, J.-G., M.-Z. Lai, T. J. Briner, J. A. Smith, and M. L. Gefter. 1986. Interaction of peptide antigens and Class II major histocompatibility complex antigens. *Nature* 324:260–262.

Haber, W. A., and G. W. Frankie. 1989. A tropical hawkmoth community: Costa Rican dry forest Sphingidae. *Biotropica* 21:155–172.

Hacker, S. D., and M. D. Bertness. 1996. Trophic consequences of a positive plant interaction. *American Naturalist* 148:559–575.

Hacker, S. D., and S. D. Gaines. 1997. Some implications of direct positive interactions for community species diversity. *Ecology* 78:1990–2003.

Haldane, J. B. S. 1949. Disease and evolution. *La Ricerca Scientifica* (suppl.) 19:68–76.

Hamilton, W. D. 1980. Sex versus non-sex versus parasite. *Oikos* 35:282–290.

Hamilton, W. D., R. Axelrod, and R. Tanese. 1990. Sexual reproduction as an adaptation to resist parasites (a review). *Proceedings of the National Academy of Sciences USA* 87:3566–3573.

Hamilton, W. D., and R. M. May. 1977. Dispersal in stable habitats. *Nature* 269:578–581.

Hamilton, W. D., and M. Zuk. 1982. Heritable true fitness and bright birds: A role for parasites? *Science* 218:384–387.

Hamrick, J. L., and M. J. W. Godt. 1990. Allozyme diversity in plant species. Pages 43–63 in A. H. D. Brown, M. T. Clegg, A. L. Kahler, and B. S. Weir, eds., *Plant population genetics, breeding, and genetic resources.* Sinauer Associates, Sunderland, Massachusetts.

Hanifin, C. T., M. Yotsu-Yamashita, T. Yasumoto, E. D. Brodie, III, and E. D. Brodie, Jr. 1999. Toxicity of dangerous prey: Variation of tetrodotoxin levels within and among populations of the newt *Taricha granulosa. Journal of Chemical Ecology* 25:2161–2175.

Hansen, T. F., W. S. Armbruster, and L. Antonsen. 2000. Comparative analysis of character displacement and spatial adaptations as illustrated by the evolution of *Dalechampia* blossoms. *American Naturalist* 156:S17–S34.

Hanski, I. 1999. *Metapopulation ecology.* Oxford University Press, Oxford.

———. 2003. Biology of extinctions in butterfly metapopulations. Pages 577–602 in C. L. Boggs, W. B. Watt, and P. R. Ehrlich, eds., *Butterflies: Ecology and evolution taking flight.* University of Chicago Press, Chicago.

Hanski, I., and M. E. Gilpin, eds. 1997. *Metapopulation biology: Ecology, genetics, and evolution.* Academic Press, San Diego, CA.

Hare, J. D., E. Elle, and N. M. van Dam. 2003. Costs of glandular trichomes in *Datura wrightii:* A three-year study. *Evolution* 57:793–805.

Harley, C. D. G. 2003. Species importance and context: Spatial and temporal variation in species interactions. Pages 44–68 in P. Kareiva and S. A. Levin, eds., *The importance of species: Perspectives on expendability and triage.* Princeton University Press, Princeton, New Jersey.

Harris, D. R., ed. 1996. *The origins and spread of agriculture and pastoralism in Eurasia.* Smithsonian Institution Press, Washington, D.C.

Harrison, R. G, ed. 1993. *Hybrid zones and the evolutionary process.* Oxford University Press, New York.

Harrison, S. 1999. Native and alien species diversity at the local and regional scales in a grazed California grassland. Oecologia 121:99–106.

Hartl, D. L. 2004. The origin of malaria: Mixed messages from genetic diversity. *Nature Reviews Microbiology* 2:15–22.

Harvell, C. D., K. Kim, J. M. Burkholder, R. R. Colwell, P. R. Epstein, D. J. Grimes, E. E. Hofmann, E. K. Lipp, A. D. M. E. Osterhaus, R. M. Overstreet, J. W. Porter, G. W. Smith, and G. R. Vasta. 1999. Emerging marine diseases—Climate links and anthropogenic factors. *Science* 285:1505–1510.

Harvey, P. H. 1996. Phylogenies for ecologists. *Journal of Animal Ecology* 65:255–263.

Hastings, A. 2003. Metapopulation persistence with age-dependent disturbance or succession. *Science* 301:1525–1526.

Hastings, A., and S. Harrison. 1994. Metapopulation dynamics and genetics. *Annual Review of Ecology and Systematics* 25:167–188.

Heath, B. D., R. D. J. Butcher, W. G. F. Whitfield, and S. F. Hubbard. 1999. Horizontal transfer of *Wolbachia* between phylogenetically distant insect species by a naturally occurring mechanism. *Current Biology* 9:313–316.

Hedderson, T. A., and R. E. Longton. 1996. Life history variation in mosses: Water relations, size and phylogeny. *Oikos* 77:31–43.

Hedrick, A. V., and S. F. Riechert. 1989. Genetically-based variation between two spider populations in foraging behavior. *Oecologia* 80:533–539.

Hedrick, P. W. 2002. Pathogen resistance and genetic variation at MHC loci. *Evolution* 56:1902–1908.

Hedrick, P. W., and E. King. 1996. Genetics and the environment in interspecific competition: A study using the sibling species *Drosophila melanogaster* and *Drosophila simulans. Oecologia* 108:72–78.

Heil, M., and D. McKey. 2003. Protective ant-plant interactions as model systems in ecological and evolutionary research. *Annual Review of Ecology and Systematics* 34:425–453.

Hendrix, S. D. 1979. Compensatory reproduction in a biennial herb *Pastinaca sativa* following insect defloration. *Oecologia* 42:107–118.

Hendry, A. P., and M. T. Kinnison. 1999. The pace of modern life: Measuring rates of contemporary microevolution. *Evolution* 53:1637–1653.

Henter, H. J. 1995. The potential for coevolution in a host-parasitoid system. II. Genetic variation within a population of wasps in the ability to parasitize an aphid host. *Evolution* 49:439–445.

Henter, H. J., and S. Via. 1995. The potential for coevolution in a host-parasitoid system. I. Genetic variation within an aphid population in susceptibility to a parasitic wasp. *Evolution* 49:427–438.

Herre, E. A. 1996. An overview of studies on a community of Panamanian figs. *Journal of Biogeography* 23:593–607.

Herre, E. A., N. Knowlton, U. G. Mueller, and S. A. Rehner. 1999. The evolution of mutualisms: Exploring the paths between conflict and cooperation. *Trends in Ecology and Evolution* 14:49–53.

Herre, E. A., C. A. Machado, E. Bermingham, J. D. Nason, D. M. Windsor, S. S. McCafferty, W. Van Houten, and K. Bachman. 1996. Molecular phylogenies of figs and their pollinator wasps. *Journal of Biogeography* 23:521–530.

Herrera, C. M. 1995. Plant-vertebrate seed dispersal systems in the Mediterranean: Ecological, evolutionary, and historical determinants. *Annual Review of Ecology and Systematics* 26:705–727.

———. 2002. Seed dispersal by vertebrates. Pages 185–208 in C. M. Herrera and O. Pellmyr, eds., *Plant-animal interactions: An evolutionary approach.* Blackwell, Oxford.

Herrera, C., M. Medrano, P. J. Rey, A. M. Sánchez-Lafuente, M. B. García, J. Guitián, and A. J. Manzaneda. 2002. Interaction of pollinators and herbivores on plant fitness suggests a pathway for correlated evolution of mutualism- and antagonism-related traits. *Proceedings of the National Academy of Sciences USA* 99:16823–16828.

Hewitt, G. M. 2001. Speciation, hybrid zones and phylogeography—or seeing genes in space and time. *Molecular Ecology* 10:537–549.

Hibbett, D. S., L.-B. Gilbert, and M. J. Donoghue. 2000. Evolutionary instability of ectomycorrhizal symbioses in basidiomycetes. *Nature* 407:506–508.

Hibbett, D. S., E. M. Pine, E. Langer, G. Langer, and M. J. Donoghue. 1997. Evolution of gilled mushrooms and puffballs inferred from ribosomal DNA sequences. *Proceedings of the National Academy of Sciences USA* 94:12002–12006.

Hickman, J. C., ed. 1993. *The Jepson manual: Higher plants of California.* University of California Press, Berkeley.

Hill, A. V. S. 1998. The immunogenetics of human infectious diseases. *Annual Review of Immunology* 16:593–617.

Hill, J. K., Y. C. Collingham, C. D. Thomas, D. S. Blakeley, R. Fox, D. Moss, and B. Huntley. 2001. Impacts of landscape structure on butterfly range expansion. *Ecology Letters* 4:313–321.

Hill, J. K., C. D. Thomas, and B. Huntley. 2003. Modelling present and potential future ranges of European butterflies using climate response surfaces. Pages 149–167 in C. L. Boggs, W. B. Watt, and P. R. Ehrlich, eds., *Butterflies: Ecology and evolution taking flight.* University of Chicago Press, Chicago.

Hoarau, G., A. D. Rijnsdorp, W. Van Der Veer, W. T. Stam, and J. L. Olsen. 2002. Population structure of plaice (*Pleuronectes platessa* L.) in northern Europe: Microsatellites revealed large-scale spatial and temporal homogeneity. *Molecular Ecology* 11:1165–1176.

Hochberg, M. E., and M. van Baalen. 1998. Antagonistic coevolution over productivity gradients. *American Naturalist* 152:620–634.

Hochberg, M. E., R. Gomulkiewicz, R. D. Holt, and J. N. Thompson. 2000. Weak sinks could cradle mutualistic symbioses—strong sources should harbour parasitic symbioses. *Journal of Evolutionary Biology* 13:213–222.

Hodge, A., C. D. Campbell, and A. H. Fitter. 2001. An arbuscular mycorrhizal fungus accelerates decomposition and acquires nitrogen directly from organic material. *Nature* 413:297–299.

Hoeksema, J. D. 1999. Investigating the disparity in host specificity between AM and EM fungi: Lessons from theory and better-studied systems. *Oikos* 84:327–332.

Hoeksema, J. D., and E. M. Bruna. 2000. Pursuing the big questions about interspecific mutualism: A review of theoretical approaches. *Oecologia* 125:321–330.

Hoeksema, J. D., and M. Kummel. 2003. Ecological persistence of the plant-mycorrhizal mutualism: A hypothesis from species coexistence theory. *American Naturalist* 162:S40–S50.

Hoekstra, H. E., J. M. Hoekstra, D. Berrigan, S. N. Vignieri, A. Hoang, C. E. Hill, P. Beerli, and J. G. Kingsolver. 2001. Strength and tempo of directional selection in the wild. *Proceedings of the National Academy of Sciences USA* 98:9157–9160.

Hoffman, A. A., M. Turelli, and G. M. Simmons. 1986. Unidirectional incompatibility between populations of *Drosophila simulans*. *Evolution* 40:692–701.

Hoffman, S. L., G. M. Subramanian, F. H. Collins, and J. C. Venter. 2002. *Plasmodium*, human and *Anopheles* genomics and malaria. *Nature* 415:702–709.

Holland, J. N., and T. H. Fleming. 2002. Co-pollinators and specialization in the pollinating seed-consumer mutualism between senita cacti and senita moths. *Oecologia* 133:534–540.

Hölldobler, B., and E. O. Wilson. 1990. *The ants.* Harvard University Press, Cambridge, Massachusetts.

Holt, R. D., A. P. Dobson, M. Begon, R. G. Bowers, and E. M. Schauber. 2003. Parasite establishment in host communities. *Ecology Letters* 6:837–842.

Holt, R. D., and J. H. Lawton. 1994. The ecological consequences of shared natural enemies. *Annual Review of Ecology and Systematics* 25:495–520.

Hood, M. E. 2003. Dynamics of multiple infection and within-host competition by the anther-smut pathogen. *American Naturalist* 162:122–133.

Horner-Devine, M. C., K. M. Carney, and B. J. M. Bohannan. 2004. An ecological perspective on bacterial biodiversity. *Proceedings of the Royal Society of London Biological Sciences Series B* 271:113–122.

Horner-Devine, M. C., M. A. Leibold, V. H. Smith, and B. J. M. Bohannan. 2003. Bacterial diversity patterns along a gradient of primary productivity. *Ecology Letters* 6:613–622.

Horton, T. R., and T. D. Bruns. 1998. Multiple-host fungi are the most frequent and abundant ectomycorrhizal types in a mixed stand of Douglas fir *(Pseudotsuga menziesii* D. Don) and bishop pine *(Pinus muricata* D. Don). *New Phytologist* 139:331–339.

Hougen-Eitzman, D., and M. D. Rausher. 1994. Interactions between herbivorous insects and plant-insect coevolution. *American Naturalist* 143:677–697.

Howard, R. S., and C. M. Lively. 1994. Parasitism, mutation accumulation and the maintenance of sex. *Nature* 367:554–557.

———. 1998. The maintenance of sex by parasitism and mutation accumulation under epistatic fitness functions. *Evolution* 52:604–610.

———. 2002. The ratchet and the Red Queen: The maintenance of sex in parasites. *Journal of Evolutionary Biology* 15:648–656.

———. 2003. Opposites attract? Mate choice for parasite evasion and the evolutionary stability of sex. *Journal of Evolutionary Biology* 16:681–689.

Hubbell, S. P. 2001. *The unified neutral theory of biodiversity and biogeography.* Princeton University Press, Princeton, New Jersey.

Huber, K., L. Le Loan, T. H. Hoang, S. Ravel, F. Rodhain, and A.-B. Failloux. 2002. Genetic differentiation of the dengue vector, *Aedes aegypti* (Ho Chi Minh City, Vietnam) using microsatellite markers. *Molecular Ecology* 11:1629–1635.

Huelsenbeck, J. P., B. Rannala, and B. Larget. 2003. A statistical perspective for reconstructing the history of host-parasite associations. Pages 93–119 in R. D. M. Page, ed., *Tangled trees: Phylogeny, cospeciation, and coevolution.* University of Chicago, Chicago.

Hufbauer, R. A. 2001. Pea aphid–parasitoid interactions: Have parasitoids adapted to differential resistance? *Ecology* 82:717–725.

Hufbauer, R. A., and S. Via. 1999. Evolution of an aphid-parasitoid interaction: Variation in resistance to parasitism among aphid populations specialized on different plants. *Evolution* 53:1435–1445.

Hughes, A. L., and M. Nei. 1992. Maintenance of MHC polymorphism. *Nature* 355:402–403.

Hughes, A. L., and F. Verra. 2001. Very large long-term effective population size in the virulent human malaria parasite *Plasmodium falciparum. Proceedings of the Royal Society of London Biological Sciences Series B* 268:1855–1860.

Hughes, J. B., G. C. Daily, and P. R. Ehrlich. 1997. Population diversity: Its extent and extinction. *Science* 278:689–692.

Husband, B. C. 2000. Constraints on polyploid evolution: A test of the minority cytotype exclusion principle. *Proceedings of the Royal Society of London Biological Sciences Series B* 267:217–223.

Husband, B. C., and S. C. H. Barrett. 1996. A metapopulation perspective in plant population biology. *Journal of Ecology* 84:461–469.

Husband, R., E. A. Herre, S. L. Turner, R. Gallery, and J. P. W. Young. 2002. Molecular diversity of arbuscular mycorrhizal fungi and patterns of host association over time and space in a tropical forest. *Molecular Ecology* 11:2669–2678.

Hutchinson, G. E. 1959. Homage to Santa Rosalia, or Why are there so many kinds of animals? *American Naturalist* 93:145–159.

Iwao, K., and M. D. Rausher. 1997. Evolution of plant resistance to multiple herbivores: Quantifying diffuse coevolution. *American Naturalist* 149:316–335.

Iwasa, Y., F. Michor, and M. A. Nowak. 2003. Evolutionary dynamics of escape from biomedical intervention. *Proceedings of the Royal Society of London Biological Sciences Series B* 270:2573–2578.

Jablonski, D. 1998. Geographic variation in the molluscan recovery from the end-Cretaceous extinction. *Science* 279:1327–1330.

Jablonski, D., and J. J. Sepkoski. 1996. Paleobiology, community ecology, and scales of ecological pattern. *Ecology* 77:1367–1378.

Jackman, T. R., A. Larson, K. de Queiroz, and J. B. Losos. 1999. Phylogenetic relationships and tempo of early diversification in *Anolis* lizards. *Systematic Biology* 48:254–285.

Jackson, J. B. C., M. X. Kirby, W. H. Berger, K. A. Bjorndal, L. W. Botsford, B. J. Bourque, R. H. Bradbury et al. 2001. Historical overfishing and the recent collapse of coastal ecosystems. *Science* 293:629–638.

Jaenike, J. 1978. An hypothesis to account for the maintenance of sex within populations. *Evolutionary Theory* 3:191–194.

Jain, R., M. C. Rivera, J. E. Moore, and J. A. Lake. 2003. Non-clonal evolution of microbes. *Biological Journal of the Linnean Society* 79:27–32.

James, A. C., and J. W. O. Ballard. 2000. Expression of cytoplasmic incompatibility in *Drosophila simulans* and its impact on infection frequencies and distribution of *Wolbachia pipientis. Evolution* 54:1661–1672.

James, A. C., M. D. Dean, M. E. McMahon, and J. W. O. Ballard. 2002. Dynamics of double and single *Wolbachia* infections in *Drosophila simulans* from New Caledonia. *Heredity* 88: 182–189.

Jamnongluk, W., P. Kittayapong, V. Baimai, and S. L. O'Neill. 2002. *Wolbachia* infections of tephritid fruit flies: Molecular evidence for five distinct strains in a single host species. *Current Microbiology* 45:255–260.

Janz, N. 1998. Sex-linked inheritance of host-plant specialization in a polyphagous butterfly. *Proceedings of the Royal Society of London Biological Sciences Series B* 265:1675–1678.

———. 2003. Sex linkage of host plant use in butterflies. Pages 229–239 in C. L. Boggs, W. B. Watt, and P. R. Ehrlich, eds., *Butterflies: Ecology and evolution taking flight.* University of Chicago Press, Chicago.

Janz, N., and S. Nylin. 1998. Butterflies and plants: A phylogenetic study. *Evolution* 52: 486–502.

Janz, N., and J. N. Thompson. 2002. Plant polyploidy and host expansion in an insect herbivore. *Oecologia* 130:570–575.

Janzen, D. H. 1980. When is it coevolution? *Evolution* 34:611–612.

———. 1984. The most coevolutionary animal of them all. *Crafoord Lectures of the Royal Swedish Academy of Sciences* 3:2–20.

———. 1999. Gardenification of tropical conserved wildlands: Multitasking, multicropping, and multiusers. *Proceedings of the National Academy of Sciences USA* 96:5987–5994.

Janzen, D. H., and P. S. Martin. 1982. Neotropical anachronisms: The fruits the gomphotheres ate. *Science* 215:19–27.

Janzen, F. J., J. G. Krenz, T. S. Haselkorn, E. D. Brodie, Jr., and E. D. Brodie, III. 2002. Molecular phylogeography of common garter snakes (*Thamnophis sirtalis*) in western North America: Implications for regional historical forces. *Molecular Ecology* 11:1739–1751.

Jarosz, A. M., and J. J. Burdon. 1991. Host-pathogen interactions in natural populations of *Linum marginale* and *Melampsora lini.* II. Local and regional variation in pattern of resistance and racial structure. *Evolution* 45:1618–1627.

———. 1992. Host-pathogen interactions in natural populations of *Linum marginale* and *Melampsora lini.* III. Influence of pathogen epidemics on host survivorship and flower production. *Oecologia* 89:53–61.

Jeffries, R. L. 1999. Herbivores, nutrients and trophic cascades in terrestrial environments Pages 301–330 in H. Olff, V. K. Brown, and R. H. Drent, eds., *Herbivores: Between plants and predators.* Blackwell Scientific, Oxford.

Jeyaprakash, A., and M. A. Hoy. 2000. Long PCR improves *Wolbachia* DNA amplification: *wsp* sequences found in 76% of sixty-three arthropods. *Insect Molecular Biology* 9:393–405.

Jiggins, C. D., W. O. McMillan, P. King, and J. Mallet. 1997. The maintenance of species differences across a *Heliconius* hybrid zone. *Heredity* 79:495–505.

Jiggins, F. M., J. K. Bentley, M. E. N. Majerus, and G. D. D. Hurst. 2002. Recent changes in phenotype and patterns of host specialization in *Wolbachia* bacteria. *Molecular Ecology* 11: 1275–1283.

Jiggins, F. M., G. D. D. Hurst, and M. E. N. Majerus. 2000. Sex ratio distorting *Wolbachia* causes sex-role reversal in its butterfly host. *Proceedings of the Royal Society of London Biological Sciences Series B* 267:69–73.

Jiggins, F. M., G. D. D. Hurst, and Z. Yang. 2002. Host-symbiont conflicts: Positive selection on an outer membrane protein of parasitic but not mutualistic Rickettsiaceae. *Molecular Biology and Evolution* 19:1341–1349.

Jokela, J., and C. M. Lively. 1995. Parasites, sex, and early reproduction in a mixed population of freshwater snails. *Evolution* 49:1268–1271.

Jokela, J., C. M. Lively, M. F. Dybdahl, and J. A. Fox. 2003. Genetic variation in sexual and clonal lineages of a freshwater snail. *Biological Journal of the Linnean Society* 79:165–181.

Jokela, J., C. M. Lively, J. A. Fox, and M. F. Dybdahl. 1997. Flat reaction norms and "frozen" phenotypic variation in clonal snails (*Potamopyrgus antipodarum*). *Evolution* 51:1120–1129.

Jones, B., C. Gliddon, and J. E. G. Good. 2001. The conservation of variation in geographically peripheral populations: *Lloydia serotina* (Liliaceae) in Britain. *Biological Conservation* 101:147–156.

Jordano, P. 1995. Angiosperm fleshy fruits and seed dispersers: A comparative analysis of adaptation and constraints in plant-animal interactions. *American Naturalist* 145:163–191.

Jordano, P., J. Bascompte, and J. M. Olesen. 2003. Invariant properties in coevolutionary networks of plant-animal interactions. *Ecology Letters* 6:69–81.

Joshi, A., and J. N. Thompson. 1995. Alternative routes to the evolution of competitive ability in two competing species of *Drosophila*. *Evolution* 49:616–625.

———. 1996. Evolution of broad and specific competitive ability in novel versus familiar environments in *Drosophila* species. *Evolution* 50:188–194.

———. 1997. Adaptation and specialization in a two-resource environment in *Drosophila* species. *Evolution* 51:846–855.

Judson, O. P., and B. B. Normark. 1996. Ancient asexual scandals. *Trends in Ecology and Evolution* 11:41–46.

Juenger, T., and J. Bergelson. 1998. Pairwise versus diffuse natural selection and the multiple herbivores of scarlet gilia, *Ipomopsis aggregata*. *Evolution* 52:1583–1592.

Kaltz, O., S. Gandon, Y. Michalakis, and J. A. Shykoff. 1999. Local maladaptation in the anther-smut fungus *Microbotryum violaceum* to its host plant *Silene latifolia:* Evidence from a cross-inoculation experiment. *Evolution* 53:395–407.

Kaltz, O., and J. C. Koella. 2003. Host growth conditions regulate the plasticity of horizontal and vertical transmission in *Holospora undulata,* a bacterial parasite of the protozoan *Paramecium caudatum*. *Evolution* 57:1535–1542.

Kaltz, O., and J. A. Shykoff. 1998. Local adaptation in host-parasite systems. *Heredity* 81:361–370.

———. 2002. Within- and among-population variation in infectivity, latency, and spore production in a host-pathogen system. *Journal of Evolutionary Biology* 15:850–860.

Kamp, C., C. O. Wilke, C. Adami, and S. Bornholdt. 2003. Viral evolution under the pressure of an adaptive immune system: Optimal mutation rates for viral escape. *Complexity* 8:28–33.

Kapan, D. D. 2001. Three-butterfly system provides a field test of Müllerian mimicry. Nature 409:338–340.

Karban, R., and I. T. Baldwin. 1997. *Induced responses to herbivory.* University of Chicago Press, Chicago.

Karban, R., and J. S. Thaler. 1999. Plant phase change and resistance to herbivory. *Ecology* 80:510–517.

Kato, M., A. Takimura, and A. Kawakita. 2003. An obligate pollination mutualism and reciprocal diversification in the tree genus *Glochidion* (Euphorbiaceae). *Proceedings of the National Academy of Sciences USA* 100:5264–5267.

Kaustuv, R., D. Jablonski, and J. W. Valentine. 2001. Climate change, species range limits and body size in marine bivalves. *Ecology Letters* 4:366–370.

Kernaghan, G., P. Widden, Y. Bergeron, S. Légaré, and D. Paré. 2003. Biotic and abiotic factors affecting ectomycorrhizal diversity in boreal mixed-woods. *Oikos* 102:497–504.

Kiers, E. T., R. A. Rousseau, S. A. West, and R. F. Denison. 2003. Host sanctions and the legume-rhizobium mutualism. *Nature* 425:78–81.

Kingsolver, J. G. 1995. Viability selection on seasonally polyphenic traits: Wing melanin pattern in western white butterflies. *Evolution* 49:932–941.

Kingsolver, J. G., H. E. Hoekstra, J. M. Hoekstra, D. Berrigan, S. N. Vignieri, C. E. Hill, A. Hoang, P. Gilbert, and P. Beerl. 2001. The strength of phenotypic selection in natural populations. *American Naturalist* 157:245–261.

Kirkpatrick, M., and N. H. Barton. 1997. Evolution of a species' range. *American Naturalist* 150:1–23.

Klassen, G. J. 1992. Coevolution: A history of the macroevolutionary approach to studying host-parasite associations. *Journal of Parasitology* 78:573–587.

Klironomos, J. N. 2003. Variation in plant response to native and exotic arbuscular mycorrhizal fungi. *Ecology* 84:2292–2301.

Knowles, L. L., A. Levy, J. M. McNellis, K. P. Greene, and D. J. Futuyma. 1999. Tests of inbreeding effects on host-shift potential in the phytophagous beetle *Ophraella communa*. *Evolution* 53:561–567.

Koella, J. C., P. Agnew, and Y. Michalakis. 1998. Coevolutionary interactions between host life histories and parasite life cycles. *Parasitology* 116:S47–S55.

Koella, J. C., and O. Restif. 2001. Coevolution of parasite virulence and host life history. *Ecology Letters* 4:207–214.

Koenig, W. D. 1999. Spatial autocorrelation of ecological phenomena. *Trends in Ecology and Evolution* 14:22–26.

Koivisto, R. K. K., and H. R. Braig. 2003. Microorganisms and parthenogenesis. *Biological Journal of the Linnean Society* 79:43–58.

Kondo, N., N. Ijichi, M. Shimada, and T. Fukatsu. 2002. Prevailing triple infection with *Wolbachia* in *Callosobruchus chinensis* (Coleoptera: Bruchidae). *Molecular Ecology* 11:167–180.

Koptur, S., V. Rico-Gray, and M. Palacios-Rios. 1998. Ant protection of the nectaried fern *Polypodium plebeium* in central Mexico. *American Journal of Botany* 85:736–739.

Koske, R. E., J. N. Gemma, and T. Flynn. 1992. Mycotrophy in Hawaiian angiosperms: A survey with implications for the origin of the native flora. *American Journal of Botany* 79:853–862.

Koskela, T., V. Salonen, and P. Mutikainen. 2000. Local adaptation of a holoparasitic plant, *Cuscuta europea:* Variation among populations. *Journal of Evolutionary Biology* 13:749–755.

Kraaijeveld, A. R., and H. C. J. Godfray. 1999. Geographic patterns in the evolution of resistance and virulence in *Drosophila* and its parasitoids. *American Naturalist* 153:S61–S74.

———. 2001. Is there local adaptation in *Drosophila*-parasitoid interactions? *Evolutionary Ecological Research* 3:107–116.

Krause, A. E., K. A. Frank, D. M. Mason, R. E. Ulanowicz, and W. W. Taylor. 2003. Compartments revealed in food-web structure. *Nature* 426:282–285.

Kremer, K., D. van Soolingen, R. Frothingham, W. H. Haas, P. W. M. Hermans, C. Martin, P. Palittapongarnpim, B. B. Plikaytis, L. W. Riley, M. A. Yakrus, J. M. Musser, and J. D. A. Van Emden. 1999. Comparison of methods based on different molecular epidemiological markers for typing of *Mycobacterium tuberculosis* complex strains: Interlaboratory study of discriminatory power and reproducibility. *Journal of Clinical Microbiology* 37:2607–2618.

Kuhn, G., M. Hijri, and I. R. Sanders. 2001. Evidence for the evolution of multiple genomes in arbuscular mycorrhizal fungi. *Nature* 414:745–748.

Kurland, C. G., B. Canback, and O. G. Berg. 2003. Horizontal gene transfer: A critical view. *Proceedings of the National Academy of Sciences USA* 100:9658–9662.

Kurtz, J., M. Kalbe, P. B. Aeschlimann, M. A. Häberli, K. M. Wegner, T. B. H. Reusch, and M. Milinski. 2004. Major histocompatibility complex diversity influences parasite resistance and innate immunity in sticklebacks. *Proceedings of the Royal Society of London Biological Sciences Series B* 271:197–204.

Labandeira, C. 1998. Early history of arthropod and vascular plant associations. *Annual Review of Earth and Planetary Sciences* 26:329–377.

———. 2002. Paleobiology of middle Eocene plant-insect associations from the Pacific Northwest: A preliminary report. *Rocky Mountain Geology* 37:31–59.

Labandeira, C. C., D. L. Dilcher, D. R. Davis, and D. L. Wagner. 1994. Ninety-seven million years of angiosperm-insect association: Paleobiological insights into the meaning of coevolution. *Proceedings of the National Academy of Sciences USA* 91:12278–12282.

Lajeunesse, M. J., and M. R. Forbes. 2002. Host range and local parasite adaptation. *Proceedings of the Royal Society of London Biological Sciences Series B* 269:703–710.

Lammi, A., P. Siikamaki, and V. Salonen. 1999. The role of local adaptation in the relationship between an endangered root hemiparasite *Euphrasia rostkoviana,* and its host, *Agrostis capillaris. Ecography* 22:145–152.

Lande, R., and S. J. Arnold. 1983. The measurement of selection on correlated characters. *Evolution* 37:1210–1226.

Lanfranco, L., M. Delpero, and P. Bonfante. 1999. Intrasporal variability of ribosomal sequences in the endomycorrhizal fungus *Gigaspora margarita. Molecular Ecology* 8:37–45.

Lankford, T. E. Jr., J. M. Billerbeck, and D. O. Conover. 2001. Evolution of intrinsic growth and energy acquisition rates. II. Trade-offs with vulnerability to predation in *Menidia menidia. Evolution* 55:1873–1881.

Law, R. 1985. Evolution in a mutualistic environment. Pages 145–170 in D. H. Boucher, ed., *The biology of mutualisms: Ecology and evolution.* Oxford University Press, Oxford.

Law, R., J. L. Bronstein, and R. Ferriere. 2001. On mutualists and exploiters: Plant-insect coevolution in pollinating seed-parasite systems. *Journal of Theoretical Biology* 212:373–389.

Lazzaro, B. P., B. K. Sceurman, and A. G. Clark. 2004. Genetic basis of natural variation in *D. melanogaster* antibacterial immunity. *Science* 303:1873–1876.

Leebens-Mack, J., O. Pellmyr, and M. Brock. 1998. Host specificity and the genetic structure of two yucca moth species in a yucca hybrid zone. *Evolution* 52:1376–1382.

Lenski, R. E., and R. M. May. 1994. The evolution of virulence in parasites and pathogens: Reconciliation between two competing hypotheses. *Journal of Theoretical Biology* 169:253–265.

Lenski, R. E., and M. Travisano. 1994. Dynamics of adaptation and diversification: A 10,000 generation experiment with bacterial populations. *Proceedings of the National Academy of Sciences USA* 91:6808–6814.

Leonard, K. J. 1998. Modelling gene frequency dynamics. Pages 211–230 in I. R. Crute, E. B. Holub, and J. J. Burdon, eds., *The gene-for-gene relationship in plant-parasite interactions.* CAB International, New York.

Levin, B. R., M. Lipsitch, and S. Bonhoeffer. 1999. Population biology, evolution, and infectious disease: Convergence and synthesis. *Science* 283:806–809.

Levin, S. A., and V. Andreasen. 1999. Disease transmission dynamics and the evolution of antibiotic resistance in hospitals and communal settings. *Proceedings of the National Academy of Sciences USA* 96:800–801.

Levin, S. A., B. Grenfell, A. Hastings, and A. S. Perelson. 1997. Mathematical and computational challenges in population biology and ecosystems science. *Science* 275:334–343.

Lewis, G. C., C. Ravel, W. Naffaa, C. Astier, and G. Charmet. 1997. Occurrence of *Acremonium* endophytes in wild populations of *Lolium* spp. in European countries and a relationship between level of infection and climate in France. *Annals of Applied Biology* 130:227–238.

Lewontin, R. C., L. R. Ginsburg, and S. D. Tuljapurkar. 1978. Heterosis as an explanation for large amounts of genic polymorphism. *Genetics* 88:149–170.

Lieberman, B. S., and N. Eldredge. 1996. Trilobite biogeography in the Middle Devonian: Geological processes and analytical methods. *Paleobiology* 22:66–79.

Limpens, E., C. Franken, P. Smit, J. Willemse, T. Bisseling, and R. Geurts. 2003. LysM domain receptor kinases regulating rhizobial nod factor–induced infection. *Science* 302:630–633.

Linhart, Y. B., and M. C. Grant. 1996. Evolutionary significance of local genetic differentiation in plants. *Annual Review of Ecology and Systematics* 27:237–277.

Linhart, Y. B., and J. D. Thompson. 1995. Terpene-based selective herbivory by *Helix aspersa* (Mollusca) on *Thymus vulgaris* (Labiatae). *Oecologia* 102:126–132.

Lively, C. M. 1987. Evidence from a New Zealand snail for the maintenance of sex by parasitism. *Nature* 328:519–521.

———. 1992. Parthenogenesis in a freshwater snail: Reproductive assurance versus parasite release. *Evolution* 46:907–913.

———1993. Rapid evolution by biological enemies. *Trends in Ecology and Evolution* 8:345–346.

———. 1999. Migration, virulence, and the geographic mosaic of adaptation by parasites. *American Naturalist* 153:S34–S47.

Lively, C. M., C. Craddock, and R. C. Vrijenhoek. 1990. Red Queen hypothesis supported by parasitism in sexual and clonal fish. *Nature* 344:864–866.

Lively, C. M., and M. F. Dybdahl. 2000. Parasite adaptation to locally common host genotypes. *Nature* 405:679–681.

Lively, C. M., and J. Jokela. 2002. Temporal and spatial distributions of parasites and sex in a freshwater snail. *Evolutionary Ecology Research* 4:219–226.

Lloyd, D. G. 1980. Benefits and handicaps of sexual reproduction. *Evolutionary Biology* 13:69–111.

Lockhart, A. B., and P. H. Thrall. 1996. Sexually transmitted diseases in animals: Ecological and evolutionary implications. *Biological Reviews of the Cambridge Philosophical Society* 71:415–471.

Lodwig, E. M., A. H. F. Hosie, A. Bourdès, K. Findlay, D. Allaway, R. Karunakaran, J. A. Downie et al. 2003. Amino-acid cycling drives nitrogen fixation in the legume-*Rhizobium* symbiosis. *Nature* 422:722–726.

Long, C. A., and S. L. Hoffman. 2002. Malaria—from infants to genomics to vaccines. *Science* 297:345–347.

Losos, J. B. 1994. Integrative approaches to evolutionary ecology: *Anolis* lizards as model systems. *Annual Review of Ecology and Systematics* 25:467–493.

———. 1995. Community evolution in Greater Antillean *Anolis* lizards: Phylogenetic patterns and experimental tests. *Philosophical Transactions of the Royal Society of London Biological Sciences Series B* 349:69–75.

———. 1996. Phylogenetic perspectives on community ecology. *Ecology* 77:1344–1354.

Losos, J. B., T. R. Jackman, A. Larson, K. de Queiroz, and L. Rodriguez-Schettino. 1998. Contingency and determinism in replicated adaptive radiations of island lizards. *Science* 279:2115–2118.

Losos, J. B., M. Leal, R. E. Glor, K. de Queiroz, P. E. Hertz, L. R. Schettino, A. C. Lara, T. R. Jackman, and A. Larson. 2003. Niche lability in the evolution of a Caribbean lizard community. *Nature* 424:542–545.

Losos, J. B., T. W. Schoener, K. I. Warheit, and D. Creer. 2001. Experimental studies of adaptive differentiation in Bahamian *Anolis* lizards. *Genetica* 112–113:399–415.

Losos, J. B., and D. A. Spiller. 1999. Differential colonization success and asymmetrical interaction between two lizard species. *Ecology* 80:252–258.

Losos, J. B., K. I. Warheit, and T. W. Schoener. 1997. Adaptive differentiation following experimental island colonization in *Anolis* lizards. *Nature* 387:70–73.

Lovelock, C. E., K. Andersen, and J. B. Morton. 2003. Arbuscular mycorrhizal communities in tropical forests are affected by host tree species and environment. *Oecologia* 135:268–279.

Lu, W., and P. Logan. 1994. Geographic variation in larval feeding acceptance and performance of *Leptinotarsa decemlineata* (Coleoptera, Chrysomelidae). *Annals of the Entomological Society of America* 87:460–469.

Lutzoni, F., M. Pagel, and V. Reeb. 2001. Major fungal lineages are derived from lichen symbiotic ancestors. *Nature* 411:937–940.

Lynch, M., J. Conery, and R. Burger. 1995. Mutational meltdown in sexual populations. *Evolution* 49:1067–1080.

Lythgoe, K. A. 2000. The coevolution of parasites with host-acquired immunity and the evolution of sex. *Evolution* 54:1142–1156.

Machado, C. A., E. Jousselin, F. Kjellberg, S. G. Compton, and E. A. Herre. 2001. Phylogenetic relationships, historical biogeography and character evolution of fig-pollinating wasps. *Proceedings of the Royal Society of London Biological Sciences Series B* 268:685–694.

Maddox, G. D., and R. B. Root. 1990. Structure of the encounter between goldenrod (*Solidago altissima*) and its diverse insect fauna. *Ecology* 71:2115–2124.

Madsen, E. B., L. H. Madsen, M. Radutoiu, M. Olbryt, M. Rakwalska, K. Szczyglowski, S. Sato, T. Kaneko, S. Tabata, N. Sandal, and J. Stougaard. 2003. A receptor kinase gene of the LysM type is involved in legume perception of rhizobial signals. *Nature* 425:637–640.

Magurran, A. E., B. H. Seghers, P. W. Shaw, and G. R. Carvalho. 1995. The behavioural diversity and evolution of guppy (*Poecilia reticulata*) populations in Trinidad. Pages 155–202 in P. J. B. Slater, ed., *Advances in the study of behavior. Volume 24,* Academic Press.

Majerus, M. E. N. 1998. *Melanism: Evolution in action.* Oxford University Press, Oxford.

Malinowski, D. P., and D. P. Belesky. 1999. *Neotyphodium coeniophialum*–endophyte infection affects the ability of tall fescue to use sparingly available phosphorus. *Journal of Plant Nutrition* 22:835–853.

Mallet, J. 1986. Hybrid zones of *Heliconius* butterflies in Panama and the stability and movement of warning colour clines. *Heredity* 56:191–202.

———. 1999. Causes and consequences of a lack of coevolution in Müllerian mimicry. *Evolutionary Ecology* 13:777–806.

Mallet, J., N. Barton, G. Lamas M., J. Santisteban C., M. Muedas M., and H. Eeley. 1990. Estimates of selection and gene flow from measures of cline width and linkage disequilibrium in *Heliconius* hybrid zones. *Genetics* 124:921–936.

Mallet, J., and L. E. Gilbert. 1995. Why are there so many mimicry rings? Correlations between habitat, behaviour and mimicry in *Heliconius* butterflies. *Biological Journal of the Linnean Society* 55:159–180.

Marak, H. B., A. Biere, and J. M. M. Van Damme. 2003. Fitness costs of chemical defense in *Plantago lanceolata* L: Effects of nutrient and competition stress. *Evolution* 57:2519–2530.

Marchetti, K., H. Nakamura, and H. L. Gibbs. 1998. Host-race formation in the common cuckoo. *Science* 282:471–472.

Margulis, L. 2000. Symbiosis and the origin of protists. Pages 141–157 in A. Haselton, ed., *Environmental evolution: Effects of the origin and evolution of life on planet earth.* MIT Press, Cambridge, Massachusetts.

Margulis, L., and D. Sagan. 2002. *Acquiring genomes: A theory of the origins of species.* Basic Books, New York.

Margulis, L., and R. Fester. 1991. Symbiosis as a source of evolutionary innovation: Speciation and morphogenesis. MIT Press, Cambridge, Massachusetts.

Mark Welch, D. B., and M. Meselson. 2003. Oocyte nuclear DNA content and GC proportion in rotifers of the anciently asexual Class Bdelloidea. *Biological Journal of the Linnean Society* 79:85–91.

Marr, D., L., M. T. Brock, and O. Pellmyr. 2001. Coexistence of mutualists and antagonists: Exploring the impact of cheaters on the yucca – yucca moth mutualism. *Oecologia* 128: 454–463.

Marrow, P., U. Dieckmann, and R. Law. 1996. Evolutionary dynamics of predator-prey systems: An ecological perspective. *Journal of Mathematical Biology* 34:556–578.

Marrow, P., R. Law, and C. Cannings. 1992. The coevolution of predator-prey interactions: ESSs and Red Queen dynamics. *Proceedings of the Royal Society of London B* 250: 133–141.

Martel, C., A. Réjasse, F. Rousset, M.-T. Bethenod, and D. Bourguet. 2003. Host-plant-associated genetic differentiation in Northern French populations of the European corn borer. *Heredity* 90:141–149.

Martin, G., O. Sorokine, M. Moniatte, P. Bulet, C. Hetru, and A. Van Dorsselaer. 1999. The structure of a glycosylated protein hormone responsible for sex determination in the isopod, *Armadillidium vulgare. European Journal of Biochemistry* 262:727–736.

Martin, M. P., X. Gao, J.-H. Lee, G. W. Nelson, R. Detels, J. J. Goedert, S. Buchbinder et al. 2002. Epistatic interaction between *KIR3DS1* and *HLA-B* delays the progression to AIDS. *Nature Genetics* 31:429–434.

Martin, T. E. 1993. Nest predation and nest sites: New perspectives on old patterns. *Bioscience* 43:523–532.

Martin, W. 2003. Gene transfer from organelles to the nucleus: Frequent and in big chunks. *Proceedings of the National Academy of Sciences USA* 100:8612–8614.

Martinsen, G. D., and T. G. Whitham. 1994. More birds nest in hybrid cottonwood trees. *Wilson Bulletin* 106:474–481.

Masui, S., S. Kamoda, T. Sasaki, and H. Ishikawa. 2000. Distribution and evolution of bacteriophage WO in *Wolbachia,* the endosymbiont causing sexual alterations in arthropods. *Journal of Molecular Evolution* 51:491–497.

Masui, S., H. Kuroiwa, T. Sasaki, M. Inui, T. Kuroiwa, and H. Ishikawa. 2001. Bacteriophage WO and virus-like particles in *Wolbachia,* an endosymbiont of arthropods. *Biochemical and Biophysical Research Communications* 283:1099–1104.

Matsuda, H., and T. Namba. 1991. Food web graph of a coevolutionarily stable community. *Ecology* 72:267–276.

Matsuura, K. 2003. Symbionts affecting termite behavior. Pages 131–143 in T. A. Miller, ed., *Insect symbiosis.* CRC Press, New York.

Matsuura, K., C. Tanaka, and T. Nishida. 2000. Symbiosis of a termite and a sclerotium-forming fungus: Sclerotia mimic termite eggs. *Ecological Research* 15:405–414.

Mauricio, R., E. A. Stahl, T. Korves, D. Tian, M. Kreitman, and J. Bergelson. 2003. Natural selection for polymorphism in the disease resistance gene *Rps2* of *Arabidopsis thaliana*. *Genetics* 163:735–746.

May, R. M. 1973. *Stability and complexity in model ecosystems*. Princeton University Press, Princeton, New Jersey.

May, R. M., and R. M. Anderson. 1983. Epidemiology and genetics in the coevolution of parasites and hosts. *Proceedings of the Royal Society of London Biological Sciences Series B* 219:281–313.

————. 1990. Parasite-host coevolution. *Parasitology* 100:S89–S102.

May, R. M., and M. A. Nowak. 1994. Superinfection, metapopulation dynamics, and the evolution of diversity. *Journal of Theoretical Biology* 170:95–114.

————. 1995. Coinfection and the evolution of parasite virulence. *Proceedings of the Royal Society of London Biological Sciences Series B* 261:209–215.

Maynard Smith, J. 1978. *The evolution of sex*. Cambridge University Press, Cambridge.

Maynard Smith, J., and E. Szathmáry. 1995. *The major transitions in evolution*. W. H. Freeman, San Francisco.

McDonald, R. A. 2002. Resource partitioning among British and Irish mustelids. *Journal of Animal Ecology* 71:185–200.

McGovern, P. E., D. L. Glusker, L. J. Exner, and M. M. Voigt. 1996. Neolithic resinated wine. *Nature* 381:480–481.

McGovern, T. M., and M. E. Hellberg. 2003. Cryptic species, cryptic endosymbionts, and geographic variation in chemical defences in the bryozoan *Bugula neritina*. *Molecular Ecology* 12:1207–1215.

McKey, D. 1984. Interaction of the ant-plant *Leonardoxa africana* (Caesalpiniaceae) with its obligate inhabitants in a rain forest in Cameroon. *Biotropica* 16:81–99.

McMillan, W. O., L. A. Weigt, and S. R. Palumbi. 1999. Color pattern evolution, assortative mating, and genetic differentiation in brightly colored butterflyfishes (Chaetondontidae). *Evolution* 53:247–260.

McPeek, M. A. 1998. The consequences of changing the top predator in a food web: A comparative experimental approach. *Ecological Monographs* 68:1–23.

————. 1999. Biochemical evolution associated with antipredator adaptation in damselflies. *Evolution* 53:1835–1845.

————. 2000. Predisposed to adapt? Clade-level differences in characters affecting swimming performance in damselflies. *Evolution* 54:2072–2080.

McPeek, M. A., and J. M. Brown. 2000. Building a regional species pool: Diversification of the *Enallagma* damselflies in eastern North America. *Ecology* 81:904–920.

McPhail, J. D. 1992. Ecology and evolution of sympatric sticklebacks (*Gasterosteus*): Evidence for a species-pair in Paxton Lake, Texada Island, British Columbia. *Canadian Journal of Zoology* 70:361–369.

Melville, J. 2002. Competition and character displacement in two species of scincid lizards. *Ecology Letters* 5:386–393.

Meyer, D., and G. Thomson. 2001. How selection shapes variation of the human major histocompatibility complex: A review. *Annals of Human Genetics* 65:1–26.

MHC Sequencing Consortium. 1999. Complete sequence and gene map of a human major histocompatibility complex. *Nature* 401:921–923.

Miao, E. A., and S. I. Miller. 1999. Bacteriophages in the evolution of pathogen-host interactions. *Proceedings of the National Academy of Sciences USA* 96:9452–9454.

Miklashevichs, E., H. Röhrig, J. Schell, and J. Schmidt. 2001. Perception and signal transduction of rhizobial NOD factors. *Critical Reviews in Plant Sciences* 20:373–394.

Miles, D. B., and A. E. Dunham. 1996. The paradox of the phylogeny: Character displacement of analyses of body size in island *Anolis*. *Evolution* 50:594–603.

Miller, L. H., and B. Greenwood. 2002. Malaria—a shadow over Africa. *Science* 298:121–122.

Miller, L. H., D. I. Baruch, K. Marsh, and O. K. Doumbo. 2002. The pathogenic basis of malaria. *Nature* 415:673–679.

Miller, T. E., and J. Travis. 1996. The evolutionary role of indirect effects in communities. *Ecology* 77:1329–1335.

Mitchell, C. E., and A. G. Power. 2003. Release of invasive plants from fungal and viral pathogens. *Nature* 421:625–627.

Mitchell-Olds, T., and D. Bradley. 1996. Genetics of *Brassica rapa*. III. Costs of disease resistance to three fungal pathogens. *Evolution* 50:1859–1865.

Mitchell-Olds, T., D. Siemens, and D. Pedersen. 1996. Physiology and costs of resistance to herbivory and disease in *Brassica*. *Entomologia Experimentalis et Applicata* 80:231–237.

Mittermeier, R. A., C. G. Mittermeier, T. M. Brooks, J. D. Pilgrim, W. R. Konstant, G. A. B. da Fonseca, and C. Kormos. 2003. Wilderness and biodiversity conservation. *Proceedings of the National Academy of Sciences USA* 100:10309–10313.

Mitton, J. B. 1997. *Selection in natural populations*. Oxford University Press, Oxford.

Miyasaka, S. C., M. Habte, and D. T. Matsuyama. 1993. Mycorrhizal dependency of two Hawaiian endemic tree species: Koa and mamane. *Journal of Plant Nutrition* 16:1339–1356.

Mode, C. J. 1958. A mathematical model for the co-evolution of obligate parasites and their hosts. *Evolution* 12:158–165.

Molbo, D., C. A. Machado, J. G. Sevenster, L. Keller, and E. A. Herre. 2003. Cryptic species of fig-pollinating wasps: Implications for the evolution of the fig-wasp mutualism, sex allocation, and precision of adaptation. *Proceedings of the National Academy of Sciences USA* 100:5867–5872.

Molina, R., H. Massicotte, and J. M. Trappe. 1992. Specificity phenomena in mycorrhizal symbioses: Community-ecological consequences and practical implications. Pages 357–423 in M. F. Allen, ed., *Mycorrhizal functioning: An integrative plant-fungal process*. Chapman and Hall, New York.

Mooney, H. A., and E. E. Cleland. 2001. The evolutionary impact of invasive species. *PNAS* 98:5446–5451.

Mopper, S. 1996. Adaptive genetic structure in phytophagous insect populations. *Trends in Ecology and Evolution* 11:235–238.

Mopper, S., and S. Y. Strauss, eds. 1998. *Genetic structure and local adaptation in natural insect populations: Effects of ecology, life history, and behavior*. Chapman and Hall, London.

Morales, M. A. 2002. Ant-dependent oviposition in the membracid *Publilia concava*. *Ecological Entomology* 27:247–250.

Morales, M. A., and E. R. Heithaus. 1998. Food from seed dispersal mutualism shifts sex ratios in colonies of the ant, *Aphaenogaster rudis*. *Ecology* 79:734–739.

Moran, N. A. 1996. Accelerated evolution and Muller's ratchet in endosymbiotic bacteria. *Proceedings of the National Academy of Sciences USA* 93:2873–2878.

———. 2002. The ubiquitous and varied role of infection in the lives of animals and plants. *American Naturalist* 160:S1–8.

Moran, N. A., and P. Baumann. 2000. Bacterial endosymbionts in animals. *Current Opinion in Microbiology* 3:270–275.

Moran, N. A., M. A. Munson, P. Baumann, and H. Ishikawa. 1993. A molecular clock in endosymbiotic bacteria is calibrated using the insect hosts. *Proceedings of the Royal Society of London B* 253:167–171.

Moran, N. A., and J. J. Wernegreen. 2000. Lifestyle evolution in symbiotic bacteria: Insights from genomics. *Trends in Ecology and Evolution* 15:321–326.

Morand, S., S. D. Manning, and M. E. J. Woolhouse. 1996. Parasite-host coevolution and geographic patterns of parasite infectivity and host susceptibility. *Proceedings of the Royal Society of London Biological Sciences Series B* 263:119–128.

Moreau, J., A. Bertin, Y. Caubet, and T. Rigaud. 2001. Sexual selection in an isopod with *Wolbachia*-induced sex reversal: Males prefer real females. *Journal of Evolutionary Biology* 14:388–394.

Moreau, J., and T. Rigaud. 2000. Operational sex ratio in terrestrial isopods: Interaction between potential rate of reproduction and *Wolbachia*-induced sex ratio distortion. *Oikos* 91:477–484.

Morin, P. J. 2003. Community ecology and the genetics of interacting species. *Ecology* 84:577–580.

Moritz, C., and D. P. Faith. 1998. Comparative phylogeography and the identification of genetically divergent areas for conservation. *Molecular Ecology* 7:419–429.

Morris, R. J., O. T. Lewis, and H. C. J. Godfray. 2004. Experimental evidence for apparent competition in a tropical forest food web. *Nature* 428:310–313.

Morrow, P. A., T. G. Whitham, B. M. Potts, P. Ladiges, D. H. Ashton, and J. B. Williams. 1994. Gall-forming insects concentrate on hybrid phenotypes of *Eucalyptus*. Pages 121–134 in P. W. Price, W. J. Mattson, and Y. N. Baranchikov, eds., *The ecology and evolution of gall-forming insects*. USDA Forest Service North Central Forest Experiment Station, St. Paul, Minnesota.

Motychak, J. E., E. D. Brodie, Jr., and E. D. Brodie, III. 1999. Evolutionary response of predators to dangerous prey: Preadaptation and the evolution of tetrodotoxin resistance in garter snakes. *Evolution* 53:1528–1535.

Mu, J., J. Duan, K. D. Makova, D. A. Joy, C. Q. Huynh, O. H. Branch, W.-H. Li, and X. Su. 2002. Chromosome-wide SNPs reveal an ancient origin for *Plasmodium falciparum*. *Nature* 418:323–324.

Mueller, U. G., S. A. Rehner, and T. R. Schultz. 1998. The evolution of agriculture in ants. *Science* 281:2034–2038.

Mueller, U. G., T. R. Schultz, C. R. Currie, R. M. M. Adams, and D. Malloch. 2001. The origin of the attine ant-fungus mutualism. *Quarterly Review of Biology* 76:169–197.

Murdoch, W. W., C. J. Briggs, and R. M. Nisbet. 2003. *Consumer-resource dynamics*. Princeton University Press, Princeton, New Jersey.

Murray, B. R., P. H. Thrall, and M. J. Woods. 2001. Acacia species and rhizobial interactions: Implications for restoration of native vegetation. *Ecological Management and Restoration* 2:213–219.

Mutikainen, P., V. Salonen, S. Puustinen, and T. Koskela. 2000. Local adaptation, resistance, and virulence in a hemiparasitic plant–host plant interaction. *Evolution* 54:433–440.

Myers, N., and A. H. Knoll. 2001. The biotic crisis and the future of evolution. *Proceedings of the National Academy of Sciences USA* 98:5389–5392.

Nei, M. 1973. Analysis of gene diversity in subdivided populations. *Proceedings of the National Academy of Sciences USA* 70:3321–3323.

Neuhauser, C., D. A. Andow, G. E. Heimpel, G. May, R. G. Shaw, and S. Wagenius. 2003. Community genetics: Expanding the synthesis of ecology and genetics. *Ecology* 84:545–558.

Newsham, K. K., A. H. Fitter, and A. R. Watkinson. 1995. Multi-functionality and biodiversity in arbuscular mycorrhizas. *Trends in Ecology and Evolution* 10:407–411.

Niaré, O., K. Markianos, J. Volz, F. Oduol, A. Touré, M. Bagayoko, D. Sangaré et al. 2002. Genetic loci affecting resistance to human malaria parasites in a West African mosquito vector population. *Science* 298:213–216.

Nielsen, J. K. 1996. Intraspecific variability in adult flea beetle behaviour and larval performance on an atypical host plant. *Entomologia Experimentalis et Applicata* 80:160–162.

———. 1997a. Variation in defences of the plant *Barbarea vulgaris* and in counteradaptations by the flea beetle *Phyllotreta nemorum*. *Entomologia Experimentalis et Applicata* 82:25–35.

———. 1997b. Genetics of the ability of *Phyllotreta nemorum* larvae to survive in an atypical host plant, *Barbarea vulgaris* ssp. *arcuata*. *Entomologia Experimentalis et Applicata* 82: 37–44.

———. 1999. Specificity of a Y-linked gene in the flea beetle *Phylotreta nemorum* for defences in *Barbarea vulgaris*. *Entomologia Experimentalis et Applicata* 91:359–368.

Nielson, M., K. Lohman, and J. Sullivan. 2001. Phylogeography of the tailed frog (*Ascaphus truei*): Implications for the biogeography of the Pacific Northwest. *Evolution* 55:147–160.

Nijhout, H. F. 1991. *The development and evolution of butterfly wing patterns*. Smithsonian Institution Press, Washington, D.C.

Nishiguchi, M. K., E. G. Ruby, and M. J. McFall-Ngai. 1998. Competitive dominance among strains of luminous bacteria provides an unusual form of evidence for parallel evolution in sepiolid squid-vibrio symbioses. *Applied Environmental Microbiology* 64:3209–3213.

Normark, B. B., O. P. Judson, and N. A. Moran. 2003. Genomic signatures of ancient asexual lineages. *Biological Journal of the Linnean Society* 79:69–84.

Nowak, M. A., K. Tarczy-Hornoch, and J. M. Austyn. 1992. The optimal number of major histocompatibility complex molecules in an individual. *Proceedings of the National Academy of Sciences USA* 89:10896–10899.

Nuismer, S. L., R. Gomulkiewicz, and M. T. Morgan. 2003. Coevolution in temporally variable environments. *American Naturalist* 162:195–204.

Nuismer, S. L., and M. Kirkpatrick. 2003. Gene flow and the coevolution of parasite range. *Evolution* 57:746–754.

Nuismer, S. L., and J. N. Thompson. 2001. Plant polyploidy and non-uniform effects on insect herbivores. *Proceedings of the Royal Society of London Biological Sciences Series B* 268:1937–1940.

Nuismer, S. L., J. N. Thompson, and R. Gomulkiewicz. 1999. Gene flow and geographically structured coevolution. *Proceedings of the Royal Society of London Biological Sciences Series B* 266:605–609.

———. 2000. Coevolutionary clines across selection mosaics. *Evolution* 54:1102–1115.

———. 2003. Coevolution between hosts and parasites with partially overlapping geographic ranges. *Journal of Evolutionary Biology* 16:1337–1345.

Nuzhdin, S. V., and D. A. Petrov. 2003. Transposable elements in clonal lineages: Lethal hangover from sex. *Biological Journal of the Linnean Society* 79:33–41.

Nylin, S., and N. Janz. 1999. The ecology and evolution of host plant range: Butterflies as a model group. Pages 31–54 in H. Olff, V. K. Brown, and R. H. Drent, eds., *Herbivores: Between plants and predators*. Blackwell Scientific, Oxford.

Nylin, S., N. Janz, and N. Wedell. 1996. Oviposition plant preference and offspring performance in the comma butterfly: Correlations and conflicts. *Entomologia Experimentalis et Applicata* 80:141–144.

Ohgushi, T., and H. Sawada. 1997. A shift toward early reproduction in an introduced herbivorous ladybird. *Ecological Entomology* 22:90–96.

Olesen, J. M., and P. Jordano. 2002. Geographic patterns in plant/pollinator mutualistic networks. *Ecology* 83:2416–2424.

Oliveira, P. S., V. Rico-Gray, C. Diaz-Castelazo, and C. Castillo-Guevara. 1999. Interaction between ants, extrafloral nectaries and insect herbivores in Neotropical coastal sand dunes: Herbivore deterrence by visiting ants increases fruit set in *Opuntia stricta* (Cactaceae). *Functional Ecology* 13:623–631.

Oliver, K. M., J. A. Russell, N. A. Moran, and M. S. Hunter. 2003. Facultative bacterial symbionts in aphids confer resistance to parasitic wasps. *Proceedings of the National Academy of Sciences USA* 100:1803–1807.

Olivera, B. M. 2002. *Conus* venom peptides: Reflections from the biology of clades and species. *Annual Review of Ecology and Systematics* 33:25–47.

Orr, H. A., and S. Irving. 1997. The genetics of adaptation: The genetic basis of resistance to wasp parasitism in *Drosophila melanogaster*. *Evolution* 51:1877–1885.

O'Steen, S., A. J. Cullum, and A. F. Bennett. 2002. Rapid evolution of escape ability in Trinidadian guppies (*Poecilia reticulata*). *Evolution* 56:776–784.

Otero, J. T., J. D. Ackerman, and P. Bayman. 2002. Diversity and host specificity of endophytic *Rhizoctonia*-like fungi from tropical orchids. *American Journal of Botany* 89:1852–1858.

Otto, S. P., and Y. Michalakis. 1998. The evolution of recombination in changing environments. *Trends in Ecology and Evolution* 13:145–151.

Page, R. D. M. 2003a. Introduction. Pages 1–21 in R. D. M. Page, ed., *Tangled trees: Phylogeny, cospeciation, and coevolution*. University of Chicago Press, Chicago.

———, ed. 2003b. *Tangled trees: Phylogeny, cospeciation, and coevolution*. University of Chicago Press, Chicago.

Palmer, T. M., M. L. Stanton, and T. P. Young. 2003. Competition and coexistence: Exploring mechanisms that restrict and maintain diversity within mutualist guilds. *American Naturalist* 162:S63–S79.

Palumbi, S. R. 1985. Spatial variation in an algal-sponge commensalism and the evolution of ecological interactions. *American Naturalist* 126:267–274.

———. 1994. Genetic divergence, reproductive isolation, and marine speciation. *Annual Review of Ecology and Systematics* 25:547–572.

———. 2001. Humans are the world's greatest evolutionary force. *Science* 293:1786–1790.

Pandolfi, J. M., and J. B. C. Jackson. 2001. Community structure of Pleistocene coral reefs of Curaçao, Netherlands Antilles. *Ecological Monographs* 71:49–67.

Pandolfi, J. M., C. E. Lovelock, and A. F. Budd. 2002. Character release following extinction in a Caribbean reef coral species complex. *Evolution* 56:479–501.

Parchman, T. L., and C. W. Benkman. 2002. Diversifying coevolution between crossbills and black spruce on Newfoundland. *Evolution* 56:1663–1672.

Parham, P. 2003. Innate immunity: The unsung heroes. *Nature* 423:20.

Parker, I. M. 1997. Pollinator limitation of *Cytisus scoparius* (Scotch broom), an invasive exotic shrub. *Ecology* 78:1457–1470.

Parker, M. A. 1994. Pathogens and sex in plants. *Evolutionary Ecology* 8:560–584.

———. 1996. Cryptic species within *Amphicarpaea bracteata* (Leguminosae): Evidence from isozymes, morphology, and pathogen specificity. *Canadian Journal of Botany* 74:1640–1650.

————. 1999a. Mutualism in metapopulations of legumes and rhizobia. *American Naturalist* 153:S48–S60.

————. 1999b. Relationships of bradyrhizobia from the legumes *Apios americana* and *Desmodium glutinosum. Applied and Environmental Microbiology* 65:4914–4920.

————. 2003. Genetic markers for analyzing symbiotic relationships and lateral gene transfer in neotropical bradyrhizobia. *Molecular Ecology* 12:2447–2455.

Parker, M. A., B. Lafay, J. J. Burdon, and P. van Berkum. 2002. Conflicting phylogeographic patterns in rRNA and *nifD* indicate regionally restricted gene transfer in *Bradyrhizobium. Microbiology* 148:2557–2565.

Parker, M. A., and H. H. Wilkinson. 1997. A locus controlling nodulation specificity in *Amphicarpaea bracteata* (Leguminosae). *Journal of Heredity* 88:449–453.

Parmesan, C., N. Ryrholm, C. Stefanescu, J. K. Hill, C. D. Thomas, H. Descimon, B. Huntley, L. Kaila, J. Kullberg, T. Tammaru, W. J. Tennent, J. A. Thomas, and M. Warren. 1999. Poleward shifts in geographic ranges of butterfly species associated with regional warming. *Nature* 399:579–583.

Paul, R. E. L., F. Ariey, and V. Robert. 2003. The evolutionary ecology of *Plasmodium. Ecology Letters* 6:866–880.

Paulsrud, P., J. Rikkinen, and P. Linblad. 2000. Spatial patterns of photobiont diversity in some *Nostoc*-containing lichens. *New Phytologist* 146:291–299.

————. 2001. Field investigations on cyanobacterial specificity in *Peltigera aphthosa. New Phytologist* 152:177–123.

Pawlowska, T. E., and J. W. Taylor. 2004. Organization of genetic variation in individuals of arbuscular mycorrhizal fungi. *Nature* 427:733–737.

Peek, A. S., R. A. Feldman, R. A. Lutz, and R. C. Vrijenhoek. 1998. Cospeciation of chemoautotrophic bacteria and deep sea clams. *Proceedings of the National Academy of Sciences USA* 95:9962–9966.

Peichel, C. L., K. S. Nereng, K. A. Ohgi, B. L. E. Cole, P. F. Colosimo, C. A. Buerkle, D. Schluter et al. 2001. The genetic architecture of divergence between threespine stickleback species. *Nature* 414:901–905.

Pellmyr, O. 1992. The phylogeny of a mutualism: Evolution and coadaptation between *Trollius* and its seed-parasitic pollinators. *Biological Journal of the Linnean Society* 47:337–365.

————. 1999. Systematic revision of the yucca moths in the *Tegeticula yuccasella* complex (Lepidoptera: Prodoxidae) north of Mexico. *Systematic Entomology* 24:243–270.

————. 2003. Yuccas, yucca moths, and coevolution: A review. *Annals of the Missouri Botanical Garden* 90:35–55.

Pellmyr, O., and M. Balcázar-Lara. 2000. Systematics of the yucca moth genus Parategeticula (Lepidoptera: Prodoxidae), with description of three Mexican species. *Annals of the Entomological Society of America* 93:432–439.

Pellmyr, O., and C. J. Huth. 1994. Evolutionary stability of mutualism between yuccas and yucca moths. *Nature* 372:257–260.

Pellmyr, O., and H. W. Krenn. 2002. Origin of a complex key innovation in an obligate insect-plant mutualism. *Proceedings of the National Academy of Sciences USA* 99:5498–5502.

Pellmyr, O., and J. Leebens-Mack. 2000. Reversal of mutualism as a mechanism for adaptive radiation in yucca moths. *American Naturalist* 156:S62–S76.

Pellmyr, O., J. Leebens-Mack, and C. J. Huth. 1996. Non-mutualistic yucca moths and their evolutionary consequences. *Nature* 380:155–156.

Pellmyr, O., J. H. Leebens-Mack, and J. N. Thompson. 1998. Herbivores and molecular clocks as tools in plant biogeography. *Biological Journal of the Linnean Society* 63:367–378.

Pellmyr, O., and J. N. Thompson. 1992. Multiple occurrences of mutualism in the yucca moth lineage. *Proceedings of the National Academy of Sciences* 89:2927–2929.

———. 1996. Sources of variation in pollinator contribution within a guild: The effects of plant and pollinator factors. *Oecologia* 107:595–604.

Pellmyr, O., J. N. Thompson, J. M. Brown, and R. G. Harrison. 1996. Evolution of pollination and mutualism in the yucca moth lineage. *American Naturalist* 148:827–847.

Pennings, S. C., E. L. Siska, and M. D. Bertness. 2001. Latitudinal differences in plant palatability in Atlantic coast salt marshes. *Ecology* 82:1344–1359.

Perlman, S. J., and J. Jaenike. 2003a. Evolution of multiple components of virulence in *Drosophila*-nematode associations. *Evolution* 57:1543–1551.

———. 2003b. Infection success in novel hosts: An experimental and phylogenetic study of *Drosophila*-parasitic nematodes. *Evolution* 57:544–557.

Perret, X., C. Staehelin, and W. J. Broughton. 2000. Molecular basis of symbiotic promiscuity. *Microbiology and Molecular Biology Reviews* 64:180–201.

Peters, A. D., and C. M. Lively. 1999. The Red Queen and fluctuating epistasis: A population genetic analysis of antagonistic coevolution. *American Naturalist* 154:393–405.

Pfennig, D. W. 1992. Polyphenism in spadefoot toad tadpoles as a locally adjusted evolutionarily stable strategy. *Evolution* 46:1408–1420.

Pfennig, D. W., and P. J. Murphy. 2000. Character displacement in polyphenic tadpoles. *Evolution* 54:1738–1749.

———. 2002. How fluctuating competition and phenotypic plasticity mediate species divergence. *Evolution* 56:1217–1228.

———. 2003. A test of alternative hypotheses for character divergence between coexisting species. *Ecology* 84:1288–1297.

Pierce, N. E. 1987. The evolution and biogeography of associations between lycaenid butterflies and ants. *Oxford Surveys in Evolutionary Biology* 4:89–116.

Pierce, N. E., M. F. Braby, A. Heath, D. J. Lohman, J. Mathew, D. B. Rand, and M. A. Travassos. 2002. The ecology and evolution of ant association in the Lycaenidae (Lepidoptera). *Annual Review of Entomology* 47:733–771.

Pigliucci, M. 2003. Selection in a model system: Ecological genetics of flowering time in *Arabidopsis thaliana*. *Ecology* 84:1700–1712.

Pilson, D. 1996. Two herbivores and constraints on selection for resistance in *Brassica rapa*. *Evolution* 50:1492–1500.

Pimm, S. L. 1982. *Food webs*. Chapman and Hall, New York.

Pogson, G. H. 2003. Natural selection and the genetic differentiation of coastal and arctic populations of the Atlantic cod in northern Norway: A test involving nucleotide sequence variation at the pantophysin (PanI) locus. *Molecular Ecology* 12:63–74.

Poinsot, D., S. Charlat, and H. Merçot. 2003. On the mechanism of *Wolbachia*-induced cytoplasmic incompatibility: Confronting the models with the facts. *Bioessays* 25:259–265.

Polis, G. A., and D. R. Strong. 1996. Food web complexity and community dynamics. *American Naturalist* 147:813–846.

Potter, M. A., and K. A. De Jong. 2000. Cooperative coevolution: An architecture for evolving coadapted subcomponents. *Evolutionary Computation* 8:1–29.

Poulsen, M., A. N. M. Bot, C. R. Currie, M. G. Nielsen, and J. J. Boomsma. 2003. Within-colony transmission and the cost of a mutualistic bacterium in the leaf-cutting ant *Acromyrmex octospinosus*. *Functional Ecology* 17:260–269.

Power, M. E., D. Tilman, J. A. Estes, B. Menge, W. J. Bond, L. S. Mills, G. C. Daily, J. C. Castilla,

J. Lubchenco, and R. T. Paine. 1996. Challenges in the quest for keystones. *Bioscience* 46:609–620.

Price, P. W. 1980. *Evolutionary biology of parasites.* Princeton University Press, Princeton, New Jersey.

———. 2003. *Macroevolutionary theory on macroecological patterns.* Cambridge University Press, Cambridge.

Price, P. W., and T. Ohgushi. 1995. Preference and performance linkage in a *Phyllocopa* sawfly on the willow, *Salix miyabeana,* on Hokkaido. *Researches on Population Ecology* 37:23–28.

Pritchard, J. R., and D. Schluter. 2001. Declining interspecific competition during character displacement: Summoning the ghost of competition past. *Evolutionary Ecology Research* 3:209–220.

Provorov, N. A., A. Y. Borisov, and I. A. Tikhonovich. 2002. Developmental genetics and evolution of symbiotic structures in nitrogen-fixing nodules and arbuscular mycorrhiza. *Journal of Theoretical Biology* 214:215–232.

Qian, J., S.-W. Kwon, and M. A. Parker. 2003. rRNA and *nifD* phylogeny of *Bradyrhizobium* from sites across the Pacific Basin. *FEMS Microbiology Letters* 219:159–165.

Radtkey, R. R., S. M. Fallon, and T. J. Case. 1997. Character displacement in some *Cnemidophorus* lizards revisited: A phylogenetic analysis. *Proceedings of the National Academy of Sciences USA* 94:9740–9745.

Radtkey, R. R., and M. C. Singer. 1995. Repeated reversals of host-preference evolution in a specialist insect herbivore. *Evolution* 49:351–359.

Radutoiu, S., L. H. Madsen, E. B. Madsen, H. H. Felle, Y. Umehara, M. Grønlund, S. Sato, Y. Nakamura, S. Tabata, N. Sandal, and J. Stougaard. 2003. Plant recognition of symbiotic bacteria requires two LysM receptor-like kinases. *Nature* 425:585–592.

Rannala, B., and Y. Michalakis. 2003. Population genetics and cospeciation: From process to pattern. Pages 120–143 in R. D. M. Page, ed., *Tangled trees: Phylogeny, cospeciation, and coevolution.* University of Chicago Press, Chicago.

Rasgon, J. L., L. M. Styer, and T. W. Scott. 2003. *Wolbachia*-induced mortality as a mechanism to modulate pathogen transmission by vector arthropods. *Journal of Medical Entomology* 40:125–132.

Rausher, M. D. 2001. Co-evolution and plant resistance to natural enemies. *Nature* 411:857–864.

Raymond, J., and R. E. Blankenship. 2003. Horizontal gene transfer in eukaryotic algal evolution. *Proceedings of the National Academy of Sciences USA* 100:7419–7420.

Read, A. F., and L. H. Taylor. 2001. The ecology of genetically diverse infections. *Science* 292:1099–1102.

Redecker, D., R. Kodner, and L. E. Graham. 2000. Glomalean fungi from the Ordovician. *Science* 289:1920–1921.

Remold, S. K., and R. E. Lenski, E. 2001. Contribution of individual random mutations to genotype-by-environment interactions in Escherichia coli. *Proceedings of the National Academy of Sciences USA* 98:11388–11393.

Remy, W., T. N. Taylor, H. Hass, and H. Herp. 1995. Four-hundred-million-year-old vesicular arbuscular mycorrhizae. *Proceedings of the National Academy of Sciences USA* 91:11841–11843.

Renwick, J. A. A., and F. S. Chew. 1994. Oviposition behavior In Lepidoptera. *Annual Review of Entomology* 39:377–400.

Restif, O., M. E. Hochberg, and J. C. Koella. 2001. Virulence and age of reproduction: New insights into host-parasite coevolution. *Journal of Evolutionary Biology* 14:967–979.

Restif, O., and J. C. Koella. 2003. Shared control of epidemiological traits in a coevolutionary model of host-parasite interactions. *American Naturalist* 161:827–836.

Reznick, D., M. J. Butler, and H. Rodd. 2001. Life-history evolution in guppies. VII. The comparative ecology of high- and low-predation environments. *American Naturalist* 157:126–140.

Reznick, D. N., F. H. Shaw, F. H. Rodd, and R. G. Shaw. 1997. Evaluation of the rate of evolution in natural populations of guppies (*Poecilia reticulata*). *Science* 275:1934–1937.

Rice, W. R. 1994. Degeneration of a nonrecombining chromosome. *Science* 263:230–232.

Richie, T. L., and A. Saul. 2002. Progress and challenges for malaria vaccines. *Nature* 415:694–701.

Ricketts, T. H., G. C. Daily, and P. R. Ehrlich. 2001. Countryside biogeography of moths in a fragmented landscape: Biodiversity in native and agricultural habitats. *Conservation Biology* 15:378–388.

Ricklefs, R. E., and E. Bermingham. 2001. Nonequilibrium diversity dynamics of the Lesser Antillean avifauna. *Science* 294: 1522–1524.

———. 2002. The concept of the taxon cycle in biogeography. *Global ecology and biogeography* 11:353–361.

Ricklefs, R. E., and G. C. Cox. 1972. Taxon cycles in the West Indian avifauna. *American Naturalist* 106:195–219.

Ricklefs, R. E., and D. Schluter, eds. 1993. *Species Diversity in Ecological Communities.* University of Chicago Press, Chicago.

Rico-Gray, V., J. G. García-Franco, M. Palacios-Rios, D. Díaz-Castelazo, V. Parra-Tabla, and J. A. Navarro. 1998. Geographical and seasonal variation in the richness of ant-plant interactions in Mexico. *Biotropica* 30:190–200.

Ridenhour, B. J., E. D. Brodie, Jr., and E. D. I. Brodie. 1999. Effects of repeated injection of tetrodotoxin on growth and resistance to tetrodotoxin in the garter snake *Thamnophis sirtalis. Copeia* 1999:531–535.

Rieseberg, L. H. 1997. Hybrid origins of plant species. *Annual Review of Ecology and Systematics* 28:359–389.

Rieseberg, L. H., O. Raymond, D. M. Rosenthal, Z. Lai, K. Livingstone, T. Nakazato, J. L. Durphy, A. E. Schwarzback, L. A. Donovan, and C. Lexer. 2003. Major ecological transitions in wild sunflowers facilitated by hybridization. *Science* 301:1211–1216.

Rieseberg, L. H., C. Van Fossen, and A. M. Desrochers. 1995. Hybrid speciation accompanied by genomic reorganization in wild sunflowers. *Nature* 375:313–316.

Rikkinen, J., I. Oksanen, and K. Lohtander. 2002. Lichen guilds share related cyanobacterial symbionts. *Science* 297:357.

Riley, C. V. 1889. The parsnip webworm. *Insect Life* 1:94–98.

Ristaino, J. B., C. T. Groves, and G. R. Parra. 2001. PCR amplification of the Irish potato famine pathogen from historic specimens. *Nature* 411:695–697.

Ritland, D. B. 1995. Comparative unpalatability of mimetic viceroy butterflies (*Limenitis archippus*) from four south-eastern United States populations. *Oecologia* 103:327–336.

Robinson, J., M. J. Waller, P. Parham, N. de Groot, R. Bontrop, L. J. Kennedy, P. Stoehr et al. 2003. IMGT/HLA and IMGT/MHC: Sequence databases for the study of the major histocompatibility complex. *Nucleic Acids Research* 31:311–314.

Rode, A. L., and B. S. Lieberman. 2002. Phylogenetic and biogeographic analysis of Devonian phyllocarid crustaceans. *Journal of Paleontology* 76:271–286.

Rodriguez-Lanetty, M., and O. Hoegh-Guldberg. 2002. The phylogeography and connectivity of the latitudinally widespread scleractinian coral *Plesiastrea versipora* in the Western Pacific. *Molecular Ecology* 11:1177–1189.

Rogers, A. R. 1988. Three components of genetic drift in subdivided populations. *American Journal of Physical Anthropology* 77:435–449.

Rogers, D. J., S. E. Randolph, R. W. Snow, and S. I. Hay. 2002. Satellite imagery in the study and forecast of malaria. *Nature* 415:710–715.

Romeike, J., T. Friedl, G. Helms, and S. Ott. 2002. Genetic diversity of algal and fungal partners in four species of *Umbilicaria* (lichenized Ascomycetes) along a transect of the Antarctic peninsula. *Molecular Biology and Evolution* 19:1209–1217.

Ronce, O., and M. Kirkpatrick. 2001. When sources become sinks: Migrational meltdown in heterogeneous habitats. *Evolution* 55:1520–1531.

Ronquist, F. 2003. Parsimony analysis of coevolving species associations. Pages 22–64 in R. D. M. Page, ed., *Tangled trees: Phylogeny, cospeciation, and coevolution.* University of Chicago Press, Chicago.

Roughgarden, J. 1979. *Theory of population genetics and evolutionary ecology: An introduction.* Macmillan, New York.

———. 1995. *Anolis lizards of the Caribbean: Ecology, evolution, and plate tectonics.* Oxford University Press, Oxford.

Roughgarden, J., and S. Pacala. 1989. Taxon cycles among *Anolis* lizard populations: Review of the evidence. Pages 403–432 in D. Otte and J. A. Endler, eds., *Speciation and its consequences.* Sinauer Associates, Sunderland, Massachusetts.

Rousset, F., H. R. Braid, and S. L. O'Neill. 1999. A stable triple *Wolbachia* infection in *Drosophila* with nearly additive incompatibility effects. *Heredity* 82:620–627.

Rowan, R., and N. Knowlton. 1995. Intraspecific diversity and ecological zonation in coralalgal symbiosis. *Proceedings of the National Academy of Sciences USA* 92:2850–2853.

Rowan, R., N. Knowlton, A. Baker, and J. Jara. 1997. Landscape ecology of algal symbiont communities creates variation in episodes of coral bleaching. *Nature* 388:265–269.

Roy, B. A., and J. W. Kirchner. 2000. Evolutionary dynamics of pathogen resistance and tolerance. *Evolution* 54:51–63.

Roy, B. A., J. W. Kirchner, C. E. Christian, and L. E. Rose. 2000. High disease incidence and apparent disease tolerance in a North American Great Basin plant community. *Evolutionary Ecology* 14:421–438.

Rudgers, J. A., J. M. Koslow, and K. Clay. 2004. Endophytic fungi alter relationships between diversity and ecosystem properties. *Ecology Letters* 7:42–51.

Rummel, J. D., and J. Roughgarden. 1983. Some differences between invasion-structured and coevolution-structured competitive communities: A preliminary theoretical analysis. *Oikos* 41:477–486.

Russell, C. A., D. L. Smith, L. A. Waller, J. E. Childs, and L. A. Real. 2004. *A priori* prediction of disease invasion dynamics in a novel environment. *Proceedings of the Royal Society of London Biological Sciences Series B* 271:21–25.

Russell, J. A., A. Latorre, B. Sabater-Muñoz, A. Moya, and N. A. Moran. 2003. Side-stepping secondary symbionts: Widespread horizontal transfer across and beyond the Aphidoidea. *Molecular Ecology* 12:1061–1075.

Ruwende, C., and A. Hill. 1998. Glucose-6-phosphate dehydrogenase and malaria. *Journal of Molecular Medicine* 76:581–588.

Ruwende, C., S. C. Khoo, R. W. Snow, S. N. R. Yates, D. Kwiatkowski, S. Gupta, P. Warn, C. E. M. Allsopp, S. C. Gilbert, N. Peschu, C. I. Newbold, B. M. Greenwood, K. Marsh, and

A. V. S. Hill. 1995. Natural selection of hemi- and heterozygotes for G6PD deficiency in Africa by resistance to severe malaria. *Nature* 376:246–249.

Sachs, J. D. 2002. A new global effort to control malaria. *Science* 298:122–124.

Sachs, J., and P. Malaney. 2002. The economic and social burden of malaria. *Nature* 415:680–685.

Saffo, M. B. 2002. Themes from variation: Probing the commonalities of symbiotic associations. *Integrative and Comparative Biology* 42:291–294.

Saikkonen, K., M. Helander, S. H. Faeth, F. Shulthess, and D. Wilson. 1999. Endophyte-grass-herbivore interactions: The case of *Neotyphodium* endophyte in Arizona fescue populations. *Oecologia* 121:411–420.

Saikkonen, K., S. H. Faeth, M. Helander, and T. J. Sullivan. 1998. Fungal endophytes: A continuum of interactions with host plants. *Annual Review of Ecology and Systematics* 29:319–343.

Saint, K. M., N. French, and P. Kerr. 2001. Genetic variation in Australian isolates of myxoma virus: An evolutionary and epidemiological study. *Archives of Virology* 146:1105–1123.

Sakai, A. K., F. W. Allendorf, J. S. Holt, D. M. Lodge, J. Molofsky, K. A. With, S. Baughman et al. 2001. The population biology of invasive species. *Annual Review of Ecology and Systematics* 32:305–332.

Salamini, F., H. Özkan, A. Brandolini, R. Schäfer-Pregl, and W. Martin. 2002. Genetics and geography of wild cereal domestication in the Near East. *Nature Reviews Genetics* 3:429–444.

Sanders, I. R. 2002. Ecology and evolution of multigenomic arbuscular mycorrhizal fungi. *American Naturalist* 160:S128–S141.

Sanders, I. R., M. Alt, K. Groppe, T. Boller, and A. Wiemken. 1995. Identification of ribosomal DNA polymorphisms among and within spores of the Glomales: Application to studies on the genetic diversity of arbuscular mycorrhizal fungal communities. *New Phytologist* 130:419–427.

Sanford, E., M. S. Roth, G. C. Johns, J. P. Wares, and G. N. Somero. 2003. Local selection and latitudinal variation in a marine predator-prey interaction. *Science* 300:1135–1137.

Santos, S. R., C. Gutiérrez-Rodriguez, H. R. Lasker, and M. A. Coffroth. 2003. *Symbiodinium* sp. associations in the gorgonian *Pseudopterogorgia elisabethae* in the Bahamas: High levels of genetic variability and population structure in symbiotic dinoflagellates. *Marine Biology* 143:111–120.

Sasaki, A., and H. C. J. Godfray. 1998. A model for the coevolution of resistance and virulence in coupled host-parasitoid interactions. *Proceedings of the Royal Society of London Biological Sciences Series B* 266:455–463.

Sasaki, A., I. Kawaguchi, and A. Yoshimori. 2002. Spatial mosaic and interfacial dynamics in a Müllerian mimicry system. *Theoretical Population Biology* 61:49–71.

Scalzo, A. A. 2002. Successful control of viruses by NK cells—a balance of opposing forces? *Trends in Microbiology* 10:470–474.

Schaller, G. B. 1972. *The Serengeti lion: A study of predator-prey relations.* University of Chicago Press, Chicago.

Schemske, D. W., and H. D. Bradshaw, Jr. 1999. Pollinator preference and the evolution of floral traits in monkeyflowers (*Mimulus*). *Proceedings of the National Academy of Sciences USA* 96:11910–11915.

Schluter, D. 1994. Experimental evidence that competition promotes divergence in adaptive radiation. *Science* 266:798–801.

———. 1996a. Ecological causes of adaptive radiation. *American Naturalist* 148:S40–S64.

———. 1996b. Ecological speciation in postglacial fishes. *Philosophical Transactions of the Royal Society of London Biological Sciences Series B* 351:807–814.

———. 2000a. *The ecology of adaptive radiation.* Oxford University Press, Oxford.

———. 2000b. Ecological character displacement in adaptive radiation. *American Naturalist* 156:S4–S16.

———. 2003. Frequency dependent natural selection during character displacement in sticklebacks. *Evolution* 57:1142–1150.

Schluter, D., and J. D. McPhail. 1992. Ecological character displacement and speciation in sticklebacks. *American Naturalist* 140:85–108.

Schoener, T. W., and A. Schoener. 1983. The time to extinction of a colonizing propagule of lizards increases with island area. *Nature* 302:332–334.

Schoener, T. W., and D. A. Spiller. 1999. Indirect effects in an experimentally staged invasion by a major predator. *American Naturalist* 153:347–358.

Schoener, T. W., D. A. Spiller, and J. B. Losos. 2003. Predation on a common *Anolis* lizard: Can the food-web effects of a devastating predator be reversed? *Ecological Monographs* 72:383–407.

Schulthess, F. M., and S. H. Faeth. 1998. Distribution, abundances, and associations of the endophytic fungal community of Arizona fescue (*Festuca arizonica*). *Mycologica* 90:569–578.

Schwartz, A., and J. C. Koella. 2001. Trade-offs, conflicts of interest and manipulation in *Plasmodium*-mosquito interactions. *Trends in Parasitology* 17:189–194.

Scott, T. W., W. Takken, B. G. J. Knols, and C. Boëte. 2002. The ecology of genetically modified mosquitoes. *Science* 298:117–119.

Scriber, J. M. 1996. A new "cold pocket" hypothesis to explain local host preference shifts in *Papilio canadensis*. *Entomologia Experimentalis et Applicata* 80:315–319.

Scriber, J. M., R. H. Hagen, and R. C. Lederhouse. 1996. Genetics of mimicry in the tiger swallowtail butterflies, *Papilio glaucus* and *P. canadensis* (Lepidoptera: Papilionidae). *Evolution* 50:222–236.

Seger, J. 1988. Dynamics of some simple host-parasite models with more than two genotypes in each species. *Philosophical Transactions of the Royal Society of London Biological Sciences Series B* 319:541–555.

———. 1992. Evolution of exploiter-victim relationships. Pages 3–25 in M. Crawley, ed., *The population biology of predators, parasites and diseases.* Blackwell, Oxford.

Segraves, K. A., and J. N. Thompson. 1999. Plant polyploidy and pollination: Floral traits and insect visits to diploid and tetraploid *Heuchera grossulariifolia*. *Evolution* 53:1114–1127.

Segraves, K. A., J. N. Thompson, P. S. Soltis, and D. E. Soltis. 1999. Multiple origins of polyploidy and the geographic structure of *Heuchera grossulariifolia*. *Molecular Ecology* 8:253–262.

Sessitsch, A., J. G. Howieson, X. Perret, H. Antoun, and E. Martínez-Romero. 2002. Advances in *Rhizobium* research. *Critical Reviews in Plant Sciences* 21:323–378.

Sezer, M., and R. K. Butlin. 1998. The genetic basis of oviposition preference differences between sympatric host races of the brown planthopper (*Nilaparvata lugens*). *Proceedings of the Royal Society of London Biological Sciences Series B* 265:2399–2405.

Sharakhov, I. V., A. C. Serazin, O. G. Grushko, A. Dana, N. Lobo, M. E. Hillenmeyer, R. Westerman, J. Romero-Severson, C. Constantini, N. Sagnon, F. H. Collins, and N. J. Besansky. 2002. Inversions and gene order shuffling in *Anopheles gambiae* and *A. funestus*. *Science* 298:182–185.

Sheppard, P. M., J. R. G. Turner, K. S. Brown, W. W. Benson, and M. C. Singer. 1985. Genetics and the evolution of Muellerian mimicry in *Heliconius* butterflies. *Philosophical Transactions of the Royal Society of London Biological Sciences Series B* 308:433–610.

Siepielski, A. M., and C. W. Benkman. 2004. Interactions among moths, crossbills, squirrels, and lodgepole pine in a geographic selection mosaic. *Evolution* 58:95–101.

Silvertown, J., M. Dodd, and D. J. G. Gowing. 2001. Phylogeny and the niche structure of meadow plant communities. *Journal of Ecology* 89:428–435.

Silvertown, J., M. Franco, and J. L. Harper, eds. 1997. *Plant life histories: Ecology, phylogeny, and evolution.* Cambridge University Press, Cambridge.

Simmons, R. B., and S. J. Weller. 2002. What kind of signals do mimetic tiger moths send? A phylogenetic test of wasp mimicry systems (Lepidoptera: Arctiidae: Euchromiini). *Proceedings of the Royal Society of London Biological Sciences Series B* 269:983–990.

Simms, E. L. 1996. The evolutionary genetics of plant-pathogen systems. *BioScience* 46:136–145.

Simms, E. L., and M. D. Rausher. 1993. Patterns of selection on phytophage resistance in *Ipomoea purpurea. Evolution* 47:970–976.

Simms, E. L., and D. L. Taylor. 2002. Partner choice in nitrogen-fixation mutualisms of legumes and rhizobia. *Integrative and Comparative Biology* 42:369–380.

Simon, J.-C., F. Delmotte, C. Rispe, and T. Crease. 2003. Phylogenetic relationships between parthenogens and their sexual relatives: The possible routes to parthenogenesis in animals. *Biological Journal of the Linnean Society* 79:151–163.

Simpson, J. M., S. A. Kocherginskaya, R. I. Aminov, L. T. Skerlos, T. M. Bradley, R. I. Mackie, and B. A. White. 2002. Comparative microbial diversity in the gastrointestinal tracts of food animal species. *Integrative and Comparative Biology* 42:327–331.

Sinervo, B., and K. R. Zamudio. 2001. The evolution of alternative reproductive strategies: Fitness differential, heritability, and genetic correlation between the sexes. *Journal of Heredity* 92:198–205.

Singer, M. C. 2003. Spatial and temporal patterns of checkerspot butterfly–host plant association: The diverse roles of oviposition preference. Pages 207–228 in C. L. Boggs, W. B. Watt, and P. R. Ehrlich, eds., *Butterflies: Ecology and evolution taking flight.* University of Chicago Press, Chicago.

Singer, M. C., and C. D. Thomas. 1996. Evolutionary responses of a butterfly metapopulation to human- and climate-caused environmental variation. *American Naturalist* 148:S9–S39.

Sironi, M., C. Bandi, L. Sacchi, B. Di Sacco, G. Darniani, and C. Genchi. 1995. Molecular evidence for a close relative of the arthropod endosymbiont *Wolbachia* in a filarial worm. *Molecular and Biochemical Parasitology* 74:223–227.

Slatkin, M., and J. Maynard Smith. 1979. Models of coevolution. *Quarterly Review of Biology* 54:233–263.

Smith, C. C. 1968. The adaptive nature of social organization in the genus of tree squirrels *Tamiasciurus. Ecological Monographs* 38:31–63.

———. 1970. The coevolution of pine squirrels (*Tamiasciurus*) and conifers. *Ecological Monographs* 40:349–371.

———. 1981. The indivisible niche of *Tamiasciurus:* An example of nonpartitioning of resources. *Ecological Monographs* 51:343–363.

Smith, D. L., L. Ericson, and J. J. Burdon. 2003. Epidemiological patterns at multiple spatial scales: An 11-year study of a *Triphragmium ulmariae–Filipendula ulmaria* metapopulation. *Journal of Ecology* 91:890–903.

Smith, S. E., and D. J. Read. 1997. *Mycorrhizal symbiosis.* Second edition. Academic Press, San Diego, California.

Sniegowski, P. D., P. J. Gerrish, and R. E. Lenski. 1997. Evolution of high mutation rates in experimental populations of *Escherichia coli*. *Nature* 387:703–705.

Soler, J. J., J. G. Martinez, M. Soler, and A. P. Møller. 1999. Genetic and geographic variation in rejection behavior of cuckoo eggs by European magpie populations: An experimental test of rejecter-gene flow. *Evolution* 53:947–956.

Soler, J. J., and M. Soler. 2000. Brood-parasite interactions between great spotted cuckoos and magpies: A model system for studying coevolutionary relationships. *Oecologia* 125:309–320.

Soltis, D. E., M. A. Gitzendanner, D. D. Strenge, and P. S. Soltis. 1997. Chloroplast DNA intraspecific phylogeography of plants from the Pacific Northwest of North America. *Plant Systematics and Evolution* 206:353–373.

Soltis, D. E., and P. S. Soltis. 1995. The dynamic nature of polyploid genomes. *The Proceedings of the National Academy of Sciences USA* 92:8089–8091.

———. 1999. Polyploidy: Recurrent formation and genome evolution. *Trends in Ecology and Evolution* 14:348–352.

Soltis, P. S., G. M. Plunkett, S. J. Novak, and D. E. Soltis. 1995. Genetic variation in *Tragopogon* species: Additional origins of the allotetraploids *T. mirus* and *T. miscellus* (Compositae). *American Journal of Botany* 82:1329–1341.

Sorenson, M. D., and R. B. Payne. 2002. Molecular genetic perspectives on avian brood parasitism. *Integrative and Comparative Biology* 42:388–400.

Sork, V. L., K. A. Stowe, and C. Hochwender. 1993. Evidence for local adaptation in closely adjacent subpopulations of northern red oak (*Quercus rubra* L.) expressed as resistance to leaf herbivores. *American Naturalist* 142:928–936.

Sotka, E. E., and M. E. Hay. 2002. Geographic variation among herbivore populations in tolerance for a chemically rich seaweed. *Ecology* 83:2721–2735.

Sotka, E. E., J. P. Wares, and M. E. Hay. 2003. Geographic and genetic variation in feeding preferences for chemically defended seaweeds. *Evolution* 57:2262–2276.

Speed, M. P., N. J. Alderson, C. Hardman, and G. D. Ruxton. 2001. Testing Müllerian mimicry: An experiment with wild birds. *Proceedings of the Royal Society of London B* 267:725–731.

Speed, M. P., and J. R. G. Turner. 1999. Learning and memory in mimicry: II. Do we understand the mimicry spectrum? *Biological Journal of the Linnean Society* 67:281–312.

Spiller, D. A., J. B. Losos, and T. W. Schoener. 1998. Impact of a catastrophic hurricane on island populations. *Science* 281:695–697.

Sprent, J. 2003. Mutual sanctions. *Nature* 422:672–674.

Stachowicz, J. J., and M. E. Hay. 2000. Geographic variation in camouflage specialization by a decorator crab. *American Naturalist* 156:59–71.

Stahl, E. A., G. Dwyer, R. Mauricio, M. Kreitman, and J. Bergelson. 1999. Dynamics of disease resistance polymorphism at the *Rpm1* locus of *Arabidopsis*. *Nature* 400:667–671.

Stanton, M. L. 2003. Interacting guilds: Moving beyond the pairwise perspective on mutualisms. *American Naturalist* 162:S10–S23.

Stanton, M. L., T. M. Palmer, and T. P. Young. 2002. Competition-colonization trade-offs in a guild of African acacia-ants. *Ecological Monographs* 72:347–363.

Stanton, M. L., T. M. Palmer, T. P. Young, A. Evans, and M. L. Turner. 1999. Sterilization and canopy modification of a swollen thorn acacia tree by a plant-ant. *Nature* 401:578–581.

Stearns, S. C., ed. 1999. *Evolution in health and disease*. Oxford University Press, Oxford.

Steinberg, P. D., J. A. Estes, and F. C. Winter. 1995. Evolutionary consequences of food chain length in kelp forest communities. *Proceedings of the National Academy of Sciences USA* 92:8145–8148.

Stevens, L., R. Giordano, and R. Fialho. 2001. Male-killing, nematode infections, bacteriophage infection, and virulence of cytoplasmic bacteria in the genus *Wolbachia. Annual Review of Ecology and Systematics* 32:519–545.

Stewart, F. M., R. Antia, B. R. Levin, M. Lipsitch, and J. E. Mittler. 1998. The population genetics of antibiotic resistance. II: Analytic theory for sustained populations of bacteria in a community of hosts. *Theoretical Population Biology* 53:152–165.

Stone, A. R., T. Dayan, and D. S. Simberloff. 2000. On desert rodents, favored states, and unresolved issues: Scaling up and down regional assemblages and local communities. *American Naturalist* 156:322–328.

Storfer, A., and A. Sih. 1998. Gene flow and ineffective antipredator behavior in a stream-breeding salamander. *Evolution* 52:558–565.

Stouthamer, R., J. A. J. Breeuwer, and G. D. D. Hurst. 1999. *Wolbachia pipientis:* Microbial manipulator of arthropod reproduction. *Annual Review of Microbiology* 53:71–102.

Stouthamer, R., M. van Tilborg, L. Nunney, and R. F. Luck. 2001. Selfish element maintains sex in natural populations of a parasitoid wasp. *Proceedings of the Royal Society of London Biological Sciences Series B* 268:617–622.

Strauss, S. Y. 1994. Levels of herbivory and parasitism in host hybrid zones. *Trends in Ecology and Evolution* 9:209–214.

———. 1997a. Floral characters link herbivores, pollinators, and plant fitness. *Ecology* 78:1640–1645.

———. 1997b. Lack of evidence for local adaptation to individual plant clones or site by a mobile specialist herbivore. *Oecologia* 110:77–85.

Strauss, S. Y., and W. S. Armbruster. 1997. Linking herbivory and pollination: New perspectives on plant and animal ecology and evolution. *Ecology* 78:1617–1618.

Strauss, S. Y., J. K. Conner, and S. L. Rush. 1996. Foliar herbivory affects floral characters and plant attractiveness to pollinators: Implications for male and female plant fitness. *American Naturalist* 147:1098–1107.

Strauss, S. Y., J. A. Rudgers, J. A. Lau, and R. E. Irwin. 2002. Direct and ecological costs of resistance to herbivory. *Trends in Ecology and Evolution* 17:278–285.

Strauss, S. Y., D. H. Siemens, M. B. Decher, and T. Mitchell-Olds. 1999. Ecological costs of plant resistance to herbivores in the currency of pollination. *Evolution* 53:1105–1113.

Streitwolf-Engel, R., M. G. A. van der Heijden, A. Wiemken, and I. R. Sanders. 2001. The ecological significance of arbuscular mycorrhizal fungal effects on clonal reproduction in plants. *Ecology* 82:2846–2859.

Strong, D. R. 1999. Predator control in terrestrial ecosystems: The underground food chain of bush lupine. Pages 577–602 in H. Olff, V. K. Brown, and R. H. Drent, eds., *Herbivores: Between plants and predators.* Blackwell Scientific, Oxford.

Stuart-Fox, D. M., C. J. Schneider, C. Moritz, and P. J. Couper. 2001. Comparative phylogeography of three rainforest-restricted lizards from mid-east Queensland. *Australian Journal of Zoology* 49:119–127.

Suzuki, Y., and M. Nei. 2002. Origin and evolution of influenza virus hemagglutinin genes. *Molecular Biology and Evolution* 19:501–509.

Switkes, J. M., and M. E. Moody. 2001. Coevolutionary interactions between a haploid species and a diploid species. *Journal of Mathematical Biology* 42:175–194.

Tamas, I., L. Klasson, B. Canbäck, A. K. Näslund, A.-S. Eriksson, J. J. Wernegreen, J. P. Sandström, N. A. Moran, and S. G E. Andersson 2002. Fifty million years of genomic stasis in endosymbiotic bacteria. *Science* 296:2376–2379.

Taper, M. L., and T. J. Case. 1985. Quantitative genetic models for the coevolution of character displacement. *Ecology* 66:355–371.

———. 1992. Models of character displacement and the theoretical robustness of taxon cycles. *Evolution* 46:317–333.

Taylor, D. L. 2000. A new dawn—the ecological genetics of mycorrhizal fungi. *New Phytologist* 147:236–239.

Taylor, D. L., and T. D. Bruns. 1997. Independent, specialized invasions of ectomycorrhizal mutualism by two nonphotosynthetic orchids. *Proceedings of the National Academy of Sciences USA* 94:4510–4515.

———. 1999. Population, habitat and genetic correlates of mycorrhizal specialization in the "cheating" orchids *Corallorhiza maculata* and *C. mertensiana. Molecular Ecology* 8:1719–1732.

Taylor, D. L., T. D. Bruns, and S. A. Hodges. 2004. Evidence for mycorrhizal races in a cheating orchid. *Proceedings of the Royal Society of London Biological Sciences Series B* 271:35–43.

Taylor, D. L., T. D. Bruns, T. M. Szaro, and S. A. Hodges. 2003. Divergence in mycorrhizal specialization within *Hexalectris spicata* (Orchidaceae), a nonphotosynthetic desert orchid. *American Journal of Botany* 90:1168–1179.

Taylor, E. B., and J. D. McPhail. 1999. Evolutionary history of an adaptive radiation in species pairs of threespine sticklebacks (*Gasterosteus*): Insights from mitochondrial DNA. *Biological Journal of the Linnean Society* 66.

———. 2000. Historical contingency and ecological determinism interact to prime speciation in sticklebacks, *Gasterosteus. Proceedings of the Royal Society of London Biological Sciences Series B* 267:2375–2384.

Taylor, L. H., S. M. Latham, and M. E. J. Woolhouse. 2001. Risk factors for human disease emergence. *Philosophical Transactions of the Royal Society of London B* 356:983–989.

Taylor, M. S., and M. E. Hellberg. 2003. Genetic evidence for local retention of pelagic larvae in a Caribbean reef fish. *Science* 299:107–109.

Telschow, A., P. Hammerstein, and J. H. Werren. 2002. Effects of *Wolbachia* on genetic divergence between populations: Mainland-island model. *Integrative and Comparative Biology* 42:340–351.

Temeles, E. J., and W. J. Kress. 2003. Adaptation in a plant-hummingbird association. *Science* 300:630–633.

Templeton, A. R., R. J. Robertson, J. Brisson, and J. Strasburg. 2001. Disrupting evolutionary processes: The effect of habitat fragmentation on collared lizards in the Missouri Ozarks. *Proceedings of the National Academy of Sciences USA* 98:5426–5432.

Terborgh, J., L. Lopez, P. Nuñez V., M. Rao, G. Shahabuddin, G. Orihuela, M. Riveros, R. Ascanio, G. H. Adler, T. D. Lambert, and L. Balbas. 2001. Ecological meltdown in predator-free forest fragments. *Science* 294:1923–1926.

Termonia, A., T. H. Hsiao, J. M. Pasteels, and M. C. Milinkovitch. 2001. Feeding specialization and host-derived chemical defense in Chrysomeline leaf beetles did not lead to an evolutionary dead end. *Proceedings of the National Academy of Sciences USA* 98:3909–3914.

Terry, A., G. Bucciarelli, and G. Bernardi. 2000. Restricted gene flow and incipient speciation in disjunct Pacific Ocean and Sea of Cortez populations of a reef fish species, *Girella nigricans. Evolution* 54:652–659.

Thomas, C. D., E. J. Bodsworth, R. J. Wilson, A. D. Simmons, Z. G. Davies, M. Musche, and L. Conradt. 2001. Ecological and evolutionary processes at expanding range margins. *Nature* 411:577–581.

Thomas, J. A. 1995. The ecology and conservation of *Maculinea arion* and other European species of large blue butterfly in A. S. Pullin, ed., *Ecology and conservation of butterflies.* Chapman & Hall, London.

Thomas, J. A., G. W. Elmes, J. C. Wardlaw, and M. Woyciechowski. 1989. Host specificity among *Maculinea* butterflies in *Myrmica* ant nests. *Oecologia* 79:452–457.

Thomas, J. A., M. G. Telfer, D. B. Roy, C. D. Preston, J. J. D. Greenwood, J. Asher, R. Fox, R. T. Clarke, and J. H. Lawton. 2004. Comparative losses of British butterflies, birds, and plants and the global extinction crisis. *Science* 303:1879–1881.

Thompson, J. N. 1978. Within-patch structure and dynamics in *Pastinaca sativa* and resource availability to a specialized herbivore. *Ecology* 59:443–448.

———. 1982. *Interaction and coevolution.* Wiley, New York.

———. 1986a. Constraints on arms races in coevolution. *Trends in Ecology and Evolution* 1:105–107.

———. 1986b. Oviposition behaviour and searching efficiency in a natural population of a braconid parasitoid. *Journal of Animal Ecology* 55:351–360.

———. 1986c. Patterns in coevolution. Pages 119–143 in A. R. Stone and D. J. Hawksworth, eds., *Coevolution and systematics.* Clarendon Press, Oxford.

———. 1987a. Symbiont-induced speciation. *Biological Journal of the Linnean Society* 32:385–393.

———. 1987b. Variance in number of eggs per patch: Oviposition behaviour and population dispersion in a seed parasitic moth. *Ecological Entomology* 12:311–320.

———. 1988a. Evolutionary genetics of oviposition preference in swallowtail butterflies. *Evolution* 42:1223–1234.

———. 1988b. Variation in interspecific interactions. *Annual Review of Ecology and Systematics* 19:65–87.

———. 1989. Concepts of coevolution. *Trends in Ecology and Evolution* 4:179–183.

———. 1993. Preference hierarchies and the origin of geographic specialization in host use in swallowtail butterflies. *Evolution* 47:1585–1594.

———. 1994. *The coevolutionary process.* University of Chicago Press, Chicago.

———. 1997. Evaluating the dynamics of coevolution among geographically structured populations. *Ecology* 78:1619–1623.

———. 1998a. Coping with multiple enemies: 10 years of attack on *Lomatium dissectum* plants. *Ecology* 79:2550–2554.

———. 1998b. The evolution of diet breadth: Monophagy and polyphagy in swallowtail butterflies. *Journal of Evolutionary Biology* 11:563–578.

———. 1998c. Rapid evolution as an ecological process. *Trends in Ecology and Evolution* 13:329–332.

———. 1999a. Coevolution and escalation: Are ongoing coevolutionary meanderings important? *American Naturalist* 153:S92–S93.

———. 1999b. Specific hypotheses on the geographic mosaic of coevolution. *American Naturalist* 153:S1–S14.

———. 1999c. The evolution of species interactions. *Science* 284:2116–2118.

———. 1999d. The raw material for coevolution. *Oikos* 84:5–16.

———. 1999e. What we know and do not know about coevolution: Insect herbivores and plants as a test case. Pages 7–30 in H. Olff, V. K. Brown, and R. H. Drent, eds., *Herbivores: Between plants and predators*. Blackwell Scientific, Oxford.

———. 2001. The geographic dynamics of coevolution. Pages 331–343 in C. W. Fox, D. A. Roff, and D. J. Fairbairn, eds., *Evolutionary Ecology: Concepts and Case Studies*. Oxford University Press, Oxford.

Thompson, J. N., and J. J. Burdon. 1992. Gene-for-gene coevolution between plants and parasites. *Nature* 360:121–125.

Thompson, J. N., and R. Calsbeek. 2004. Molecular differentiation of species and species interactions across large geographic regions: California and the Pacific Northwest. In J. Rolff and M. D. E. Fellowes, eds., *Evolutionary Ecology of Insects*. Blackwell Scientific, Cambridge (in press).

Thompson, J. N., and B. M. Cunningham. 2002. Geographic structure and dynamics of coevolutionary selection. *Nature* 417:735–738.

Thompson, J. N., B. M. Cunningham, K. A. Segraves, D. M. Althoff, and D. Wagner. 1997. Plant polyploidy and insect/plant interactions. *American Naturalist* 150:730–743.

Thompson, J. N., S. L. Nuismer, and R. Gomulkiewicz. 2002. Coevolution and maladaptation. *Integrative and Comparative Biology* 42:381–387.

Thompson, J. N., S. L. Nuismer, and K. Merg. 2004. Plant polyploidy and the evolutionary ecology of plant/animal interactions. *Biological Journal of the Linnean Society* (in press).

Thompson, J. N., and O. Pellmyr. 1992. Mutualism with pollinating seed parasites amid co-pollinators: Constraints on specialization. *Ecology* 73:1780–1791.

Thompson, J. N., and P. W. Price. 1977. Plant plasticity, phenology and herbivore dispersion: Wild parsnip and the parsnip webworm. *Ecology* 58:1112–1119.

Thomson, J. 2003. When is it mutualism? *American Naturalist* 162:S1–S9.

Thomson, J. D., P. Wilson, M. Valenzuela, and M. Malzone. 2000. Pollen presentation and pollination syndromes, with special reference to *Penstamon*. *Plant Species Biology* 15:11–29.

Thrall, P. H., A. Biere, and J. Antonovics. 1993. Plant life-history and disease susceptibility—the occurrence of *Ustilago violacea* on different species within the Caryophyllaceae. *Journal of Ecology* 81:489–498.

Thrall, P. H., and J. J. Burdon. 2002. Evolution of gene-for-gene systems in metapopulations: The effect of spatial scale of host and pathogen dispersal. *Plant Pathology* 51:169–184.

———. 2003. Evolution of virulence in a plant host-pathogen metapopulation. *Science* 299:1735–1737.

Thrall, P. H., J. J. Burdon, and J. D. Bever. 2002. Local adaptation in the *Linum marginale–Melampsora lini* host-pathogen interaction. *Evolution* 56:1340–1351.

Thrall, P. H., J. J. Burdon, and M. J. Woods. 2000. Variation in the effectiveness of symbiotic associations between native rhizobia and temperate Australian legumes: Interactions within and among genera. *Journal of Applied Ecology* 37:52–65.

Thrall, P. H., J. J. Burdon, and A. Young. 2001. Variation in resistance and virulence among demes of a plant host-pathogen metapopulation. *Journal of Ecology* 89:736–748.

Tian, D., M. B. Traw, J. Q. Chen, M. Kreitman, and J. Bergelson. 2003. Fitness costs of R-gene-mediated resistance in *Arabidopsis thaliana*. *Nature* 423:74–77.

Tibbets, T. M., and S. H. Faeth. 1999. *Neotyphodium* endophytes in grasses: Deterrents or promoters of herbivory by leaf-cutting ants? *Oecologia* 118:297–305.

Tishkoff, S. A., R. Varkonyi, N. Cahinhinan, S. Abbes, G. Argyropoulos, G. Destro-Bisol, A. Drousiotou et al. 2001. Haplotype diversity and linkage disequilibrium at human G6PD: Recent origin of alleles that confer malarial resistance. *Science* 293:455–462.

Tofts, R., and J. Silvertown. 2000. A phylogenetic approach to community assembly from a local species pool. *Proceedings of the Royal Society of London B* 267:363–369.

Torchin, M. E., K. D. Lafferty, A. P. Dobson, V. J. McKenzie, and A. M. Kuris. 2003. Introduced species and their missing parasites. *Nature* 421:628–630.

Torre, A. della, C. Costantini, N. J. Besansky, A. Caccone, V. Petrarca, J. R. Powell, and M. Coluzzi. 2002. Speciation within *Anopheles gambiae*—the glass is half full. *Science* 298:115–117.

Travis, J. 1996. The significance of geographical variation in species interactions. *The American Naturalist* 148:S1–S8.

Trussell, G., P. J. Ewanchuk, and M. D. Bertness. 2003. Trait-mediated effects in rocky intertidal food chains: Predator risk cues alter prey feeding rates. *Ecology* 84:629–640.

Tsuchida, T., R. Koga, H. Shibao, T. Matsumoto, and T. Fukatsu. 2002. Diversity and geographic distribution of secondary endosymbiotic bacteria in natural populations of the pea aphid, *Acyrthosiphon pisum*. *Molecular Ecology* 11:2123–2135.

Turelli, M., and A. A. Hoffmann. 1991. Rapid spread of an inherited incompatibility factor in California *Drosophila*. *Nature* 353:440–442.

Turner, J. R. G., and J. L. B. Mallet. 1996. Did forest islands drive the diversity of warningly coloured butterflies? Biotic drift and the shifting balance. *Philosophical Transactions of the Royal Society of London Biological Sciences Series B* 351:835–845.

Turner, S. L., and J. P. W. Young. 2000. The glutamine synthetases of rhizobia: Phylogenetics and evolutionary implications. *Molecular Biology and Evolution* 17:309–319.

Vamosi, S. M., and D. Schluter. 2002. Impacts of trout predation on fitness of sympatric sticklebacks and their hybrids. *Proceedings of the Royal Society of London Biological Sciences Series B* 269:923–930.

———. 2004. Character shifts in the defensive armor of sympatric sticklebacks. *Evolution* 58:376–385.

van Baalen, M., and M. W. Sabelis. 1995. The dynamics of multiple infection and the evolution of virulence. *American Naturalist* 146:881–910.

van der Heijden, M. G. A., T. Boller, A. Wiemken, and I. R. Sanders. 1998. Different arbuscular mycorrhizal fungal species are potential determinants of plant community structure. *Ecology* 79:2082–2091.

van der Heijden, M. G. A., and I. R. Sanders, eds. 2002. *Mycorrhizal ecology.* Springer, Berlin.

van der Laan, J. D., and P. Hogeweg. 1995. Predator-prey coevolution: Interactions across different timescales. *Proceedings of the Royal Society of London Biological Sciences Series B* 259:35–42.

van Ommeren, R. J., and T. G. Whitham. 2002. Changes in interactions between juniper and mistletoe mediated by shared avian frugivores: Parasitism to potential mutualism. *Oecologia* 130:281–288.

van Oppen, M. J. H., F. P. Palstra, A. M.-T. Piquet, and D. J. Miller. 2001. Patterns of coral-dinoflagellate associations in *Acropora*: Significance of local availability and physiology of *Symbiodinium* strains and host-symbiont selectivity. *Proceedings of the Royal Society of London Biological Sciences Series B* 268:1759–1767.

Van Veghel, M. L. J., D. F. R. Cleary, and R. P. M. Back. 1996. Interspecific interactions and competitive ability of the polymorphic reef-building coral *Montastrea annularis*. *Bulletin of Marine Science* 58:792–803.

Van Zandt, P. A., and S. Mopper. 1998. A meta-analysis of adaptive deme formation in phytophagous insect populations. *American Naturalist* 152:595–604.

Vásquez, D. P., and M. A. Aizen. 2003. Null model analyses of specialization in plant-pollinator interactions. *Ecology* 84:2493–2501.

Vavre, F., F. Fleury, D. Lepetit, P. Fouillet, and M. Boulétreau. 1999. Phylogenetic evidence for horizontal transmission of *Wolbachia* in host-parasitoid associations. *Molecular Biology and Evolution* 16:1711–1723.

Vermeij, G. J. 1978. *Biogeography and adaptation: Patterns of marine life.* Harvard University Press, Cambridge, Massachusetts.

———. 1987. *Evolution and escalation: An ecological history of life.* Princeton University Press, Princeton, New Jersey.

———. 1994. The evolutionary interaction among species: Selection, escalation, and coevolution. *Annual Review of Ecology and Systematics* 25:219–236.

———. 2002a. Evolution in the consumer age: Predators and the history of life. Pages 375–393 in M. Kowalewski and P. H. Kelley, eds., *The fossil record of predation.* Paleontological Society Papers 8, New Haven, Connecticut.

———. 2002b. Characters in context: Molluscan shells and the forces that mold them. *Paleobiology* 28:41–54.

Vermeij, G. J., and D. R. Lindberg. 2000. Delayed herbivory and the assembly of marine benthic ecosystems. *Paleobiology* 26:419–430.

Via, S. 1994. Population structure and local adaptation in a clonal herbivore. Pages 58–85 in L. A. Real, ed., *Ecological Genetics.* Princeton University Press, Princeton, New Jersey.

———. Reproductive isolation between sympatric races of pea aphids. I. Gene flow restriction and habitat choice. *Evolution* 53:1446–1457.

Via, S., A. C. Bouck, and S. Skillman. 2000. Reproductive isolation between divergent races of pea aphids on two hosts. II. Selection against migrants and hybrids in the parental environments. *Evolution* 54:1626–1637.

Via, S., and D. J. Hawthorne. 2002. The genetic architecture of ecological specialization: Correlated gene effects on host use and habitat choice in pea aphids. *American Naturalist* 159:S76–S88.

Vitousek, P. M., H. A. Mooney, J. Lubchenco, and J. M. Melillo. 1997. Human domination of earth's ecosystems. *Science* 277:494–499.

Volis, S., S. Mendlinger, and N. Orlovsky. 2000. Variability in phenotypic traits in core and peripheral populations of wild barley *Hordeum spontaneum* Koch. *Hereditas* 133:235–247.

Wade, M. J. 2002. A gene's eye view of epistasis, selection and speciation. *Journal of Evolutionary Biology* 15:337–346.

———. 2003. Community genetics and species interactions. *Ecology* 84:583–585.

Wade, M. J., and C. J. Goodnight. 1998. The theories of Fisher and Wright in the context of metapopulations: When nature does many small experiments. *Evolution* 52:1537–1553.

Wade, M. J., and J. R. Griesemer. 1998. Population heritability: Empirical studies of evolution in metapopulations. *American Naturalist* 151:135–147.

Wainwright, P. C. 1996. Ecological explanation through functional morphology: The feeding biology of sunfishes. *Ecology* 77:1336–1343.

Walker, D., and J. C. Avise. 1998. Principles of phylogeography as illustrated by freshwater and terrestrial turtles in the southeastern United States. *Annual Review of Ecology and Systematics* 29:23–58.

Warren, M. S., J. K. Hill, J. A. Thomas, J. Asher, R. Fox, B. Huntley, D. B. Roy, M. G. Teller, S. Jeffcoate, P. Harding, G. Jeffcoate, S. G. Willis, J. N. Greatorex-Davies, D. Moss, and C. D. Thomas. 2001. Rapid responses of British butterflies to opposing forces of climate and habitat change. *Nature* 414:65–69.

Warren, P. H. 1996. Structural constraints on food web assembly. Pages 142–161 in M. E. Hochberg, J. Clobert, and R. Barbault, eds., *Aspects of the Genesis and Maintenance of Biological Diversity.* Oxford University Press, Oxford.

Waser, N. M., L. Chittka, M. V. Price, N. M. Williams, and J. Ollerton. 1996. Generalization in pollination systems, and why it matters. *Ecology* 77:1043–1060.

Watt, W. B., C. W. Wheat, E. H. Meyer, and J.-F. Martin. 2003. Adaptation at specific loci. VII. Natural selection, dispersal and the diversity of molecular-functional variation patterns among butterfly species complexes (*Colias:* Lepidoptera, Pieridae). *Molecular Ecology* 12:1265–1275.

Weatherall, D. J., and A. B. Provan. 2000. Red cells: I: Inherited anemias. *Lancet* 355:1169–1175.

Webb, C. O. 2000. Exploring the phylogenetic structure of ecological communities: An example for rain forest trees. *American Naturalist* 156:145–155.

Webb, C. O., D. D. Ackerly, M. A. McPeek, and M. J. Donoghue. 2002. Phylogenies and community ecology. *Annual Review of Ecology and Systematics* 33:475–505.

Webster, J. P., and M. E. J. Woolhouse. 1998. Selection and strain specificity of compatibility between snail intermediate hosts and their parasitic schistosomes. *Evolution* 52:1627–1634.

Weeks, A. R., R. Velten, and R. Stouthamer. 2003. Incidence of a new sex-ratio-distorting endosymbiotic bacterium among arthropods. *Proceedings of the Royal Society of London Biological Sciences Series B* 270:1857–1865.

Wegner, K. M., M. Kalbe, J. Kurtz, T. B. H. Reusch, and M. Milinski. 2003. Parasite selection for immunogenetic optimality. *Science* 301:1343.

Wegner, K. M., T. B. H. Reusch, and M. Kalbe. 2003. Multiple parasites are driving major histocompatibility complex in the wild. *Journal of Evolutionary Biology* 16:224–232.

Wehling, W. F., and J. N. Thompson. 1997. Evolutionary conservatism of oviposition preference in a widespread polyphagous insect herbivore, *Papilio zelicaon. Oecologia* 111:209–215.

Weiblen, G. D. 2000. Phylogenetic relationships of functionally dioecious *Ficus* (Moraceae) based on ribosomal DNA sequences and morphology. *American Journal of Botany* 87:1342–1357.

———. 2001. Phylogenetic relationships of fig wasps pollinating functionally dioecious figs based on mitochondrial DNA sequences and morphology. *Systematic Biology* 50:243–267.

Weiblen, G. D., and G. L. Bush. 2002. Speciation in fig pollinators and parasites. *Molecular Ecology* 11:1573–1578.

Weiher, E., and P. Keddy, eds. 1999. *Ecological assembly rules: Perspectives, advances, retreats.* Cambridge University Press, Cambridge.

Wellems, T. E. 2002. *Plasmodium* chloroquine resistance and the search for a replacement antimalarial drug. *Science* 298:124–126.

Wernegreen, J. J. 2002. Genome evolution in bacterial endosymbionts of insects. *Nature Reviews Genetics* 3:850–861.

Werner, E. E., and J. F. Gilliam. 1984. The ontogenetic niche and species interactions in size-structured populations. *Annual Review of Ecology and Systematics* 15:393–426.

Werren, J. H. 1998. *Wolbachia* and speciation. Pages 245–269 in D. J. Howard and S. H. Berlocher, eds., *Endless forms: Species and speciation.* Oxford University Press, Oxford.

Werren, J. H., and D. M. Windsor. 2000. *Wolbachia* infection frequencies in insects: Evidence of a global equilibrium? *Proceedings of the Royal Society of London Biological Sciences Series B* 267:1277–1285.

West, K., and A. Cohen. 1996. Shell microstructure of gastropods from Lake Tanganyika, Africa: Adaptation, convergent evolution, and escalation. *Evolution* 50:672–681.

West, K., A. Cohen, and M. Baron. 1991. Morphology and behavior of crabs and gastropods from Lake Tanganyika, Africa: Implications for lacustrine predator-prey coevolution. *Evolution* 45:589–607.

West, S. A., A. W. Gemmill, A. Graham, M. E. Viney, and A. F. Read. 2001. Immune stress and facultative sex in a parasitic nematode. *Journal of Evolutionary Biology* 14:333–337.

West, S. A., E. T. Kiers, E. L. Simms, and R. F. Denison. 2002. Sanctions and mutualism stability: Why do rhizobia fix nitrogen? *Proceedings of the Royal Society of London Biological Sciences Series B* 269:685–694.

West, S. A., C. M. Lively, and A. F. Read. 1999. A pluralist approach to sex and recombination. *Journal of Evolutionary Biology* 12:1003–1012.

Western, D. 2001. Human-modified ecosystems and future evolution. *Proceedings of the National Academy of Sciences USA* 98:5458–5465.

Whitham, T. G. 1981. Individual trees as heterogeneous environments: Adaptation to herbivory or epigenetic noise? Pages 9–27 in R. F. Denno and H. Dingle, eds., *Insect life history patterns: Habitat and geographic variation.* Springer-Verlag, New York.

Whitham, T. G., G. D. Martinsen, K. D. Floate, H. S. Dungey, B. M. Potts, and P. Keim. 1999. Plant hybrid zones affect biodiversity: Tools for a genetic-based understanding of community structure. *Ecology* 80:416–428.

Whitham, T. G., P. A. Morrow, and B. M. Potts. 1994. Plant hybrid zones as centers of biodiversity: The herbivore community of two endemic Tasmanian eucalypts. *Oecologia* 97:481–490.

Whitham, T. G., W. P. Young, G. D. Martinsen, C. A. Gehring, J. A. Schweitzer, S. M. Shuster, G. M. Wimp et al. 2003. Community and ecosystem genetics: A consequence of the extended phenotype. *Ecology* 84:559–573.

Wilf, P., C. C. Labandeira, K. R. Johnson, P. D. Coley, and A. D. Cutter. 2001. Insect herbivory, plant defense, and early Cenozoic climate change. *Proceedings of the National Academy of Sciences USA* 98:6221–6226.

Wilf, P., C. C. Labandeira, W. J. Kress, C. L. Staines, D. M. Windsor, A. L. Allen, and K. R. Johnson. 2000. Timing the radiations of leaf beetles: Hispines on gingers from latest Cretaceous to recent. *Science* 289:291–294.

Wilkinson, H. H., J. M. Spoerke, and M. A. Parker. 1996. Divergence in symbiotic compatibility in a legume-*Bradyrhizobium* mutualism. *Evolution* 50:1470–1477.

Williams, B. L., E. D. Brodie, Jr., and E. D. Brodie, III. 2003. Coevolution of deadly toxins and predator resistance: Self assessment of resistance by garter snakes leads to behavioral rejection of toxic newt prey. *Herpetologica* 59:155–163.

Williams, E. E. 1969. The ecology of colonization as seen in the zoogeography of anoline lizards on small islands. *Quarterly Review of Biology* 44:345–389.

Williams, G. C. 1975. *Sex and evolution.* Princeton University Press, Princeton, New Jersey.

Wilson, D. S., and W. Swenson. 2003. Community genetics and community selection. *Ecology* 84:586–588.

Wilson, E. O. 1959. Adaptive shift and dispersal in a tropical ant fauna. *Evolution* 13:122–144.

———. 1961. The nature of the taxon cycle in the Melanesian ant fauna. *American Naturalist* 95:169–193.

Wilson, P., M. C. Castellanos, J. N. Hogue, J. D. Thomson, and W. S. Armbruster. 2004. A multivariate search for polllination syndromes in penstemons. *Oikos* 104:345–361.

Wilson, W. G., W. F. Morris, and J. L. Bronstein. 2003. Coexistence of mutualists and exploiters on spatial landscapes. *Ecological Monographs* 73:397–413.

Wirth, R., H. Herz, R. J. Ryel, W. Beyschlag, and B. Hölldobler. 2003. *Herbivory of leaf-cutting ants: A case study on Atta colombica in the tropical rainforest of Panama.* Springer, Berlin.

Wolf, P. G., D. E. Soltis, and P. S. Soltis. 1990. Chloroplast-DNA and allozymic variation in diploid and autotetraploid *Heuchera grossulariifolia* (Saxifragaceae). *American Journal of Botany* 77:232–244.

Woodruff, D. S. 2001. Declines of biomes and biotas and the future of evolution. *Proceedings of the National Academy of Sciences USA* 98:5471–5476.

Wooton, J. C., X. Feng M. T. Ferdig, R. A. Cooper, J. Mu, D. I. Baruch, A. J. Magill, and X. Su. 2002. Genetic diversity and chloroquine selective sweeps in *Plasmodium falciparum. Nature* 418:320–323.

Wootton, J. T. 1994. The nature and consequences of indirect effects in ecological communities. *Annual Review of Ecology and Systematics* 25:443–466.

Wright, S. 1951. The genetical structure of populations. *Annals of Eugenics* 15:323–354.

———. 1965. The interpretation of population structure by F-statistics with special regard to systems of mating. *Evolution* 19:395–420.

Wright, S., and D. Finnegan. 2001. Genome evolution: Sex and the transposable element. *Current Biology* 11:296–299.

Yu, D. W., and D. W. Davidson. 1997. Experimental studies of species-specificity in *Cecropia*-ant relationships. *Ecological Monographs* 67:273–294.

Yu, D. W., H. B. Wilson, and N. E. Pierce. 2001. An empirical model of species coexistence in a spatially structured environment. *Ecology* 82:1761–1771.

Zabinski, C. A., L. Quinn, and R. M. Callaway. 2002. Phosphorus uptake, not carbon transfer, explains arbuscular mycorrhizal enhancement of *Centaurea maculosa* in the presence of native grassland species. *Functional Ecology* 16:758–765.

Zangerl, A. R., and M. R. Berenbaum. 1990. Furanocoumarin induction in wild parsnip: Genetics and population variation. *Ecology* 71:1933–1940.

———. 1993. Plant chemistry, insect adaptations to plant chemistry, and host plant utilization patterns. *Ecology* 74:47–54.

———. 1997. Cost of chemically defending seeds: Furanocoumarins and *Pastinaca sativa. American Naturalist* 150:491–504.

———. 2003. Phenotype matching in wild parsnip and parsnip webworms: Causes and consequences. *Evolution* 57:806–815.

Zhivotovsky, L. A., and M. W. Feldman. 1993. Heterogeneous selection in subdivided populations. *Journal of Mathematical Biology* 31:747–759.

Zinkernagel, R. M., and P. C. Doherty. 1974. Restriction of *in vitro* T cell–mediated cytotoxicity in lymphocytic choriomeningitis within a syngeneic or semiallogeneic system. *Nature* 248:701–702.

Index

Aanen, D. K., 248, 266
Aarssen, L. W., 318
Abbott, A., 359
Abbott, P., 277
Abbott, R. J., 358
ABO blood groups, 69
Abrams, P. A., 88, 95, 181, 241, 242, 243, 325
Acacia, 264, 270–271, 274
Acanthaster, 18
Acaulospora, 303
Acromyrmex, 257–260
Acyrthosiphon, 45–48, 217, 278–281
Adams, D. C., 316, 347
Adams, R. M. M., 259–260
adaptation, genetics, 138
adaptation: local, 9, 29, 47–48, 50–96, 128–133, 164, 188–189, 211–220, 248, 251, 252; metapopulation level, 188–189
adaptive peaks, 72, 133–134
adaptive zones, 148–153, 366
Addicott, J. F., 34, 309
Aedes, 359
Africa, 55, 66–71, 182, 231, 238, 265–266, 348
Agaonidae, 160–161
Agathis, 23
Agavaceae, 32
Agnew, P., 254
Agrawal, A., 42, 73, 356
Agrawal, A. F., 112, 180, 191
Agricultural coevolution, 342–348
Agriculture, 5, 47, 176
Aizen, M. A., 291
Aksoy, S., 275, 283
Alders, 14, 16
alfalfa, 45–48
algae, 39, 64, 153, 284
alkaloids, 42
allelic diversity, 175–220, 250, 277, 352, 360, 367

Allen, M. F., 302
allopatric *vs.* sympatric. *See* adaptation, local
Alnus, 16
Alonso, L. E., 308
Alphey, L., 348
Als, T. D., 61
alternation, coevolutionary. *See* coevolutionary alternation
Althoff, D. M., 14, 17, 23, 31
Ambystoma, 130
Amia, 14
amphibians, 4, 17, 58–59, 107–111, 142, 146, 165, 170–171, 202, 319–322, 329
Amphicarpea, 272–273
amphipod, 64
Ampithoe, 64–65
Anacardiaceae, 32
anchovy, 25
Anderson, K., 303
Anderson, R. A., 71
Anderson, R. M., 88, 112, 181, 249, 251, 252, 350–351
Andreasen, V., 351
Anolis, 316, 327–328
Anopheles, 67–68, 71, 340, 347
antagonism, attenuated, 7, 9, 43, 89, 92–93, 212, 246–269, 368
antibiotics, 5, 9, 224, 225, 348–353, 364
antibodies, 195, 219
antigens, 225
Antiv, R., 250
Antonovics, J., 63, 112, 183, 336, 357
Antonsen, L., 316
ants, 61–62, 153, 248, 257–260, 265–266, 274–275, 295–296, 300, 307–308
Aphidius, 45–48, 280–281
aphids, 45–47, 153, 217, 275–281
Aphomomyrmex, 265

Apiaceae, 31–32, 36, 198–204
Apios, 264
Apocryptophagus, 160–161
aquaculture, 5
Arabidopsis, 182
arbitrary fragment length polymorphisms, 21
Ariey, F., 66
Arkhipova, I., 209
Armadillidium, 222
Armbruster, W. S., 168, 228, 243, 309, 316, 326
arms races, 227–245. *See also* coevolutionary escalation
Arnold, A. E., 42, 244, 267
Arnold, S. J., 58, 170, 240
artificial intelligence, 242
Ascomycota, 284
asexual lineages, 205–226
Asia, 32, 55, 58–59, 61, 64, 67, 69, 83, 160–161, 182, 235, 274, 278–279, 342, 359
Asobara, 51–54, 224
assembly rules, 288–313, 331–332
asymmetries, in interactions, 288–313
Atmar, W., 291
Atta, 257–260
Austin, D. J., 350–352
Australia, 64, 67, 84, 157, 184–191, 217–218, 270–271, 275, 291
Austyn, J. M., 193
autopolyploidy, 21–23
Avise, J. C., 13, 14, 15
Ayala, F. J., 66
Ayres, D. R., 358
Azorhizobium, 261
Azzolini, D., 302

Back, R. P. M., 334
bacteria, 5, 24, 35, 36, 62, 84, 147–149, 153, 182–184, 195, 218, 225, 246, 252, 253, 254, 259, 261–264, 275–277, 283, 353
bacteriophage, 225, 230
Baker, A. C., 268, 269
Balcázar-Lara, M., 153
Baldwin, B. G., 168
Ballard, J. W. O., 224
Ball-Ilosera, N. S., 25
Bandi, C., 225, 275
Barbarea, 59–61
Baron, M., 238
Barone, J. A., 63, 65
Barrett, S. C. H., 12
Barrio, E., 66
Barton, N. H., 21, 127, 336
Bascompte, J., 291–293

Baumann, P., 248, 277
Bazzaz, F. A., 353
Bdelloidea, 206
bears, 14, 16
Beattie, A. J., 48, 363
Becerra, J. X., 150
bees, 76, 138, 289, 297, 357
beetle, 59, 153, 253, 277
Bekkevold, D., 260
Beldade, P., 31
Belesky, D. P., 43
Bell, G., 207, 212
Beltman, J. B., 197
Belwood, D. R., 331
Benassi, V., 52
Benkman, C. W., 51, 56, 138–147, 169, 170, 228, 229
Bennett, A. F., 56, 76, 80
Benzie, J. A. H., 19
Berenbaum, M. R., 51, 56, 76, 98, 129, 130, 131, 150, 170, 198–204, 228, 357
Berg, O. G., 210
Bergelson, J., 88, 183, 228, 243
Bergman, S. H., 81
Bermingham E., 14, 25, 332
Bernardi, G., 18, 19, 20
Bernatchez, L., 14
Bertness, M. D., 63, 240, 289
Betulaceae, 32
Bever, J. D., 56, 57, 170, 185, 256, 296, 302
Bidartondo, M. I., 303, 304
Billerbeck, J. M., 64
biogeography, 11–33
birds, 26, 58–59, 77, 138–147, 231, 232, 234–236, 288, 289, 290, 294–295, 312, 313, 325, 330
Biston, 76–77
Blackwell, M., 247
Blanford, S., 217–218, 252
Blankenship, R. E., 210
Blum, M. J., 299–300
Boag, P. T., 77
Bohannan, B. J. M., 24, 73, 147, 183, 230
Bohonak, A. J., 13
Bombus, 23
Bonfante, P., 302
Bonhoeffer, S., 220, 351
Boomsma, J. J., 61, 244, 257, 259, 260
Boots, M., 250
Borghans, J. A. M., 197
Borisov, A. Y., 263
Bossard, C. C., 356
Bot, A. N. M., 244, 257, 259, 260
Bouchon, D., 222

Bouck, A. C., 346
Boulton, W. J., 242
Bowler, P. J., 361
Boyd, C., 83, 357
Braconidae, 51–54
Bradley, D., 183, 228
Bradshaw, H. D., Jr., 138, 311
Bradyrhizobium, 261, 264, 271–274
Braid, H. R., 253
Braig, H. R., 221
Brakefield, P. M., 31, 77, 171, 239
Breeuwer, J. A. J., 221
Breznak, J. A., 267
Briggs, C. J., 13
Brock, M. T., 158, 309
Brodie, E. D., III, 51, 56, 58, 91, 98, 107–111, 165, 169, 170
Brodie, E. D., Jr., 51, 56, 58, 91, 107–111, 108, 170
Brodo, I. M., 284
Brody, A. K., 39, 228, 229
Bronstein, J. L., 34, 48, 88, 228, 247, 308, 309, 311
brood parasites, 234–236
Brooke, M. De L., 234–235
Brooks, D. R., 154, 168
Brouat, C., 265, 266, 274
Broughton, W. J., 262, 263
Brown, A. H. D., 217–218
Brown, J. H., 42, 332, 333
Brown, J. M., 15, 18, 28, 30, 33
Brown, M. T., 247
Brown, W. L., Jr., 329
Bruna, E. M., 308
Bruns, T. D., 303, 304, 309
Brunsfeld, S. J., 157
bryozoan, 64
Buchnera, 153, 275–277
Buckling, A., 147–149
Budd, A. F., 334–335
Budiansky, S., 69
bugs, soapberry, 82–83
bugs, true, 221, 260, 357
Bugula, 64
Bull, J. J., 252
Burdon, J. J., 12, 51, 56, 57, 63 98, 113, 125, 129, 170, 176, 185–191, 217–218, 264, 270–271, 344
Burger, R., 207
Burns, J. G., 250, 252
Burton, R. S., 18
Bush, G. L., 160–161
Bush, L. P., 43
Bush, R. M., 86, 194
Butler, M. J., 80

Butlin, R. K., 206, 232
butterflies, 37, 61, 62, 81, 150, 153, 232, 296–300, 308, 357

cacao, 244
Caldarelli, G., 242
California, 14, 16, 19, 20
Callaway, R. M., 302
Calloselasma, 58
Callosobruchus, 253
Calsbeek, R., 14, 15, 18, 109, 168
Campbell, C. D., 301
Campbell, D. R., 156
Canback, B., 210
Cannings, C., 181
Cardiospermum, 83
Caribbean, 295, 316, 327–329, 334–336
Carius, H. J., 48, 62, 188
Cariveau, D., 228
Carlsson-Graner, U., 354
Carney, K. M., 24
Carroll, S. P., 82–83, 357
Carton, Y., 51–54, 55, 190
Caruso, C. M., 316
Case, T. J., 88, 125, 127, 181, 318, 330, 334
Casiraghi, M., 225
Castellanos, M. C., 290
Cataulacus, 266
Cattin, M.-F., 242
Cavender-Bares, J., 33
Cecidosidae, 32
Ceratosolen, 160–161
Cercropia, 274
Chagas' disease, 352, 353
Chapela, I. H., 257
character displacement, 314–338
character release, 333–336
Charina, 16
Charlat, S., 221, 222
Charleston, M. A., 154
Charlesworth, B., 21, 209
Charlesworth, D., 21
Chase, J. M., 330, 336
cheaters in mutualism, 131, 264–265, 267–268, 308–311
Chen, D. Q., 280
Chen, X. A., 95, 275, 283, 325
Chenuil, A., 265
Chew, F. S., 232
Chiastochaeta, 306
chitons, 40
Chlamydia, 275
chloroplasts, 5, 247, 248, 326

cholera, 251
Chorthippus, 16
Christophides, G. K., 71
Chu, L.-R., 328
Citernesi, A. S., 302
cladogenesis, parallel, 150, 153–155
Clark, J. W., 275
Clark, M. A., 153
Clavicipitaceae, 42
Clay, K., 42, 43, 220
Clayton, D. H., 26
cleaner fish, 288
Cleary, D. F. R., 334
Cleland, E. E., 353
clines, 103, 118, 120–121, 124–125, 165, 299, 314, 318
clones, 205–226
clover, 45–48, 289
Cobb, N. S., 40, 42
cod, 19
codominance, 178
coevolution: applied, ix, 9, 339–364; central problems, 3, 6, 11; classes of dynamics, 88–96, 173–338 passim; constraints, 228; definition, vii, 3; diffuse, vii, 91–92, 197, 229, 243; diversifying, 136–162, 366; escape-and-radiate, 148–153, 366; forms of evidence, 163–172; gene-for-gene, 89, 111–133, 176, 179–191, 193, 249, 342; geographic mosaic theory, 97–135; hierarchical structure, 50, 111, 134, 135, 136; history of study, 4; and humans, 4–5, 9–10, 67–71, 192, 251, 339–364; misuse, 153–154, 163, 228–229; multispecific, 3, 34–49, 89–90, 95, 192–204, 227–245, 314–338, 327–333; number of traits, 167–168; and phylogeny, 136–162; protocols, 163–172; rate, 4, 72; raw materials, 6–7, 11–49
coevolutionary alternation, 7, 8, 89, 92, 176, 227–245, 255, 292, 368
coevolutionary coldspots, 7, 97–135, 138–147, 156, 255, 282, 333–336, 344
coevolutionary complementarity, 246–269
coevolutionary convergence, 7, 8, 288–313
coevolutionary displacement, 7, 9, 89, 95, 314–338
coevolutionary escalation, 7, 8, 89, 90–92, 152, 227–245, 368
coevolutionary hotspots, 7, 97–135, 136, 138–147, 156, 169, 182, 211, 213, 255, 282, 314, 315, 317, 318, 324, 327, 331, 334, 337, 340, 364, 365
coevolved traits, number of, 103
coexistence, 314–338
Cohen, A., 238
Coleman, A. W., 24

Coleman, R. M., 219
Coley, P. D., 63, 65
Collins, J. P., 336
Colpaert, J. V., 305
commensalism, 48, 106, 121, 246–269
commensalistic, 39
community context, 35, 42, 111, 135, 181, 202, 228
community organization and phylogeny, 33
comparative genomics, 176
compartmentalization, 240, 291–293, 308
competition, 36, 42, 74–76, 92, 95, 176, 252, 256, 278, 314–338, 369–370; apparent, 95, 324–326, 369
Conery, J., 207
conifers, 138–147
Conner, J. K., 76, 171, 243
Conover, D. O., 64
conservation, 5
conserving interactions, 339–364
Conus, 158
convergence, 288–313
Cook, J. M., 77, 168, 358
Cooper, V. S., 147
copepods, 17, 77
coral, 19
Corallina, 39
Corallorhiza, 304–305
corals, 5, 244, 247, 248, 268, 326, 334–336
Cordeiro, N. J., 354
Cordia, 274
Cornell, H. V., 150
cospeciation, 150, 153–155
Cowdria, 226
Cox, G. C., 332
Coyne, J. A., 336
crabs, 64, 87, 238
Craddock, C., 208
Craig, D. M., 318, 336
Craig, T. P., 232
Crawley, M. J., 348
Crespi, B. J., 128
crossbills, 138–147, 155, 166
Crow, J. F., 206
Crute, I. R., 176, 190
Cuautle, M., 307
cuckoos, 234–236
Culaea, 323
Cullum, A. J., 56, 76, 80
Cunningham, B. M., 98, 105, 106, 117, 170, 228, 229, 306, 307, 309
Currie, C. R., 159, 244, 257
Cuscuta, 130
Cushman, J. H., 48

cyanobacteria, 284–286
Cyphomyrmex, 259
cytochrome oxidase, 17
cytochrome P-450, 200–204
Cytophaga-like organisms, 221
cytoplasmic incompatibility, 221–226

Daily, G. C., 24, 355, 362, 363
Dalechampia, 325
Daltry, J. C., 58–59
damselflies, 28–30
Daphnia, 42
Darwin, C., 4, 131, 364
Darwin's finches, 77–79
Darwinulidae, 206
Davidson, D. W., 274
Davidson, S. K., 64
Davies, N. B., 234–235
Davies, S. J., 274
Davis, D. R., 31, 167, 168, 306
Davis, M. B., 25
Dawkins, R., 341
Day, T., 250, 252
Dayan, T., 330, 332
De Boer, R. J., 193, 197
De Jong, K. A., 242
De Jong, P. W., 56, 59–61
de Queiroz, K., 328
Debevec, E. M., 316, 326
defense, 40, 48, 51–55, 56, 57, 59, 64, 71, 73, 78,
    80, 89–92, 129, 130, 147, 150–153, 175–204,
    227–245, 292, 342–353
defenses, induced, 191–198, 230–231
Delpero, M., 302
Denham, T. P., 342
Depressaria, 198–204
Desrochers, A. M., 358
Diamond, J. M., 67, 331, 342
Diaptomus, 77
Dictyota, 64–64
Dieckmann, U., 181, 242
Dietl, G. P., 57, 92
Dingle, H., 13, 82–83
dinoflagellates, 247, 268–269, 334
Dioryctria, 40, 41
Dirzo, R., 355
diseases. *See* parasitism
dispersal, seed. *See* seeds
distributed outcomes, 34–49
diversification, 136–162
Dobson, S. L., 353
Doebeli, M., 88, 242, 246
Doherty, P. C., 195

Doherty, P. F., Jr., 210
domatia, 265
Donnelly, C. A., 351
Donoghue, M. J., 303
Doolittle, W. F., 210
Douglas, A. E., 247, 248, 278
dragonflies, 28–30
Drosophila, 51–55, 74, 159, 190, 222–224, 226,
    230, 250, 253
Duda, T. R., Jr., 159
Duffy, J. E., 57
Dunham, A. E., 328
dunnocks, 234–236
Dupas, S., 51, 53, 55, 190
durable resistance, 344
Dushoff, J., 351
Dwyer, G., 88
Dyall, S. D., 247
Dybdahl, M. F., 56, 128, 129, 188, 208, 213–215
dysentery, 251

Earn, D. J. D., 351
Ebert, D., 48, 56, 62, 188, 252
ecological opportunity, 31–33
ecoregions, 14
Edwards, M. E., 316, 326
Ehrlich, P. R., 24, 26, 148–154, 155, 340, 353, 355,
    363
Eickbush, T. H., 209
El Niño, 11, 77
elaiosomes, 195–196
Eldredge, N., 25
Elle, E., 183
Elliott, P. F., 138, 140
Ellison, K., 362
Ellner, S. P., 77
Embiotoca, 19, 20
Emerson, J. J., 88
Enallagma, 28–30
encapsulation, 51–55
Endler, J. A., 76, 80
endophytes, 36, 42–45, 244, 267
endosymbionts, 220–226
enemies, number of, 34–37
Eom, A.-H., 303
Epicephala, 306
epidemics, 255
epidemiology, 5, 185, 339–353; evolutionary, 250
epistasis, 133
Epperson, B. K., 13
equilibrium allele frequencies, 111–133
Ericson, L., 12, 63
Erinaceus, 16

Ernest, S. K., 333
Erynia, 217
Erythrocytes, 196
Escalante, A. A., 66
escalation. *See* coevolutionary escalation
Escherichia, 147–148, 206, 230
Escovopsis, 259–260
Estes, J. A., 63, 64–65, 355
Eucalyptus, 157
Eucosma, 142
Eulampis, 294–295
Euphydryas, 62
Europe, 14, 15, 16, 25, 32, 43, 51–54, 55, 59–62,
    67, 69, 76, 77, 131, 182, 196–197, 198, 234–
    236, 285, 291, 346, 347, 349–350, 357, 358
Evans, A. G., 66, 67, 196
evolution: asymmetry in rates, 79–80; meaning,
    87; rapid, ix, 62, 72–96, 136, 348–364; rates,
    196, 205, 226
Ewald, P. W., 43, 192, 251, 353
Ewanchuk, P. J., 240
extinction, 7, 12, 13, 25, 61, 63, 81, 99, 143–144,
    185, 206–207, 210, 212, 249, 353

Faegri, K., 290
Faeth, S. H., 43, 44
Fagan, W. F., 43, 44
Faith, D. P., 14
Fallon, S. M., 334
Falush, D., 195
Farrell, B., 150, 153
Feder, J. L., 81
Feldhaar, H., 274
Feldman, M. W., 177
Fellowes, M. D. W., 52, 53, 73, 228, 230
Felsenstein, J., 207
feminization, and interactions, 221–222
Fenner, P. J., 84
Ferguson, N. M., 86, 194, 219
Ferraz, G., 354
Ferriere, R., 88
Fester, R., 248
Festuca, 43–45
Fialho, R., 225
Fiedler, K., 61
figs, 34, 160–161, 307
fig wasps, 34, 160–161
Finlay, B. J., 24
Finnegan, D., 209
Firbank, L. G., 348
Fischer, B., 62
fish, 4, 5, 14, 20, 28, 77, 78, 159, 196–197, 288,
    322–324

Fisher, R. A., 133
fitness, 7, 34, 37, 50, 100, 101, 178, 183, 211, 222,
    228, 246, 255, 307
Fitter, A. H., 301
Flavobacteria, 275
Fleming, T. H., 306
flies, 81, 104, 253, 297, 306, 307
Floate, K. D., 156
Flor, H., H., 343
flowers, 5, 8, 93, 104–106, 138, 160–161, 167, 290
Flynn, T., 301
Foitzik, S., 62
food webs, 239–243
Forbes, D. G., 358
Forbes, M. R., 56
fossil assemblages, 26, 334–336
Fox, B. J., 332
Fox, C. W., 56, 232
fragmentation, 355
Franco, M., 26
Frank, S. A., 43, 88, 180, 181, 194, 252, 257
Frankie, G. W., 290
Frankino, W. A., 321
Freeland, W. J., 242
French, B., 239
French, N., 84
Frey, F., 52
Fritz, R. S., 157, 191
Frost, S. D., 218
fruits and frugivory, 26–27, 31, 81, 82–83, 93,
    160–161, 288, 293
Frydenberg, J., 260
F-statistics, 13
Fukatsu, T., 260, 261
fungi, 5, 35, 36, 42–45, 56, 84, 131, 153, 156–157,
    184–191, 206, 217, 244, 246, 248, 257–260,
    265–266, 267, 283–286, 301–305, 356
Funk, D. J., 154, 168, 232, 275, 277, 283
Furano, A. V., 209
furanocoumarins, 200–204
Futuyma, D. J., 26, 28, 33, 154, 168, 232, 360

G6PD deficiency, 69, 70
Gadus, 19
Gaines, S. D., 289
Galápagos, 77–79
Galvani, A. P., 86, 194, 219
Gambusia, 14
Gandon, S., 98, 115–116, 127, 128, 188, 189, 252,
    255
Ganusov, V. V., 250
Garcia-Marin, J. L., 25
García-Ramos, G., 84

Garrett, L., 192
Garrido, J. L., 296
Gasterosteus, 196–197, 322–324
Gavrilets, S., 88, 148, 181, 242
Geffeney, S., 107, 109
Gehring, C. A., 40
Gelb, M. H., 352
Gemma, J. N., 301
Gemmill, A. W., 183, 219
gene flow, 7, 12, 13, 19, 25, 63, 76, 99, 111–133, 168, 169, 171, 197, 206, 211–217, 337, 354
gene-for-gene interactions. *See* coevolution, gene-for-gene
generalists, 34–37
genes: sex-linked, 69, 232, 236; surrogate, 9, 348–354
genetic correlations, 31, 52, 60, 154
genetic drift. *See* random genetic drift
genetic elements, 225
genetic equilibrium, 74, 86
genetic handshaking, 261–267, 286
genetic modification, 9, 347
genetics of coevolution, 158–159, 175–226, 261–267, 271, 275–277
genome reduction, 275–277
genotype-by-genotype-by-environment interaction, 98, 99
geographic ranges. *See* species ranges
geographic selection mosaics, 7, 50–71, 98, 99, 100, 138–147,182, 331, 340, 364, 365
Geospiza, 77–79
Gerrish, P. J., 206
Gil, R., 277
Gilbert, L. E., 297, 298
Gilbert, L.-B., 303
Gill, D. E., 244
Gillespie, J. P., 51
Gillespie, R., 33
Gilliam, J. F., 230, 268
Gilpin, M. E., 12
Ginsburg, L. R., 178
Giordano, R., 225
Giovannetti, M., 302
Gliddon, C., 25
global warming, 11
Glochidion, 306
Glor, R. E., 328
Godfray, H. C. J., 51–54, 73, 95, 183, 228, 230, 254
Godt, M. J. W., 13
Goebel, W., 275
Goetze, E., 19
Goldin, A. L., 107

Gomulkiewicz, R., 88, 96, 98, 100, 112–114, 117–133, 179, 228, 255, 274, 281, 282, 310, 311
Good, J. E. G., 25
Good, J. M., 18
Goodnight, C. J., 49, 133, 318, 336
Gotelli, N. J., 331
Gould, S. J., 127, 361
Goulson, D., 357
Graham, L. E., 247
Grant, B. R., 77–79
Grant, B. S., 77
Grant, M. C., 56
Grant, P. R., 77–79
grasses, 42, 43, 63
grasshoppers, 14, 16, 230
Graves, G. R., 331
Gray, S. M., 324
grazers, 89, 175, 227–245
grazing, 8, 35, 246,
Green, A. M., 259
Greenwood, B., 348
Greya, 15–18, 22–23, 31, 104–106, 117, 153, 156, 166–167, 202, 307
Griesemer, J. R., 336
Gross, R., 275
Grossulariaceae, 32
Groth, J. G., 138
Groves, C. T., 346
guilds, 314–338
Guillet, J.-G., 195
guppy, 78, 80
Gupta, S., 351
gut microorganisms, 36

Haber, W. A., 290
Habte, M., 301
Hacker, S. D., 289
Haldane, J. B. S., 180
half-sib families, 45
Halichrondria, 39
Hamilton, W. D., 180, 210, 252
Hammerstein, P., 223
Hamrick, J. L., 13
Hanifin, C. T., 108
Hansen, T. F., 316
Hanski, I., 12
Hare, J. D., 183
Harley, C. D. G., 38
Harpagoxenus, 61
Harper, J. L., 26
Harris, D. R., 342
Harrison, R. G., 156
Harrison, S., 12, 356

Hartl, D. L., 67
Hartnett, D. C., 303
Harvell, C. D., 356
Harvey, P. H., 26
Hastings, A., 12, 88
Hawaii, 224, 301
Hawkins, B. A., 150
Hawthorne, D. J., 232, 346
Hay, M. E., 57, 63, 64, 240
Haygood, M. G., 64
Heath, B. D., 253
Hedderson, T. A., 26
hedgehogs, 14, 16
Hedrick, A. V., 240
Hedrick, P. W., 74, 178, 318, 336
Heil, M., 308
Heinze, J., 62
Heithaus, E. R., 307
Helicobacter, 195
Heliconia, 294–295
Heliconius, 296–300
Hellberg, M. E., 19, 64
Helleborus, 295–296
Hemagglutinin, 84, 85
Hendrix, S. D., 198
Hendry, A. P., 76
Hensel, J. L., Jr., 108
Henter, H. J., 45–47
Heracleum, 198–204
herbivores, 5, 36, 56, 63, 156–167, 228, 305
herbivory, 198–204
heritability, 45, 76, 108
Herre, E. A., 34, 160, 257
Herrera, C. M., 26, 168, 309
heterozygote advantage, 178, 181, 204
Heuchera, 21–23, 31
Hewitt, G. M., 15, 16, 156
Hibbett, D. S., 303
Hickman, J. C., 356
Higgs, P. G., 242
Hijri, M., 302
Hill, A. V. S., 69, 70, 340
Hill, J. K., 354
HIV, 62, 192, 195, 218
HLA, 69
Hoarau, G., 18
Hochberg, M. E., 125, 127, 242, 254, 282, 318
Hochwender, C., 62
Hodge, A., 301
Hodges, S. A., 304
Hoegh-Guldberg, O., 19
Hoeksema, J. D., 244, 268, 304, 308
Hoekstra, H. E., 57, 76

Hoffman, A. A., 222
Hoffman, S. L., 177, 352
Hogeweg, P., 181
Hol, W. G. J., 352
Holah, J., 42
Holimon, W. C., 56, 98, 138–142, 169, 228, 229
Holland, J. N., 306
Hölldobler, B., 259
Holospora, 252
Holt, R. D., 95, 240
Holub, E. B., 176, 190
Hood, M. E., 253, 357
horizontal transmission. See transmission,
    horizontal
Horner, J. D., 232
Horner-Devine, M. C., 24, 25
Horton, T. R., 304
Hoshovsky, M. C., 356
Hosokawa, T., 260, 261
Hossaert-McKey, M., 34
host races, 213
host shifts, 152
hotspots. See coevolutionary hotspots
Hougen-Eitzman, D., 230
Howard, R. S., 207, 208, 210, 219
Howardula, 250
Howe, H. F., 354
Hoy, M. A., 221
Hubbell, S. P., 331
Huber, K., 359
Hudson, P. J., 250
Huelsenbeck, J. P., 154
Hufbauer, R. A., 47, 56
Hughes, A. L., 66, 178
Hughes, J. B., 24
human leukocyte system, 195–198
humans. See coevolution, and humans
hummingbird, 8, 138
Huntley, B., 354
Hurst, G. D. D., 221, 222, 225, 226
Husband, B. C., 12, 346
Husband, R., 303
Hutchinson, G. E., 329
Huth, C. J., 34
hybrid inferiority, 136, 137
hybridization, 78, 169, 348, 354, 358
hybrids, 156–157
hybrid zone, 156–158, 298–300

immune system, 194–198, 219, 228, 231, 243
immunity, 71, 89, 93, 176, 348–353
indirect effects, 240, 243–244, 278, 309
infectious spread. See transmission, horizontal

influenza, 84, 85, 86, 194
insects, 26, 35, 59, 62, 138, 153, 157, 160–161, 166,
    198–204, 217, 218, 221–226, 230, 231, 232,
    244, 250, 253, 274–281, 289, 296–300, 306,
    307, 313, 340, 346, 353, 357
interaction: ecological strength, 58; gene-by-gene-
    by-environment, 7; modes, 160–162; norms,
    41–45
interaction biodiversity, 4
introduced species, 81–84, 356–358
Irving, S., 52
isolating mechanisms, 147
isopods, 221
Itami, J. K., 232
Iwao, K., 228
Iwasa, Y., 351

Jabbouri, S., 262
Jablonski, D., 25
Jackman, T. R., 327
Jackson, J. B. C., 334, 355
Jadera, 82–83
Jaenike, J., 250
Jain, R., 210
Jalmenus, 308
James, A. C., 224, 253
Jamnongluk, W., 253
Janz, N., 22, 56, 150, 232, 241
Janzen, D. H., 81, 163, 229, 341, 362
Janzen, F. J., 18, 109
Jarosz, A. M., 63, 184, 185
Jeffries, R. L., 140
Jeyaprakash, A., 221
Jiggins, F. M., 221, 222, 225, 226
Johnson, J. A., 108
Johnson, P. J., 247
Jokela, J., 208, 212, 214
Jones, B., 25
Jordano, P., 26, 27, 291
Joshi, A., 73, 74–76, 318, 336
Juchault, P., 222
Judson, O. P., 206
Juenger, T., 228, 243

Kalbe, M., 176, 193, 196–197
Kaltz, O., 56, 57, 128, 179, 188, 252
Kamp, C., 220
Kanost, M. R., 51
Kapan, D. D., 297
Karban, R., 229, 231
Kato, M., 306
Kaustuv, R., 25
Kawaguchi, I., 125, 300

Kawakita, A., 306
Keddy, P., 331
Keese, M. C., 26, 154, 168, 232
Kelley, P. H., 92
Kelp, 64
Kelt, D. A., 332
Keogh, T., 318
Kernaghan, G., 304
Kerr, P., 84
keystone mutualists, 160
Kimura, M., 206
King, E., 74, 318, 336
Kingsolver, J. G., 76
Kinnison, M. T., 76
Kirchner, J. W., 88, 254
Kirkpatrick, M., 127, 129, 354
Klassen, G. J., 153
Klassen, S. P., 82–83
Klironomos, J. N., 303
Knight, T. M., 336
Knoll, A. H., 353
Knols, B. J. G., 71
Knowles, L. L., 154
Knowlton, N., 88, 246, 268
Kodner, R., 247
Koella, J. C., 71, 252, 254, 351
Koelreuteria, 83
Koenig, W. D., 126
Kohn, A. J., 159
Koivisto, R. K. K., 221
Kondo, N., 253, 254
Koops, K., 31
Koptur, S., 307
Koske, R. E., 301
Koskela, T., 56, 130
Koslow, J. M., 42
Kotanen, P. M., 356
Kover, P., 220
Kraaijeveld, A. R., 51–54, 73,183, 228, 230
Krause, A. E., 240
Kremer, K., 195
Krenn, H. W., 32, 34, 168
Kristinsson, K. G., 350–351
Kuhn, G., 302
Kummel, M., 244, 268
Kurland, C. G., 210
Kurtz, J., 193
Kwon, S.-W., 262

Labandeira, C., 26, 154
Lajeunesse, M. J., 56
Lammi, A., 56
Lampronia, 31

Lande, R., 170
Lanfranco, L., 302
Lankford, T. E., Jr., 64
Larget, B., 154
lateral gene transfer, 210, 274
Latham, S. M., 192
Lathyrus, 280
Law, R., 43, 88, 181, 242
Lawrence, G. J., 51, 63, 98, 170, 185, 189, 344
Lawton, J. H., 95
Leebens-Mack, J., 34, 158, 168, 246, 306, 309, 310
legumes, 5, 45–48, 248, 261–264, 270–274, 342
Leibold, M. A., 330
Leishmaniasis, 352, 353
Lenski, R. E., 73, 147,183, 206, 230, 250
Leonardoxa, 265–266
Lepidoptera. *See* butterflies; moths
Lepiotaceae, 257
Lepomis, 14, 15
leprosy, 195
Leptopilina, 51–54, 190
Leptothorax, 61
Levin, B. R., 351
Levin, S. A., 351
Lewis, G. C., 43
Lewis, O. T., 95
Lewontin, R. C., 127, 178
Li, S., 275, 283
Libinia, 64
lichens, 5, 153, 247, 283–286, 301, 326
Lieberman, B. S., 25
Liebert, T. G., 77
limiting resources, 314–338
Limpens, E., 248
Linblad, P., 285, 286
Lindberg, D. R., 154
Linhart, Y. B., 39, 56
linkage disequilibrium, 205
Linum, 184–191, 217–218
lions, 231
Lipsitch, M., 351
Lithophragma, 104–106, 117, 156, 202, 307
Lively, C. M., 56, 76, 112, 125, 127, 128, 129, 180, 188, 191, 207, 208, 211–217
Lockhart, A. B., 252
Lodwig, E. M., 264
Logan, P., 232
Lohman, K., 18
Lohtander, K., 284
Lomatium, 36
Long, C. A., 352
Long-lived species, 244
Longton, R. E., 26

Losos, J. B., 33, 63, 327–328
Lotka-Volterra models, 124–145
Lovelock, C. E., 303, 334–335
Loxia, 138–147
Lu, W., 232
Lutzoni, F., 284, 285
Lynch, M., 207
Lythgoe, K. A., 219

Macaranga, 274
Machado, C. A., 153
Macrotermitidae, 265
Maculinea, 61
Madagascar, 224
Maddox, G. D., 36
Madsen, E. B., 248
Magurran, A. E., 80
Majerus, M. E. N., 77, 221
major histocompatibility complex, 195–198
maladaptation, viii, 8, 57, 103, 126–133, 171
Malaney, P., 66, 67
malaria, 66–71, 178, 192, 195, 340, 347, 352, 353
Malayan pit viper, 58
Malinowski, D. P., 43
Mallet, J., 297, 298
mammals, 4, 5, 26, 58–59, 81, 84, 138–147, 225, 231, 313, 330, 333
Manning, S. D., 112
Margulis, L., 248
marine, 18–19, 57–58, 158–159, 247, 283, 334–336, 356
Mark Welch, D. B., 206
Marr, D. L., 309
Marrow, P., 181, 242
Martel, C., 346
Martens, K., 206
Martin, M. P., 194, 222
Martin, P. S., 81
Martin, T. E., 325
Martin, W., 277
Martinsen, G. D., 158
Massicotte, H., 303
Masui, S., 225
matching alleles, 111–133, 179–181, 191
mathematical models, 100, 111–133, 208, 215–216, 219–220, 241–242, 254, 273–274, 282, 309, 318, 325, 329, 331, 336, 351, 354, 360
Matsuda, H., 181, 242
Matsuura, K., 244
Matsuyama, D. T., 301
May, R. M., 88, 112, 181, 240, 249, 250, 251, 252, 351
Maynard Smith, J., 88, 219, 248

McCabe, D. J., 331
McDonald, R. A., 331
McFall-Ngai, M. J., 283
McGovern, T. M., 64, 342
McKane, A. J., 242
McKey, D., 265, 308
McLennan, D. A., 154, 168
McMillan, W. O., 21
McPeek, M. A., 28–30, 33, 240
McPhail, J. D., 316, 317, 323
Medicago, 280
Melampsora, 184–191, 217–218
Melville, J., 316
memes, 341
Mendlinger, S., 25
Merçot, H., 221, 222
Merg, K., 346, 358
Meselson, M., 206, 209
Mesepiola, 31
Mesorhizobium, 261
metapopulations, 12, 13, 57, 63, 115, 127, 165, 171,
    185–191, 197, 217–218, 282
Meyer, D., 195
Miao, E. A., 225
Michalakis, Y., 15, 98, 128, 155, 180, 189, 254
Michor, F., 351
Microbotryum, 131, 357
Microorganisms 24, 56, 210–226, 244, 246–287.
    See also bacteria; *names of other microorganisms*
Micropterus, 14
microsatellites, 21, 69, 196–197, 268, 359
migration, 13
Miklashevichs, E., 262
Miler, S. I., 225
Miles, D. B., 327
Miller, L. H., 340, 348
Miller, T. E., 240
mimicry, 61, 288, 296–300
Mimulus, 138
mismatched traits, 8, 51, 57, 103–133, 171, 198–
    204, 296
Mitchell, C. E., 356
Mitchell-Olds, T., 183, 228, 230
mites, 221
mitochondria, 5, 222, 225, 247, 248
mitochondrial DNA, 13–25
Mitter, C., 26, 28, 33, 150, 153
Mittermeier, R. A., 362
Mitton, J. B., 177
Miyasaka, S. C., 301
mobile elements, 208–209
Mode, C., 342
Mode, C. J., 181

Molbo, D., 307
molecular differentiation, 13–25, 58
Molina, R., 303
mollusks, 158–159
Monocultures, 9
Monotropa, 305
Monotropoideae, 304–305
Montastraea, 224–336
Montllor, C. G., 280
Moody, M. E., 88
Mooney, H. A., 353
Mopper, S., 62, 218
Morales, M. A., 307, 308
Moran, N. A., 206, 207, 248, 275, 277
Morand, S., 112
Moraxella, 349–350
Moreau, J., 221
Morgan, M. T., 88, 179, 282
Morin, P. J., 329
Moritz, C., 14
Morris, R. J., 95
Morris, W. F., 228, 311
Morrow, P. A., 157, 358
Morton, J. B., 303
moths, 18, 31–32, 40, 41, 42, 76–77, 93, 142, 168,
    198–204, 231, 290, 306–307, 309
Mousseau, T. A., 56, 232
Mu, J., 66
Mueller, U. G., 248, 257, 259
Muller, W. J., 63
Muller's ratchet, 206–208
Munger, J. C., 333
Murdoch, W. W., 13
Murphy, P. J., 318–322
mussel, 58
mutation, 7, 50, 131, 133, 147, 148, 151, 177, 179,
    205–208, 219, 244, 298
Mutikainen, P., 56, 188, 130
mutualism, 8, 43, 93–94, 104–106, 112, 117, 119,
    131, 158, 175, 179, 288–313
mutualists, 35, 36, 39, 40, 42, 48, 56, 101, 160–
    161, 196, 224, 244, 246–313, 317, 324–326,
    356
Mycobacteriaum, 195
mycorrhizae, 5, 40, 228, 247, 248, 268, 300–305
Myers, N., 353
Myrmica, 61
Myrtaceae, 32
Mytilus, 58
Myxoma, 84

Namba, T., 242
Nappi, A. J., 53

Nash, D. R., 61
natural-killer cells, 194
nectar, 289, 291, 307
Nei, M., 13, 84, 85, 178
nematodes, 35, 221, 225, 250, 353
Nemoria, 231
Neotyphodium, 43–45
nestedness, 291–293, 308
networks, multispecific, 35, 92, 227–245, 257,
     288–313, 369
Neuhauser, C., 86, 344
New Caledonia, 224, 253
New Guinea 348
New Zealand 213–217
Newsham, K. K., 301
Niaré, O., 71
niches, 314–338
Nichols-Orians, C. M., 157
Nielsen, J., 56, 59–61, 232
Nielson, M., 18
Nijhout, H. F., 298
Nisbet, R. M., 13
Nishida, T., 244
Nishiguchi, M. K., 283
Nitao, J. K., 198–200
nitrogen-fixation, 248
nod factors, 248, 274
Nolinaceae, 31, 32
nonequilibrium dynamics, 86–87, 96
Normack, B. B., 206
North America, 14, 15, 17–24, 25, 28–30, 32, 40–
     49, 55, 62, 63–64, 67, 76, 77, 80, 81–83, 104–
     111, 138–147, 156, 158, 167, 170, 182, 198–
     104, 222, 224, 230, 244, 264, 271–274, 285,
     291, 294–200, 308, 309, 318–325, 334, 344,
     346, 356, 357, 358
North Atlantic oscillations, 11
Nostoc, 285–286
Nowak, M. A., 193, 351
Nucella, 19, 58
Nuismer, S. L., 22, 31, 88, 90, 96, 98, 112–114,
     117–133, 179, 228, 255, 274, 281, 282, 310,
     311, 346, 354, 358
Nuzhdin, S. V., 209
Nylin, S., 56, 150, 232

oaks, 14, 16, 62
ocean. *See* marine
Ohgushi, T., 39, 83
Oksanen, I., 284
Olesen, J. M., 291
Oliveira, P. S., 307

Oliver, K. M., 280
Olivera, B. M., 158–159
O'Neill, S. L., 253
ontogenetic shifts, 227–245, 268, 330
optimal allelic diversification, 8, 175, 176
orchids, 5, 304–305
organelle formation, 276–277
Orlovsky, N., 25
Orr, H. A., 52
O'Steen, S., 56, 76, 80
ostracods, 206
Otero, J. T., 205
otters, sea, 64
Otto, S. P., 180
overdominance, 178

Pacala, S., 332
Page, R. D. M., 154
Pagel, M., 284, 285
Palacios-Rios, M., 307
Palmer, T. M., 274, 289
Palumbi, S. R., 18, 40, 159, 362
pandemics, 194
Pandolfi, J. M., 334–336
panmixis, 13, 133, 359
Papilio, 81–82, 232
Paramecium, 252
parasites, 4, 5, 8, 35, 48, 56, 58, 61, 62, 73, 84, 89,
     101, 104–106, 112, 117, 119, 131, 150–153,
     160–161, 166, 167, 175–245, 304–305, 317,
     324–325, 340, 342–364
parasitoid, 23–24, 45–48, 51–54, 191, 224, 253,
     280–291
Parategeticula, 31, 306
Parchman, T. L., 143–146
Parham, P., 194
Parker, I. M., 356
Parker, M. A., 125, 180, 262, 264, 271–274
Parmesan, C., 354
Parra, G. R., 346
parsimony, 154
parsnip, 130, 166, 198–204, 357
parsnip webworm, 131, 166, 198–204
Partain, J., 357
parthenogenesis, and interactions, 221–226
Pastinaca, 198–204
Patterson, B. D., 291
Paul, R. E. L., 66
Paulsrud, P., 285, 286
Pawlowska, T. E., 302
Pedersen, D., 230
Peek, A. S., 247, 275

Peichel, C. L., 323
Pellmyr, O., 31, 32, 34, 104, 153, 158, 167,168, 246, 306–307, 309, 310
Pennings, S. C., 63
Penstamon, 290
Perelson, A. S., 193
Perkins, S. L., 154
Perlman, S. J., 250
Perret, X., 262, 263
Petalomyrmex, 265
Peters, A. D., 208
Petrov, D. A., 209
Pfennig, D. W., 318–322
phage, 147–149
Phyllotreta, 59–61
phylogenetic analyses, 136–172
phylogenetic conservatism, 7, 12, 26–33, 49
phylogeography, 13–25, 64, 109, 168, 176, 274, 311, 332, 344
Phytophthora, 244, 345
phytoplankton, 19
Picea, 143–146
Pierce, N. E., 61, 153, 274, 308
Pigliucci, M., 171
Pilson, D., 230
Pimm, S. L., 241
Pinaceae, 304
Pinus, 40, 41, 42, 138–147, 155, 166
Pistacia, 342
Pla, C., 25
plague, 192
plaice, 18
Planctomyces, 275
plants 31, 32, 35, 36, 40, 41, 42, 56, 76, 77–79, 81, 93, 130, 131, 138, 150–153, 156–158, 160–161, 166, 176, 177, 181, 182–191, 198–204, 206, 217, 218, 227, 230, 231, 244, 246, 247, 248, 261–268, 270–274, 280, 281, 290–296, 300–313, 326, 356–358
plasmids, 262
Plasmodium, 66–71, 178, 196, 340
pleiotropy, 133
Pleistocene, 14, 15, 25, 156, 323, 334–336, 364
Plesiastrea, 19
Pleuronectes, 18
ploidy, 21–23
Poaceae, 42
Poecilia, 78, 80
Pogson, G. H., 19
Poiriè, M., 51, 53, 55, 190
Polemonium, 76
Polis, G. A., 240

pollination, 32, 76, 104–106, 160–161, 288, 290–296, 300, 305–307, 312, 325,
pollinators, 5, 36, 158,167, 228, 246
polymorphism, 6, 7, 13, 69, 89–90, 92, 111–133, 175–226, 255, 277, 366
polyploidy, 21–23, 346, 358
population differentiation, 11–25
populations, number on earth, 24
population subdivision, 13–25
Populus, 158
Potamopyrgus, 213–217
Potter, M. A., 242
Potts, B. M., 358
Poulsen, M., 259
Power, A. G., 356
Power, M. E., 355
predation, 8, 28–30, 35, 61, 77, 78, 158–159, 246
predators, 58, 64, 89–90, 92, 117, 125, 175, 227–245, 249, 297–300, 317, 324–325
preference. *See* specialization
Price, P. W., 26, 39, 160, 175, 199
Pritchard, J. R., 323
Prodoxidae, 31–32, 168, 306
Prodoxus, 31
Prokaryotes, 210
Proteobacteria, 221, 222, 260, 275–277, 278
protozoa, 252, 352
Provorov, N. A., 263
Pseudacris, 329
Pseudomonas, 147–149, 182
Pseudotsuga, 139
Ptelea, 230
Puccinia, 36
Pungitius, 323
Purcell, A. H., 280
Purrington, C. B., 183

Qian, J., 262
Quercus, 16, 62
Quinn, L., 302

Radtkey, R. R., 232, 334
Radutoiu, S., 248, 263
Rainey, P. D., 147–149
Rana, 17
Randall, J. M., 356
random genetic drift, 7, 12, 21, 25, 76, 99, 133, 169, 177, 207, 275, 277, 278, 286, 359
Rannala, B., 154, 155
Rasgon, J. L., 353
Rausher, M. D., 228, 230
Raven, J. A., 248, 278

Raven, P. H., 26, 148–154, 155, 355
Raymond, J., 210
reaction norms, 41–45
Read, A. F., 183, 208, 211, 219, 252, 253
Read, D. J., 301
recolonization, 12, 13, 14
recombination, 179, 205–226
Red Queen, 205–226, 367–368
Redecker, D., 247
redwoods, 244
Reeb, V., 284, 285
Rehner, S. A., 257, 259, 260
Remold, S. K., 147
Remy, W., 301
Renwick, J. A. A., 232
reproduction, sexual, 89, 162, 176, 185, 205–226
reproductive character displacement, 315, 316
reptiles, 4, 58–59, 107–111, 142, 146, 165, 170–
    171, 202, 316, 327–328, 334
resistance. *See* defense
Restif, O., 254, 351
restriction fragment length polymorphisms, 21, 69
retrotransposons, 209
Reusch, T. B. H., 176, 193, 196–197
Reznick, D. N., 56, 80
Rhagoletis, 81
Rhizobia, 5, 247, 261–254, 270–274, 301
Rhizobium, 261, 263
Rice, W. R., 207
Richardson, J., 14, 15, 109
Richie, T. L., 352
Rickettsia, 196, 276, 278, 279, 280
Rickettsiaceae, 225
Ricklefs, R. E., 25, 33, 63, 332
Rico-Gray, V., 307, 308
Ridenhour, B. J., 56, 98, 107, 108, 109, 169
Riechert, S. F., 240
Rieseberg, L. H., 156, 358
Rigaud, T., 221, 222
Rikkinen, J., 284, 285, 286
Riley, C. V., 204
ripple effects, 40
Ristaino, J. B., 346
Ritland, D. B., 297
Robert, V., 66
Roberts, J. K., 185
Robinson, B. W., 324
Robinson, J., 195
Rodd, H., 80
Rode, A. L., 25
Rodríguez, D., 84
Rodriguez-Lanetty, M., 19

Rogers, A. R., 13
Rohlf, F. J., 316
Romeike, J., 153, 284
Ronce, O., 127
Ronquist, F., 154
Root, R. B., 36
Rosaceae, 32, 230
rotifers, 206, 209
Roughgarden, J., 88, 327, 329, 332
Rousset, F., 253
Rowan, R., 268
Roy, B. A., 88, 254
Rpm1 gene, 182–184
Rubus, 230
Ruby, E. G., 283
Rudgers, J. A., 42
Rummel, J. D., 329
Rush, S. L., 243
Russell, C., 360
Russell, J. A., 278
Russulaceae, 304
rust, 36
Rutaceae, 230
Ruwende, C., 69, 70

Sabelis, M. W., 252
Sachs, J. D., 66, 67
Saffo, M. B., 248
Sagan, D., 248
Saikkonen, K., 42, 43
Saint, K. M., 84
Sakai, A. K., 84
Salamini, F., 342
Salk, J., 341
Salonen, V., 56, 130
Sánchez-Azofeifa, G. A., 355
Sanders, I. R., 301, 302
Sanford, E., 19, 56, 57, 58, 240
Santos, S. R., 268
Sapindaceae, 82–83, 357
Sardine, 25
Sasaki, A., 125, 250, 254, 300
Saul, A., 352
Sawada, H., 83
Saxifragaceae, 21, 31–32, 306
Scalzo, A. A., 194
Schaller, G. B., 231
Schardl, C. L., 42, 43
Scheffer, S. J., 26, 168
Schemske, D. W., 138, 311
Schistocerca, 230
Schluter, D., 33, 63, 163, 316, 317, 322–325, 329,
    330

Schöen, I., 206
Schoener, A., 328
Schoener, T. W., 240, 327–328
Schulthess, F. M., 43
Schwartz, A., 71
Scophthalmus, 18
Scott, T. W., 347, 353
Scriber, J. M., 232
seeds, 82–83, 138–147, 184, 228, 295–296
Seger, J., 181
Segraves, K., 17, 21–23, 346
selection: artificial, 73–76, 147–148, 230, 336; asymmetries, 94–95; density-dependent, 88, 89; directional, 88, 89, 92; fluctuating, 57, 77–79, 87; frequency-dependent, 88, 89, 118, 178–191, 204, 205–219, 300; interdemic, 133, 336; and local adaptation, 50–96; mosaics (*see* geographic selection mosaics); positive, 226; sexual, 176, 206–211, 221; stabilizing, 88, 89; strength of, 76; temporally varying, 178–179, 181, 204, 294, 282; weak, 86
senescence, 48
senita cactus, 306
serial transfer, 74
Sessitsch, A., 261, 262
sex, 205–226
sex ratio, 161, 213–226
sex-linked genes, 59–6, 232, 236
sexual reproduction. *See* reproduction, sexual
Sezer, M., 232
Sharakhov, I. V., 67, 69
Sharnoff, S., 284
Sharnoff, S. D., 284
Shaw, R. G., 25
Sheppard, P. M., 298
shrews, 14, 16
shrimp, fairy, 318–322
Shultz, T. R., 257
Shykoff, J. A., 56, 57, 128, 179, 188
sickle-cell, 69, 70, 178
Siemens, D., 230
Siepielski, A. M., 142
Sih, A., 130
Siikamaki, P., 56
Silene, 131, 357
Silvertown, J., 26, 33
Simberloff, D. S., 330, 332
Simmons, G. M., 222
Simms, E. L., 176, 228, 230, 264
Simon, J.-C., 206
Simpson, J. M., 267
Sinervo, B., 176
Singer, M. C., 56, 62, 232, 241

Sinorhizobium, 261
Sironi, M., 225
Siska, E. L., 63
size ratios, 329–331
Skillman, S., 346
Slatkin, M., 88
Smith, C. C., 138, 140
Smith, D. L., 12
Smith, J. C., 138–142, 169
Smith, J. W., 56, 98, 228, 229
Smith, S. E., 301
snails, 19, 58, 213–217, 238
snakes, 17, 58, 107–111, 142, 146, 165, 170–171, 202
Sniegowski, P., 206, 209, 220
social insects, 244
Soler, J. J., 236
Soler, M., 236
Solidago, 36
Soltis, D. S., 14, 15, 22, 346
Soltis, P. S., 15, 22, 346, 358
somatic mutation, 244
Sorex, 16
Sork, V. L., 62
Sotka, E. E., 57, 63, 64, 240
South America, 32, 55, 67, 296–300
Spartina, 358
Spea, 319–322
specialization, in interactions, 12, 28, 34–49, 50–96, 72, 73, 138–147, 158–159, 168, 227–245, 288–313
speciation, 4, 6, 7, 136–162, 165, 275, 283–286, 366
species, sibling, 35
species ranges, 12, 25, 69, 103, 124–145, 155–156, 282, 310, 315, 331–332, 357
Speed, M. P., 297
Spiller, D. A., 240, 327
Spirochaetes, 275
Spiroplasma, 276, 278, 279
Spoerke, J. M., 271
sponges, 39
spruce, 143–146, 155
squid, 283
squirrels, 138–147
Stachowicz, J. J., 64
Staehelin, C., 263
Stahl, E. A., 182
Stanton, M. L., 274, 289, 309
starfish, 19
Stearns, S. C., 353
Steinberg, P. D., 63, 64–65
Stephan, W., 209

Stevens, L., 225
Stone, A. R., 332
Storfer, A., 130
Stouthamer, R., 221
Stowe, K. A., 62
Strauss, S. Y., 56, 62, 156, 183, 218, 243, 309
Streitwolf-Engel, R., 302
Streptococcus, 349–350
Strong, D. R., 240, 358
Styer, L. M., 353
Stylidium, 326
substitutions, nonsynonymous, 226
Sullivan, J., 18
Sullivan, T. J., 44
surrogate coevolution, 5, 9, 348–353
surrogate genes, 340
Suzuki, Y., 84, 85
Swenson, W., 336
Switkes, J. M., 88
Symbiodinium, 268–269
symbiont, definition, 175
symbionts, 5, 8, 35, 93–94, 246–287
sympatric *vs.* allopatric. *See* adaptation, local
sympatry, 316–338
Szathmáry, E., 219, 248

Tahiti, 224
Takimura, A., 306
Tamas, I., 257, 277
Tanaka, C., 244
tannins, 64
Taper, M. L., 88, 125, 127, 181, 318, 330
Tarczy-Hornoch, K., 193
Taricha, 107–111, 142, 165, 170–171
Tasmania, 189
taxon cycle, 331–332
Taylor, D. L., 264, 302, 303, 304, 305, 309
Taylor, E. B., 323
Taylor, J. W., 302
Taylor, L. H., 192, 252, 253
Taylor, M. S., 19
Tegeticula, 31, 306, 310
Telshow, A., 223
Templeton, A. R., 353
Terborgh, J., 355
termites, 167, 248
Termonia, A., 168
Terry, A., 18
Tetragma, 32
tetrodotoxin, 107–111, 165
Thaler, J. S., 229, 231
Thamnophis, 107–111, 165–166
Theobroma, 244

Thomas, C. D., 56, 62, 232, 354
Thomas, J. A., 61
Thomson, G., 195
Thomson, J. D., 228, 290, 309
Thorpe, R. S., 58–59
Thrall, P. H., 51, 56, 57, 63, 98, 112, 113, 125, 129, 170, 176, 183, 185–191, 217–218, 264, 344, 354
Tian, D., 182–184
Tibbets, T. M., 43
ticks, 225
Tigriopus, 16
Tikhonovich, I. A., 263
time lags, 57
Tishkoff, S. A., 70
Tofts, R., 33
Torre, A., 67, 68, 69
tradeoffs, 228
trait remixing, 7, 98–135, 168, 182, 211, 364, 366
transgenic lines, 183–185, 347
transmission: horizontal, 244, 248–287, 307, 326; vertical, 225, 248–287
transmission rates, of parasites, 92, 251–260
transposable elements, 208–209
Trappe, J. M., 303
Travis, J., 38, 51, 240
Travisano, B. J. M., 73, 147, 183, 230
Trebouxia, 284
trematodes, 213–217
Trenczek, T., 51
Trichogramma, 224
Tricholoma, 305
Trinidad, 78, 80
Trollius, 306, 307
trophic antagonists, 175–245
Trussell, G., 240
trypanosomes, 196
Tsuchida, T., 257, 278, 279
Tsuga, 139
tuberculosis, 195
Tuljapurkar, S. D., 178
turbot, 18
Turelli, M., 222, 336
Turner, J. R. G., 297, 198
Turner, S. L., 262
typhoid, 251

urbanization, 358–359
Ursus arctos, 16
Ustilago, 344

vaccines, 340, 364
Valentine, J. S., 25

Vamosi, S. M., 325
van Baalen, 125, 127, 242, 252, 318
van Dam, N. M., 183
van der Heijden, M. G. A., 301, 303
van der Laan, J. D., 181
van der Pijl, L., 290
Van Fossen, C., 358
van Ommeren, R. J., 309
van Oppen, M. J. H., 268
Van Veghel, M. L. J., 334
Van Zandt, P. A., 62, 218
Vásquez, D. P., 291
Vavre, F., 253
Velten, R., 221
venom, 58–59, 158–159
Vermeij, G. J., 63, 64, 91–92, 154, 238
Verra, F., 66
vertical transmission. *See* transmission, vertical
Via, S., 46–47, 232, 246
Viney, M. E., 219
virulence, 73, 84, 93, 147–148, 180, 185, 212, 215–216, 225, 249–269, 282
viruses, 84, 194–198, 218, 220
Vitousek, P. M., 362
Volis, S., 25
Vrijenhoek, R. C., 208

Waddell, K. J., 56, 232
Wade, M. J., 49, 133, 336
wagtails, 236
Wainwright, P. C., 239
Walker, D., 15
Wares, J. P., 57, 63, 240
Warheitt, K. I., 328
warning colors. *See* mimicry
Warren, M. S., 354
Warren, P. H., 240
Waser, N. M., 39, 289
wasps, 160–161, 221, 224, 280–281, 297, 307
Watkinson, A. R., 301
Watt, W. B., 177
Webb, C. O., 33
Webster, J. P., 56
Wedell, N., 232
Weeks, A. R., 221
Wegner, K. M., 176, 193, 196–197
Wehling, W. F., 82, 232
Weiblen, G. D., 160–161
Weiher, E., 331
Wellems, T. E., 66, 67, 196, 348
Wernegreen, J. J., 275–277
Werner, E. E., 230, 268

Werren, J. H., 221, 223, 225, 253
West, K., 238
West, S. A., 208, 211, 219
Western, D., 353
Whitham, T., 156, 157, 158, 244, 309, 336, 358
Wilczek, A., 33
wilderness, 9, 361–364
Wilf, P., 26, 154
Wilkinson, H. H., 43, 271–272
Williams, E. E., 327
Wilson, C. C., 14
Wilson, D. S., 336
Wilson, E. O., 259, 329, 331, 341
Wilson, G. W. T., 303
Wilson, H. B., 274
Wilson, P., 290
Wilson, W. G., 228, 311
Windsor, D. M., 221, 253
Winter, F. C., 63–65
Wirth, R., 259, 342
Wiseman, L. L., 77
Wolbachia, 4, 221–226, 249, 251, 253–254, 275, 353
Wolf, P. G., 22
Woodruff, D. S., 353
Woods, M. J., 264
Woolhouse, M. E. J., 56, 112, 192
Wooton, J. T., 66
Wootton, J. C., 240
Wright, S., 13, 133–134, 209
Wüster, W., 58–59

Yang, Z., 225, 226
Yoshimori, A., 125, 300
Young, A., 185
Young, J. P. W., 262
Young, T. P., 274, 289
Yu, D. W., 274
Yucca, 31–32, 34, 306, 309
yucca moths, 34, 153,167, 267

Zabinski, C. A., 302
Zamudio, K. R., 176
Zangerl, A. R., 51, 56, 76, 98, 129–130, 131, 170, 198–204, 228, 357
Zhivotovsky, L. A., 177
Zinkernagel, R. M., 195
zooanthallae, 5, 248
Zschokkek-Rohringer, C. D., 48, 62, 188
Zuk, M., 210
Zwaan, B. J., 239